Photoelectron Spectroscopy in Inorganic Materials Science

Inorganic Materials Series

Series editors:
Duncan W. Bruce, *University of York, UK*
Elizabeth A. Gibson, *Newcastle University, UK*
Dermot O'Hare, *University of Oxford, UK*
Richard I. Walton, *University of Warwick, UK*

For a list of titles in this series see:
rsc.li/inorg-mater-series

How to obtain future titles on publication:
A standing order plan is available for this series. A standing order will bring delivery of each new volume immediately on publication.

For further information please contact:
Book Sales Department, Royal Society of Chemistry, Thomas Graham House, Science Park, Milton Road, Cambridge, CB4 0WF, UK
Telephone: +44 (0)1223 420066, Fax: +44 (0)1223 420247
Email: booksales@rsc.org
Visit our website at books.rsc.org

Photoelectron Spectroscopy in Inorganic Materials Science

Edited by

Robert G. Palgrave
University College London, UK
Email: r.palgrave@ucl.ac.uk

and

Philip R. Davies
Cardiff University, UK
Email: daviespr@cardiff.ac.uk

ROYAL SOCIETY
OF **CHEMISTRY**

Inorganic Materials Series No. 19

Hardback ISBN: 978-1-83916-679-2
PDF ISBN: 978-1-83916-992-2
EPUB ISBN: 978-1-83916-993-9
Print ISSN: 2472-3819
Electronic ISSN: 2472-3827

A catalogue record for this book is available from the British Library

The Royal Society of Chemistry is a charity, registered in England and Wales, Number 207890, and a company incorporated in England by Royal Charter (Registered No. RC000524), registered office: Burlington House, Piccadilly, London W1J 0BA, UK, Telephone: +44 (0) 20 7437 8656.

For further information see our website at www.rsc.org

For general enquiries, please contact books@rsc.org

For EU product safety enquiries, please email books@rsc.org or contact Royal Society of Chemistry Worldwide (Germany) GmbH, Römischer Hof, Unter den Linden 10, 10117 Berlin.

Printed in the United Kingdom by CPI Group (UK) Ltd, Croydon, CR0 4YY, UK

Series Preface

Materials chemistry has grown hugely in the past few decades and continues to evolve rapidly. This is especially the case since the subject now finds itself central to many of the major challenges that face global society, such as those concerning energy and the environment; but applications of materials are also found in many everyday situations, with ongoing developments in technology around electronics and the miniaturisation of devices. Although the study of materials is a highly interdisciplinary topic, the role of innovative, fundamental chemistry continues to lie at the heart of the discovery of new materials, the understanding of their atomic-scale structure and the relationship of structure to a useful property.

With the *Inorganic Materials Series*, our aim is to provide a series of themed volumes, each reflecting the diversity of the topic as it continues to develop over the coming years, with many chosen to relate closely to emerging practical applications. The use of the title *Inorganic Materials* is by no means limiting, but instead emphasises the study of the chemistry of all elements in the Periodic Table and how they impact the development of new materials. The chapters in each book will have a pedagogical flavour with a target audience from first-year postgraduate student upwards, but with wide subject coverage ranging from continuous inorganic solids, to molecular and soft matter systems. We will use expert, active researchers as editors and contributors to provide an up-to-date perspective of their field. At the same time our aim is to provide an international perspective, so to reflect the diversity and interdisciplinarity of the now very broad area of the subject.

Duncan W. Bruce
Elizabeth A. Gibson
Dermot O'Hare
Richard I. Walton

Inorganic Materials Series No. 19
Photoelectron Spectroscopy in Inorganic Materials Science
Edited by Robert G. Palgrave and Philip R. Davies

Published by the Royal Society of Chemistry, www.rsc.org

Preface

Photoelectron spectroscopy has evolved from a specialist tool of atomic physics into one of the central techniques for materials characterisation. It now plays a vital role across chemistry, physics, and materials science, providing direct insight into surface elemental composition, chemical state, and electronic structure. Yet, despite this wide adoption, photoemission remains a technique that can easily be misunderstood. With modern equipment it is straightforward to collect a spectrum, but still difficult to interpret one well.

Moreover, many of those who now use photoemission spectroscopy in their research are not specialists in the technique. Gone are the days when photoemission users were united in a distinct specialist community; today photoemission is likely to be one of a plethora of analytical tools used in a modern materials research programme, by scientists with backgrounds in physics, chemistry, engineering or many other disciplines. The primary intended audience for this book is exactly these types of researchers, who need to know enough to properly interpret their spectra, without dedicating their whole research practice to it. Therefore, the opening chapters begin with an accessible introduction to the theory that underpins photoemission spectroscopy, showing how this theory connects to the interpretation of experimental data. Later chapters deal with practical use cases in several topical areas of modern inorganic and materials chemistry. We hope to guide the reader from fundamental physical ideas to practical analysis, always with reference to real materials and real experiments, and give an understanding of the capabilities and limitations of photoelectron spectroscopy. Perhaps most importantly, we hope to show how to avoid the common pitfalls and errors made in the analysis of photoelectron spectra.

Inorganic Materials Series No. 19
Photoelectron Spectroscopy in Inorganic Materials Science
Edited by Robert G. Palgrave and Philip R. Davies

Published by the Royal Society of Chemistry, www.rsc.org

The chapters that follow reflect the combined expertise of many contributors, each representing a different aspect of the modern photoemission community: from those developing the instruments themselves to those applying spectroscopic methods to catalytic, electronic, and energy materials. Their collective experience demonstrates the remarkable range and continuing evolution of photoemission as a tool for understanding the surface and near-surface chemistry of materials.

Our hope is that this book will serve as both a practical guide and a conceptual introduction to photoemission. Whether you are encountering photoemission for the first time or refining an established analytical approach, we hope it will help you to use the technique with confidence, and to interpret spectra with understanding.

R. G. Palgrave, London
P. R. Davies, Cardiff

Contents

Inorganic Materials Series No. 19
Photoelectron Spectroscopy in Inorganic Materials Science
Edited by Robert G. Palgrave and Philip R. Davies

Published by the Royal Society of Chemistry, www.rsc.org

CHAPTER 1

A Perspective on Photoelectron Spectroscopy for Materials Analysis

R. G. PALGRAVE[a,b*] AND P. R. DAVIES[b,c*]

[a]Department of Chemistry, University College London, London, WC1H 0AJ, UK; [b]HarwellXPS, EPSRC's National Research Facility in PES, R92 Research Complex at Harwell, Rutherford Appleton Laboratory, Harwell Campus, Didcot, OX11 0FA, UK; [c]Cardiff Catalysis Institute, School of Chemistry, Cardiff University, Maindy Road, Cardiff, CF24 4HQ, UK
*E-mails: r.palgrave@ucl.ac.uk, daviespr@cardiff.ac.uk

1.1 Introduction

Photoelectron spectroscopy is now a mainstream technique for materials analysis. It holds this position due to the unique information that it can provide about a sample: the chemical composition, electronic structure, chemical environments and oxidation states of elements. Such information is particularly important for both fundamental and applied work in material science and related fields. Advances in instrumentation have brought the technique to an ever-wider audience, and photoelectron spectroscopy now forms part of the standard characterisation toolkit for inorganic, and to a lesser extent, organic materials, and regularly appears in published research in general materials, chemistry, and physics journals, as well as in its own specialist publications.

Inorganic Materials Series No. 19
Photoelectron Spectroscopy in Inorganic Materials Science
Edited by Robert G. Palgrave and Philip R. Davies

Published by the Royal Society of Chemistry, www.rsc.org

Despite its widespread use, photoelectron spectroscopy is a challenging technique to master. The physical basis of photoemission has been understood from the early days of quantum mechanics, and simple spectra, especially those arising from light elements, can be interpreted satisfactorily (for most purposes) using well-known ideas of atomic structure combined with Einstein's explanation for the photoelectric effect. But success with these straightforward spectra can build overconfidence in the novice analyst. Writing in the 1980s, G. K. Wertheim commented that '*photoelectron spectroscopy is a deceptively simple technique*',[1] and while many aspects have changed in the decades since, Wertheim's comment is still true today. Understanding complex spectra, such as those from transition metal or f-block elements with open shells that give rise to fantastically intricate photoemission profiles, can be a daunting task, and is still a topic of research today. In this book we aim to equip the reader with the knowledge and resources to carry out analysis of materials using photoelectron spectroscopy.

1.1.1 Basic Physical Principle of Photoemission

Photoelectron spectroscopy (PES) involves the ejection of electrons from a sample by irradiation with photons; therefore, it can be thought of as a *photon-in-electron-out* technique. The electrons ejected directly by absorption of a photon are known as *photoelectrons*, and in photoelectron spectroscopy we measure the kinetic energy of the photoelectrons to produce a spectrum.

From measurement of the kinetic energy (E_K), and knowledge of the photon energy ($h\nu$), we can calculate the 'binding energy' (E_B) of the electron, according to eqn (1.1):

$$h\nu = E_K + E_B \tag{1.1}$$

The exact meaning of the binding energy will be explored in detail in Chapter 2, but for now we can see it as a measure of the energy needed to remove the electron from the atom, and can be usefully thought of as approximating the ionisation energy of the electron. A photoelectron spectrum is conventionally plotted as binding energy against intensity, with the binding energy scale increasing from right to left – the reason for this is that, historically, spectra were plotted using kinetic energy, which runs opposite to binding energy as seen in eqn (1.1).

An atom consists of bound electrons arranged in atomic orbitals, which are labelled 1s, 2s, 2p, 3s, 3p, 3d *etc.* These orbitals have discrete energies, which depend primarily on the nuclear charge, and so the XP spectrum contains sharp lines corresponding to the ionisations of electrons from these orbitals. Different elements, having different nuclear charges, have orbitals of characteristic energies. For example, the 1s orbital of carbon has a binding energy of around 285 eV, while the 1s orbital in nitrogen has a binding energy of 400 eV; the difference caused by the single additional proton in the nucleus of nitrogen.

These sharp peaks in the spectrum are called *core lines* – arising as they do from core atomic orbitals – and the main application of PES is in using the core lines to understand the nature of the sample under investigation. The intensity of each core line will be proportional to the amount of that element in the analysis volume, and so PES can be a quantitative method. High-resolution spectra of core lines reveal that orbitals of the same element do not always appear with exactly the same binding energy. There is a *chemical shift* that means different chemical environments of the same element can give rise to core lines at different energies. Thus, through measurement of this chemical shift, PES not only provides compositional information, but also chemical information. Each of these areas will be discussed in greater detail in Chapter 2.

1.1.2 Some Nomenclature and Definitions

Photoelectron spectroscopy can be divided into several sub-disciplines, often differentiated by the nature of the exciting radiation, but also at times the manner of photoelectron detection and the sample environment. In nomenclature of these techniques, a point of confusion is that the words 'photoelectron' and 'photoemission' can be and are used interchangeably. Thus *X-ray photoelectron spectroscopy* or *X-ray photoemission spectroscopy* may be found, but both phrases have the same meaning. In the list of acronyms below we use only one version – *photoelectron* – for brevity but note that in all cases *photoemission* can be substituted.

PES (Photoelectron spectroscopy): a general term for any *photon-in-electron-out* spectroscopy where the kinetic energy of the emerging photoelectron is measured.

XPS (X-ray photoelectron spectroscopy): PES carried out with X-ray incident radiation.

ESCA (Electron Spectroscopy for Chemical Analysis): a now largely obsolete name for XPS, especially when that technique is focused on understanding chemical environments (see Section 1.2 Historical Background).

UPS (Ultraviolet photoelectron spectroscopy): PES carried out with UV incident radiation.

APXPS/NAPXPS (Ambient Pressure/Near Atmospheric Pressure XPS): XPS carried out with the sample close to atmospheric pressure – this is unusual as most PES is done in high vacuum.

ARPES (Angle-resolved PES): despite the general sounding name, in practice this means UPS where the angle of emission of the photoelectrons is carefully measured. The angular dependence of the emission can be related to the momentum of the bound electron and so provides considerable information on the electronic structure.

ARXPS (Angle-resolved XPS): not to be confused with the previously listed technique, this is XPS where the angle of emission of the photoelectrons is measured. This gives information on the depth distribution of the emitted electrons.

HAXPES (Hard X-ray Photoelectron Spectroscopy): XPS carried out with high photon energy (typically above 2 keV, but definitions vary[2]).

SXPS (Soft XPS or Synchrotron XPS): two different meanings of this acronym are common. Soft X-rays here refer to photon energies of 2 keV or below.

Synchrotron XPS uses synchrotron light, which has the important characteristic of being able to vary the photon energy.

AES (Auger Emission Spectroscopy): a related technique that studies electrons emitted by the Auger–Meitner effect, after creation of an excited state by either X-ray or electron beam excitation.

IPES (Inverse Photoelectron Spectroscopy): an *electron-in-photon-out* technique used to probe the empty electronic states of a material.

PEM (Photoelectron Microscopy – also written PEEM): a microscopy technique that uses UV light to excite photoelectrons which are detected to form an image.

Note that phrases such as '*XPS spectrum*' are very commonly found, but once the acronym is expanded the phrase ('X-ray photoelectron spectroscopy spectrum') is seen to be awkward at best and grammatically incorrect at worst. We prefer '*XP spectrum*', where XP stands for *X-ray photoelectron*. Some may argue whether *spectroscopy* is even the right word for a technique that measures outgoing electrons rather than light. But the weight of historical usage means most now accept that the use of an incident photon beam, and the wave-like nature of the emitted electrons, classifies the techniques under discussion as forms of electron spectroscopy.

1.1.3 Gas Phase and Condensed Matter PES

One issue which is not addressed in the acronyms given above is the state of matter of the sample. Samples may either be in the gas phase, or in a condensed phase (solid or, much more rarely, liquid). This distinction between gas phase and condensed matter photoemission turns out to be quite fundamental, and effectively divides the field into two (barely interacting) camps. Gas phase PES of atoms and molecules is able to probe discrete, well-defined atomic or molecular orbitals while minimising intermolecular interactions, and so is particularly valuable for studying fundamental aspects of chemical bonding in isolated systems and benchmarking theoretical calculations.[3,4] Gas phase molecular photoelectron spectra are additionally strongly influenced by vibrational and rotational fine structure, and there are opportunities for studies of emission from excited states using time-resolved measurements. In contrast, in condensed phase samples, rotational freedom is virtually eliminated, and vibrational fine structure is usually not seen either. Solid-state molecular spectra are very strongly influenced by intermolecular interactions. In non-molecular solids, photoelectron spectra, especially of the valence electrons, often reflect collective properties like density of states or band dispersion.

The gas phase and condensed matter PES research communities have little overlap, but the term 'PES' is used by both, usually without further clarification, so the reader needs to be alert to which method is being described. The term 'XPS', however, almost always refers to the study of condensed matter samples.

1.1.4 The Focus of This Book

The techniques listed above all have utility in analysis of inorganic materials. The main focus of this book, however, will be on use of *condensed matter XPS for materials analysis*, as this is by far the most widely used photoelectron spectroscopy method across all materials classes, especially in inorganic materials. Ultraviolet photoelectron spectroscopy (UPS) has uses in understanding the workfunction and electronic structure, and these are discussed in Chapter 6 in the context of solar cell materials. We also refer the reader to several excellent texts on the UPS method.[5,6] As mentioned above, gas phase PES is essentially a separate and more specialist field and will not be discussed further here.

1.2 Historical Background

We can trace a history of photoelectron spectroscopy back to the revolutionary period in physics at the turn of the 20th century; indeed, the technique embodies many of the major discoveries of that era: X-rays, the quantised nature of light, and the quantum mechanical model of the atom. In this section we recount some of the important steps in progressing PES, especially XPS, as an analytical technique over the past 150 years.

Beginning as a tool to experimentally probe the nature of light, electrons and atomic structure, photoelectron spectroscopy was initially far less popular than other branches of X-ray science; we can attribute this in part to practical difficulties in carrying out the experiments. In XPS, electrons emerge from the sample and their kinetic energy and intensity must be measured. This is a significantly more difficult task than measuring the energy and intensity of emitted photons, as electrons are strongly attenuated by the atmosphere. Therefore, photoemission experiments must be carried out at low pressure. Perhaps as a result, X-ray absorption, emission and diffraction methods, which all measure photons emerging from the sample, and so can be carried out in air without any great loss of signal, saw much more rapid progress.

Hertz first reported the observation of what we now call the photoelectric effect in 1887, by shining UV light onto a spark gap transmitter (two metal strips with a small air gap between them, used to detect radio waves). When UV light shone onto the metal strips, the sparks between them became more frequent. The observation immediately attracted the attention of the scientific community, and intense research into the action of UV light on metal bodies was performed in the final decade of the 19th century. Most important was that of Philipp Lenard, who was the first to conduct photoemission experiments in a vacuum,[7] and can claim to have first understood the emission from the metals as 'cathode rays', or as we would say today, electrons. Wheaton provides a fascinating history of the years 1887–1905, where wildly different ideas of the nature of cathode rays, X-rays and the cause of the photoemission competed.[8] The confusion was eventually resolved by Einstein in

1905 who was able to explain the observations of Lenard by introducing the idea of quanta of light (although his ideas were not universally accepted for many years):[9]

> If each energy quantum of the incident light imparts its entire energy to a single electron, then the velocity of emission will depend only on the frequency and not on the intensity of the light.

Photoelectric experiments until this point had been carried out only with UV light. In 1907, P. D. Innes working at Cambridge, England, used X-rays from a Pt anode to emit electrons from metals, performing the first XPS experiments, and observing that the kinetic energies of the electrons were not uniform.[10] He did not link this to Einstein's work of two years prior, and his physical interpretation, coming before the nuclear model of the atom, would be seen as incorrect today.

Decisive advances were then made in Ernest Rutherford's lab, then in Manchester, where Henry Robinson and Henry Moseley were working collaboratively on X-ray spectroscopy. While Moseley focused on X-ray emission,[11] Robinson pursued XPS and published his first results in 1914.[12] His work halted due to the First World War and resumed only in the 1920s when he received a fellowship from the Royal Society to work in Cambridge, funded by the bequest of his late colleague Moseley, who had been killed in action in 1915. Robinson's paper of 1923 is monumental and is often seen as the foundation of photoelectron spectroscopy; the language used to describe both photoemission theory and experiment appears almost entirely modern, only using the term '*corpuscular rays*' for what we today term '*photoelectrons*':[13]

> [T]he method consisted in exposing a narrow strip of metal to the beam of X-rays, and using a uniform magnetic field to bend the stream of ejected electrons onto a photographic plate, the experiment being performed in a high vacuum.
>
> X-radiation, of a sufficiently high single frequency, should give rise to what may be called a 'line spectrum' of secondary corpuscular rays [photoelectrons] from the target, this line spectrum being composed of groups of electrons from the different levels of the atom.

It is notable however, that, like Innes in 1907, Robinson in his extensive and detailed paper doesn't mention Einstein's work at all, even though the latter had won the Nobel Prize for the photoelectric effect two years earlier. It's hard to understand from this vantage point whether the connection to Einstein's work was overlooked, or too obvious to need mentioning.

Robinson observed spectra of gold, lead, bismuth, iodine and several light elements using a Cu Kα X-ray source (photon energy *ca.* 8 keV), and interpreted the spectra in terms of orbital energies, foreshadowing Koopmans' later theorem. Robinson's experimental binding energies of the heavier elements differ from modern values by less than 1%, but there were greater errors in the light elements.

From 1923 till 1940, Robinson was the world's foremost (and for some periods, only) XPS researcher, moving from Cambridge to Cardiff to Queen Mary's College, London. Using ever more refined versions of his 1914 apparatus, he observed chemical shift and equated this with similar shifts seen in X-ray absorption spectra, and studied Auger–Meitner lines in detail. He recognised that the low energy resolution of his instruments prevented the fine structure of spectra from being revealed.

After the Second World War, there was renewed interest in XPS in several countries, with the aim of increasing energy resolution and utilising XPS for chemical analysis. In efforts to achieve higher resolution, Ralph Steinhardt at Lehigh University, USA, built a spectrometer using very high energy X-rays (Mo Kα, 17.5 keV), and reported the first ever quantitative XPS.[14] However, the key breakthrough was achieved by Kai Siegbahn, in Uppsala, Sweden, who developed high-resolution spectrometers, first using magnetic focusing,[15] but from 1964 moving to an electrostatic system, with the invention of the double-focussing method. This increased sensitivity enough to allow monochromated soft X-rays from Al or Mg anodes – much less intense, but offering much higher energy resolution – to be used, producing a step change in spectrometer performance. The first high-resolution angle-resolved XP spectrum was recorded on a cleaved NaCl crystal,[16] and showed variation in intensity due to photoemission diffraction; coincidentally this was the same material whose crystal structure was the first to be solved by Bragg using X-ray diffraction in 1913. The new high-resolution spectrometers in Uppsala enabled observation of the elastic photoemission peak, distinct from the inelastic background, for the first time. Siegbahn, like Steinhardt before him, recognised the importance of the chemical shift, but Siegbahn now had the tools to fully exploit the chemical characterisation potential of XPS. He transformed the method from one of niche interest for atomic and nuclear physicists, to an analytical tool capable of impacting the wide range of science we see today. The transformation came with a change of name – Siegbahn encouraged the use of the name *Electron Spectroscopy for Chemical Analysis (ESCA)* to emphasise the chemical capabilities of the method, on both solids and gases. Today, the term ESCA is rather outdated, and most modern literature uses the term 'XPS' as we do in this book.[17] The developments in Uppsala from 1945 to the late 1960s are summarised in the epoch-defining *Atomic, Molecular and Solid State Structure Studied by Means of Electron Spectroscopy*.[18] This book was many scientists' first exposure to the concept of XPS, and it served as a catalyst for a rapid expansion of the technique beyond the original pioneers. As well as solids, Siegbahn began to study XPS of gas phase molecules, with much success, essentially inventing a whole new discipline; a second book followed on gas phase molecular photoemission.[19]

The focus of XPS as a chemical analysis technique, directed by Siegbahn, somewhat downplayed the surface sensitivity of the method. Kelly recalls that around that time, it was hoped that XPS would become the '*Nuclear Magnetic Resonance (NMR) of solids*', meaning a method capable of revealing the concentrations and spatial arrangements of all chemical environments present in a solid sample, as solution phase NMR then could for organic compounds (in the event, of course, solid-state NMR became the NMR of solids).[20] Still, recommendations were made at the time to coat XPS samples with carbon or gold to improve conductivity, completely sacrificing any information on the surface layer. Perhaps because of this, the discipline of *surface science*, then an increasingly popular field concerned with the way the surfaces and adsorbates behave and key to the field of heterogeneous catalysis, was a surprisingly late adopter of XPS; the first applications began in the late 1960s.[21] But from that point it was soon established that XPS (and UPS) had a central role to play in understanding how adsorbates bond to surfaces.[22] The 1970–80s saw commercialisation of XPS instruments by a number of manufacturers, with landmark spectrometers produced by Hewlett-Packard, Scientia and DuPont.[20] As scientists worldwide gained access to XP spectrometers without having to build one themselves, applications increased rapidly. XPS became a cornerstone method in surface science, catalysis, corrosion science, and other fields. Rapid developments were made on all fronts, from fundamental theory of quantum mechanics and its application to XPS, to quantifying the probing depth of the technique and the behaviour of the emitted photoelectrons, to chemical quantification, metrology, sample preparation and handling.[23]

Today, the success of XPS and its applicability to so many fields can be seen as the hard-won prize of more than a century of work. The challenges faced in bringing the technique to this point have been considerable, from the difficulty in detecting and measuring photoelectrons, to the complex atomic physics of the excited states produced after photoemission, to the difficulty in maintaining a clean surface. While method development work continues in many areas, XPS can now be regarded as a mature technique, one that is widely available internationally, and can usefully be applied to a great many scientific problems, likely far beyond the range that it currently is.

We have barely scratched the surface of the history of XPS in this short account; the reader who wishes to delve deeper is directed to some of the more detailed historical works in the literature.[24–27]

1.3 XPS as a Surface-sensitive Technique

One of the distinguishing features of XPS as a technique is its surface sensitivity. When photoelectrons are ejected from their atoms, they must travel through the sample, escape into the spectrometer environment and eventually reach the detector. Free electrons travelling through space have a very high chance of being scattered by other atoms they encounter. Therefore, for an electron to have a high chance of escaping the sample without scattering, it must originate from an atom very near to the sample surface; those that emanate from deeper inside the sample are almost certain to be scattered by the atoms of the sample as they pass through. For this reason, XPS is a surface-sensitive technique, and it turns out that the vast majority of the primary signal (meaning the sharp peaks) measured by an XP spectrometer originates from atoms in the top few nanometres of the sample. This surface sensitivity is the key to many important uses of XPS.

In a solid cube of size 1 cm^3 there are approximately 10^{23} atoms, but fewer than 10^{15} atoms (one in a hundred million) are at the surface. Yet for many purposes, those 10^{15} surface atoms are the most important in the material; any chemical interaction, and many physical interactions, of the solid with the surroundings occurs at the surface. Furthermore, we know that the atoms at the surface will be different from those in the bulk, because their average number of nearest neighbours is lower. We cannot necessarily extrapolate from the behaviour of the bulk to understand what's going on at a surface. Isolating the signal from the 1 in 10^8 surface atoms is impossible for analytical methods that probe all atoms in a sample equally. XPS is one of a small number of techniques that can provide this surface-specific information.

1.4 Atomic Spectroscopy, Surface Science and Surface Analysis

Scientists who work with condensed matter XPS are a diverse group, with interests and research questions that bridge many of the traditionally-defined research fields. One view that we find useful is that modern XPS work can be divided into three groupings: atomic spectroscopy, surface science, and surface analysis. These groupings have large overlap, with many shared methods and paradigms, but are distinguished by the different types of research question they ask and attempt to answer.

Atomic spectroscopy is the study of interactions of light (here X-rays) and matter. Deeply rooted in the quantum description of the atom, the discipline seeks to understand atomic structure and dynamics – in practice this means mostly electronic structure and dynamics, since X-rays interact much more strongly with electrons than nucleons. A typical research question in atomic spectroscopy might be: *what is the origin of the satellite features seen in the Fe 2p XP spectrum in* Fe_2O_3*?* One possible way to answer this is to experimentally measure the Fe 2p XP spectrum from a pure sample of Fe_2O_3, and compare this with *ab initio* calculations. The spectroscopist may then attempt to interpret the experimental spectrum in terms of the theory, assigning causes such as 'charge transfer' or 'shake up' to the various spectral features seen (phenomena that are tackled in Chapter 2). They may wish to understand the effects of vibrations, atomic recoil, the lifetime of excited states, and incident photon energy on the observed spectrum, with their answers leaning towards fundamental properties of bound electrons and excited states.

Through atomic spectroscopy comes an understanding of XP spectra that underpins every other application of XPS.

Surface science is the study of physical and chemical processes that occur at the surface of a material or at the interface of two materials. It explores fundamental phenomena such as adsorption, desorption, catalysis, and surface reactions, often under highly controlled environments (ultra-high vacuum, single crystal surfaces, controlled does rates of gases and extended surface cleaning regimes). Surface scientists are usually concerned with understanding fundamental features of a surface and so often design their experiments to strip away any complications such as surface roughness, inhomogeneity and contamination, and think hard about how to observe the phenomenon of interest in as much isolation as possible. For example, a typical surface science research question may be: *how does water interact with oxygen vacancies on the* TiO_2 *(110) surface?* To answer this, the surface scientist may attempt to prepare an ideal TiO_2 (110) surface, free from contamination, and with a controlled number of vacancies, characterise this surface fully (with XPS and other methods such as electron diffraction or scanning probe microscopy), then introduce a controlled partial pressure of water, and repeat the analysis. Surface science uses knowledge from atomic spectroscopy as a tool to understand these ideal surfaces and adsorbates.

Surface analysis is the division closest to the topic of this book. It is more commonly concerned with what we might call real-world materials, investigating their surface composition, contamination, oxidation state distributions, and how these properties relate to material properties performance and processing. Surface analysts are concerned with obtaining an accurate and useful picture of the sample under study, whatever that might be. For this reason, features such as surface oxidation, contamination, and multiphase samples, which might be considered by the surface scientist or atomic spectroscopist as complications to be engineered away, may in fact be the exact features of interest to the surface analyst. That doesn't mean that the analyst can be careless as to whether they themselves contaminate, oxidise or otherwise change the surface of their sample; quite the opposite, as for the analyst any observed contamination may in fact be the object of their study, so it is especially important that they did not introduce it themselves.

Some typical surface analysis research questions are:

- *How does the surface chemistry of this functional material change during operation?*
- *What is the chemical nature of the corrosion on this material?*
- *How do changes to the preparation of my material alter the elemental composition and chemical environments of the surface?*

To answer such questions, an understanding of the physical process of spectroscopy is vital, but so too is an appreciation of the chemical nature of the surface, and both physical and chemical knowledge may have to be employed to reach a successful answer. As a bridge from the fundamental to the applied, modern surface analysis with XPS might be seen as the full flowering of the ESCA philosophy envisaged by Siegbhan, and an important way that the knowledge produced by atomic spectroscopists and surface scientists can be translated into practical technological advancements.

We find the demarcations discussed in this section useful when considering the knowledge required to successfully answer questions in each field. Of course, any one person may be an atomic spectroscopist one day, a surface scientist the next and a surface analyst the following week, without necessarily making any distinction between their activities. At the same time, many who work with XPS would place their activities exclusively in one of these categories, perhaps enjoying friendly rivalry with, or even mild indifference to, those working in the other fields. From here on, we largely consider XPS from the point of view of the *surface analyst*, but emphasise that the analyst must have knowledge of both surface science and atomic spectroscopy if they are to come to sound conclusions about their samples.

1.5 XPS as a Tool for Materials Analysis

One of the primary strengths of XPS lies in its ability to provide not only elemental analysis but also insight into the chemical states, bonding environments and electronic structure at the surface. These are often central questions in the field of materials chemistry and materials science. Specific examples of applications to current fields will be provided in Chapters 4–6, but the physical principles outlined above lead to a number of useful features of XPS for inorganic materials analysis in general:

1. *Direct information on the filled electronic states of a material*: perhaps more than any other method, XPS can give the direct measurements of electronic structure, although several caveats apply, and the physics of the photoemission process must be understood to a degree (see Chapter 2).
2. *Surface-sensitive elemental analysis*: XPS can identify and quantify elements heavier than helium in the outermost few nanometres of a sample surface.
3. *A high vacuum technique*: conventionally, XPS is carried out at very low pressure. The main reason for this is to allow photoelectrons emitted from the sample to reach the detector – a high pressure of gas would attenuate the electron signal too much. Some spectrometers permit higher pressure directly around the sample – this 'near ambient pressure XPS' is still restricted to specialist instruments.
4. *Chemical state determination*: XPS provides detailed information about the chemical environment of elements through shifts in binding energies. This allows researchers to distinguish between different oxidation states, chemical bonds, and coordination environments.

5. *Quantitative analysis*: the intensity of photoelectron peaks is proportional to the concentration of the corresponding elements, enabling semi-quantitative determination of surface composition without the need for matrix-matched standards.
6. *Depth profiling*: when combined with ion sputtering, or other ablation methods, XPS can analyse composition and chemistry as a function of depth, revealing layered structures and concentration gradients.
7. *Spatial mapping*: modern XPS instruments can map the lateral distribution of elements and chemical states across a surface, though with more limited spatial resolution compared to some other techniques.
8. *Non-destructive analysis:* XPS can be non-destructive, allowing samples to be analysed by other techniques after XPS analysis, although depth profiling with ion sputtering is destructive, and some materials are unstable to X-rays and vacuum.

1.6 Sample Suitability and Sample Mounting for XPS

The importance of surface analysis in modern science and technology is reflected in the availability of internationally agreed sample preparation methods described in the ISO 18116 (ASTM E1078) and ISO 18117 (ASTM E1829) standards. There are also several excellent articles available discussing common pitfalls when conducting and analysing photoelectron spectroscopy.[14,20,21] Samples for conventional XPS must be compatible with the spectrometer environment. For most spectrometers, this means compatibility with high or ultra-high vacuum, as well as X-ray and electron beam irradiation. Categories of sample which typically don't meet these requirements include molecular liquids and many molecular solids, on account of their high volatility. Samples of this type will evaporate during introduction to the spectrometer (potentially causing damaging contamination depending on the material). Likewise, samples that are contaminated with molecular liquids (*i.e.* wet samples) are also troublesome and are better dried before any XPS analysis is attempted. Specialist spectrometers can now operate at much higher pressure, and these allow samples to be measured that would be incompatible with conventional high vacuum, such as wet, volatile or biological samples.

Inorganic samples are often presented as monoliths, films, crystals or powders. Since most modern photoelectron spectrometers can focus on samples down to a few millimetres in size, individual crystals or film fragments can often be secured mechanically and studied directly. But for small crystals and in cases where the sample is a powder, there are really two main options: placing the sample in a recessed area of the sample plate or pressing the powder into an adhesive tape. The former has the benefit of minimising contamination from the support, but the sample is at the mercy of the inevitable vibrations as the sample plate is moved into the analyser and of gas currents as the chamber is evacuated and refilled. Magnetic materials can also be a significant risk with this approach since the use of electric or magnetic fields for focussing electrons is commonplace. Pressing samples into an adhesive tape has been the preferred method of sample mounting for many years, and as long as the analyst is alert to the possibility of contamination by the adhesive can be very successful. A modified version of the tape approach is to press the powder into a malleable metal; indium is commonly used although aluminium and copper foils are much cheaper and less likely to contaminate the sample. This also has the advantage of excellent electrical grounding and minimal contamination except from the press and on occasion from the supporting foil. Different approaches to mounting samples are compared in a recent paper by Davies and Morgan[21] and some more advanced aspects are discussed in a paper by Isaacs *et al.*[22]

1.7 Sample Cleaning

The surface sensitivity of XPS should be at the forefront of the analyst's mind when preparing a sample for study; cleanliness is paramount, a few monolayers of contamination can deposit on a sample remarkably easily and obscure most of, or all the information beneath it. Manipulation of the sample should be kept to a minimum, and where manipulation is absolutely necessary, tweezers, gloves and face masks will minimise the transfer of organic materials. Exposure to volatile substances, for example vapours from adhesives, alcohol or acetone cleaning solutions should also be avoided and in that theme, storage of the samples needs to be considered carefully; glass, metal and most plastic containers are relatively safe, but avoid plastic films in close proximity of the samples, as plasticisers diffuse readily from flexible films and can easily contaminate samples.

Despite best efforts, the surface of most samples will be contaminated to some extent by oxygen-containing organic matter. A number of procedures have been developed to remove this accumulated contamination. Heating the sample is the simplest approach, however whilst many photoelectron spectroscopy systems are equipped with heating elements, samples pressed into sticky tape cannot be heated. The spectroscopist must also ask the question of how heat might affect the sample itself. In some cases heating is used to restore surface order but in many, even though the bulk may be known to be stable, surfaces can be much more susceptible to thermal effects.

Argon ion bombardment (ion sputtering) is a common method of removing surface contamination; the kinetic energy delivered by the accelerated ions causing the desorption of unwanted (and occasionally wanted) surface species. However, the sputtering process will also affect the substrate to be studied, and in the case of inorganic materials in particular, the spectroscopist must be aware of the possibility of selective sputtering which can remove a higher percentage of one or more of the different types of ions at the surface, changing apparent surface stoichiometries. Thermal annealing can be used to try to redress the balance of surface composition after cleaning but, as mentioned above, is not always possible and can't be relied upon to recreate the original surface. An improved method that has been developed in recent years is the bombardment of surfaces with argon clusters rather than individual ions. This method, which was developed originally in association with secondary ion mass spectrometry, has been shown to be much less prone to selective ion etching, and even maintains the ratios of organic functional groups in sensitive biological samples.

1.8 Synergy with Other Techniques

While XPS alone provides extensive chemical and electronic information, its integration with complementary analytical methods significantly enhances understanding of inorganic materials. Here, we outline several techniques frequently combined with XPS, each bringing distinct yet overlapping insights.

X-ray diffraction (XRD) is perhaps the most complementary technique to XPS in the analysis of inorganic materials. XRD primarily provides long-range structural information, including crystal structure, lattice parameters, and phase identification, but lacks sensitivity to chemical states or surface-specific features. Coupling XRD with XPS allows for correlating structural data with surface chemical states, bridging bulk and surface properties.

X-ray absorption spectroscopy (XAS), including near-edge (XANES) and extended fine structure (EXAFS), complements XPS by probing local atomic structure and oxidation states around selected elements. Unlike XPS, XAS can examine both occupied and unoccupied electronic states and provides local structural detail, such as bond lengths and coordination environments. Combining XAS and XPS strengthens interpretation by providing direct correlations between local coordination environment and electronic structure derived from XPS.

Atomic force microscopy (AFM) provides high-resolution surface topography, morphology, and mechanical property information. AFM complements XPS's chemical specificity by adding spatially resolved physical characteristics, allowing correlation between surface chemistry and nanoscale morphology. This combination is particularly valuable for studying surface modification, thin films, and heterogeneities within composite materials.

Scanning electron microscopy (SEM) complements XPS by offering detailed imaging and morphological analysis at the micro to nanoscale. SEM provides visual evidence of surface structures, compositional contrasts through backscattered electron imaging, and elemental mapping when coupled with energy dispersive X-ray spectroscopy (EDS). Combining SEM and XPS allows for the direct correlation of elemental composition, chemical states, and surface morphology, especially useful in characterising heterogeneous or structured materials.

Inverse photoelectron spectroscopy (IPES) complements XPS by directly probing unoccupied electronic states, whereas XPS probes occupied states. Together, IPES and XPS can fully map the electronic structure around the Fermi level, giving comprehensive electronic band structure information critical in semiconductor, catalytic, and battery research.

Density functional theory (DFT) and other computational methods synergistically enhance XPS analysis by predicting spectral features, assisting in peak assignments, and elucidating underlying electronic structures. Computational simulations of XP spectra validate experimental observations, improve the reliability of chemical state assignments, and deepen the mechanistic understanding of observed spectral phenomena. Table 1.1 summarises the complementary strengths of these techniques relative to XPS.

Table 1.1 Comparison of techniques complementary to XPS.[a]

Technique	Primary information	Surface sensitivity	Complements XPS by
XRD	Crystal structure, phase identification, lattice parameters	Low (bulk)	Linking bulk structural data with surface chemistry
Elemental analysis	Chemical composition	Low (bulk)	Measuring accurately the bulk chemical composition, contrasting with that at the surface
XAS	Local atomic structure, oxidation state, coordination environment	Medium to low	Correlating bulk with surface oxidation states
AFM	Surface morphology, topography, mechanical properties	High (surface)	Providing spatial context to surface chemistry
SEM	Morphology, microstructure, elemental mapping (with EDS)	High to medium	Correlating morphology with surface chemistry and composition. EDS provides a way to measure more bulk composition
IPES	Unoccupied electronic states, conduction band structure	High (surface)	Completing electronic structure information (occupied and unoccupied states)
DFT	Electronic structure simulation, spectral prediction	Can model surfaces or bulk	Validating and interpreting experimental spectra and assignments

[a]The synergy of XPS with these techniques enables a comprehensive, multi-dimensional understanding of inorganic materials, crucial for both fundamental studies and technological applications.

1.9 The Challenges of Correctly Interpreting XP Spectra

The importance of the information that XPS can provide has led to widespread adoption across materials science, and as result of the ease of use of modern instruments, analysts are often less experienced in the technique than was the case 30 years ago, and this is becoming a cause for concern in the community.

Several factors complicate XPS data interpretation. Initially, experimental decisions regarding sample preparation, cleaning, and mounting significantly affect data quality. For instance, contamination or incorrect handling can introduce misleading signals. Moreover, the choice of instrument settings, such as pass energy, step size, and energy range for narrow scans, profoundly influences spectral resolution and accuracy. But more subtle issues also create pitfalls for the unwary. XP spectra often require peak fitting due to overlapping signals from different chemical states. Many software packages offer simple curve fitting methods, but errors frequently arise from incorrect peak fitting due to insufficient understanding of the underlying physics and chemistry. Selecting improper synthetic peak shapes, widths, or positions leads to inaccurate interpretations, despite the models themselves fitting the data very well.

Recent systematic assessments highlight the magnitude of erroneous interpretations in published XPS data. A review of approximately 400 articles across reputed journals found that roughly 30% contained errors severe enough to compromise the reliability of the conclusions.[28] Even more concerning, about 80% of peak-fitted spectra exhibited significant issues, reflecting widespread misunderstanding or misuse of fitting techniques. Errors ranged from minor oversights in experimental detail reporting to catastrophic misinterpretations of peak assignments and chemical states. Typically errors fall into one (or several) of the following categories:

1. *Peak overlap*: photoelectron peaks from different elements or chemical states can overlap, complicating identification and quantification. This is particularly challenging for materials with multiple chemical environments.
2. *Complex spectroscopic effects*: while some spectra are reasonably straightforward to interpret, certain elements and orbitals suffer from a variety of complicating physical effects during the photoemission process. These lead to highly complex spectra, the interpretation of which is extremely challenging.
3. *Reference data limitations*: interpretation often relies on comparison with reference spectra, but comprehensive databases for all possible chemical states and compounds do not exist, and those that do may not be applicable to every sample.
4. *Surface contamination*: the extreme surface sensitivity of XPS means that even minor contamination can significantly affect results.

5. *Quantification accuracy*: while XPS is considered quantitative, absolute accuracy is typically limited to ±10% without standards.
6. *Sample charging*: despite charge compensation techniques, analysis of highly insulating materials remains challenging and can cause uncertainty or even incorrect interpretation.
7. *Sample alteration*: the high vacuum environment and X-ray exposure can alter some materials, particularly those containing volatile components, organic compounds, easily reduced metals, or samples that normally exist in hydrated states.

The persistent occurrence of errors in published XPS data interpretation underscores the need for continuous training, rigorous reporting practices, and adherence to established standards. It is a sad fact that we cannot recommend that new users freely consult the general scientific literature for XPS knowledge, such is the level of incorrect published work. A good level of scepticism is required, but it can be hard for a beginner to ascertain the quality of a study they are reading. Seeking help from experienced colleagues or mentors is always desirable, as is reading of the many high quality accounts of common XPS pitfalls and suggestions for best practice which have appeared in recent years.[29–32] Enhanced awareness of common pitfalls, alongside targeted education and clearer reporting standards is vital for improving data quality across disciplines relying on XPS and is essential for maintaining the credibility and effectiveness of XPS as a scientific tool.

References

1. *Solid State Chemistry Techniques*, ed. A. K. Cheetham and P. Day, Clarendon Press, Oxford, 1987.
2. C. Kalha, N. K. Fernando, P. Bhatt, F. O. L. Johansson, A. Lindblad, H. Rensmo, L. Z. Medina, R. Lindblad, S. Siol, L. P. H. Jeurgens, C. Cancellieri, K. Rossnagel, K. Medjanik, G. Schönhense, M. Simon, A. X. Gray, S. Nemšák, P. Lömker, C. Schlueter and A. Regoutz, *J. Phys.: Condens. Matter*, 2021, **33**, 233001.
3. J. H. D. Eland, *Photoelectron Spectroscopy*, Butterworth-Heinemann, London, 2nd edn, 1984, pp. 1–29.
4. J. W. Kim and A. Kim, *Curr. Appl. Phys.*, 2021, **31**, 52.
5. J. E. Whitten, *Appl. Surf. Sci. Adv.*, 2023, **13**, 100384.
6. S. Krause, M. B. Casu, A. Schöll and E. Umbach, *New J. Phys.*, 2008, **10**, 085001.
7. P. Lenard, *Ann. Phys.*, 1902, **313**, 149.
8. B. R. Wheaton, *Hist. Stud. Phys. Sci.*, 1978, **9**, 299.
9. A. Einstein, *Ann. Phys.*, 1905, **322**, 132.
10. P. D. Innes, *Proc. R. Soc. London, Ser. A*, 1907, **79**, 442.
11. R. G. Egdell and E. Bruton, *Philos. Trans. R. Soc., A*, 2020, **378**, 1.
12. H. Robinson and W. F. Rawlinson, *London Edinburgh Dublin Philos. Magaz. J. Sci.*, 1914, **28**, 277.

13. H. Robinson and E. Rutherford, *Proc. R. Soc. London, Ser. A*, 1923, **104**, 455.
14. R. G. Steinhardt, Jr. and E. J. Serfass, *Anal. Chem.*, 1953, **25**, 697.
15. K. Siegbahn and K. Edvarson, *Nucl. Phys.*, 1956, **1**, 137.
16. S. Kai, G. Ulrik, S. Hans and O. Erik, *Phys. Scr.*, 1970, **1**, 272.
17. J. W. Pinder, B. Kulbacki, D. R. Baer, M. Biesinger, J. Castle, D. G. Castner, C. D. Easton, J. T. Grant, G. Greczynski, S. L. Harmer, A. E. Hughes, M. Isaacs, L. Kövér, G. H. Major, D. J. Morgan, C. J. Powell, P. M. A. Sherwood, W. M. Skinner, K. Stowers, J. Terry and M. R. Linford, *Surf. Interface Anal.*, 2025, **57**, 368–377.
18. K. Siegbahn, Atomic, molecular and solid state structure studied by means of electron spectroscopy, in *ESCA*, ed. K. Siegbahn, [and Others], Etc. [With Illustrations.], 1967.
19. K. Siegbahn, C. Nordling, G. Johansson, J. Hedman, P. F. Hedén, K. Hamrin, U. Gelius, T. Bergmark, L. O. Werme and R. Manne, *ESCA: Applied to Free Molecules*, North-Holland Publishing Company, 1969.
20. M. A. Kelly, *J. Chem. Educ.*, 2004, **81**, 1726.
21. P. R. Davies and D. J. Morgan, *J. Vac. Sci. Technol., A*, 2020, **38**, 033204.
22. M. A. Isaacs, J. Davies-Jones, P. R. Davies, S. Guan, R. Lee, D. J. Morgan and R. Palgrave, *Mater. Chem. Front.*, 2021, **5**, 7931–7963.
23. C. R. Brundle and B. V. Crist, *J. Vac. Sci. Technol., A*, 2020, **38**, 041001.
24. D. R. Baer and P. M. A. Sherwood, *Front. Anal. Sci.*, 2025, **4**, 1.
25. J. G. Jenkin, R. C. G. Leckey and J. Liesegang, *J. Electron Spectrosc. Relat. Phenom.*, 1977, **12**, 1.
26. H. P. Bonzel and C. Kleint, *Prog. Surf. Sci.*, 1995, **49**, 107.
27. F. Reinert and S. Hüfner, *New J. Phys.*, 2005, **7**, 97.
28. G. H. Major, T. G. Avval, B. Moeini, G. Pinto, D. Shah, V. Jain, V. Carver, W. Skinner, T. R. Gengenbach, C. D. Easton, A. Herrera-Gomez, T. S. Nunney, D. R. Baer and M. R. Linford, *J. Vac. Sci. Technol., A*, 2020, **38**, 100534.
29. J. W. Pinder, G. H. Major, D. R. Baer, J. Terry, J. E. Whitten, J. Čechal, J. D. Crossman, A. J. Lizarbe, S. Jafari, C. D. Easton, J. Baltrusaitis, M. A. van Spronsen and M. R. Linford, *Appl. Surf. Sci. Adv.*, 2024, **19**, 100534.
30. D. R. Baer, G. E. McGuire, K. Artyushkova, C. D. Easton, M. H. Engelhard and A. G. Shard, *J. Vac. Sci. Technol., A*, 2021, **39**, 021601.
31. G. H. Major, N. Fairley, P. M. A. Sherwood, M. R. Linford, J. Terry, V. Fernandez and K. Artyushkova, *J. Vac. Sci. Technol., A*, 2020, **38**, 061203.
32. G. Greczynski and L. Hultman, *J. Appl. Phys.*, 2022, **132**, 011101.

CHAPTER 2

Spectral Interpretation and Data Analysis

ROBERT G. PALGRAVE[a,b*], PHILIP R. DAVIES[b,c] AND DAVID J. MORGAN[b,c*]

[a]Department of Chemistry, University College London, 20 Gordon St. London, WC1H 0AJ, UK; [b]HarwellXPS, EPSRC's National Research Facility in PES, R92 Research Complex at Harwell, Rutherford Appleton Laboratory, Harwell Campus, Didcot, OX11 0FA, UK; [c]Cardiff Catalysis Institute, School of Chemistry, Cardiff University, Maindy Road, Cardiff, CF24 4HQ, UK
*E-mails: r.palgrave@ucl.ac.uk, morgandj3@cardiff.ac.uk

2.1 Introduction

In the previous chapter, we introduced the history and basic physical principles of photoemission. In this chapter, we turn to the spectra themselves, and explore how physical effects translate into the line shapes, binding energies, and chemical shifts that we observe. X-ray photoelectron spectra (XP spectra) of condensed phases contain a wealth of information on elemental ratios, chemical states and their depth distributions. Interpreting XP spectra is a challenging task, requiring the analyst to not only understand the physical processes of atomic spectroscopy and the manifestation of these effects in spectra, but in many cases to also fit the data with synthetic line shapes and backgrounds to reach a meaningful physical and chemical interpretation.

Inorganic Materials Series No. 19
Photoelectron Spectroscopy in Inorganic Materials Science
Edited by Robert G. Palgrave and Philip R. Davies

Published by the Royal Society of Chemistry, www.rsc.org

This chapter aims to introduce the physical processes of photoemission, explain how they give rise to features in the spectrum, and use this as a basis for successful data analysis.

2.2 The Anatomy of X-ray Photoelectron Spectra

The wide scan, or survey spectrum, of a sodium chloride (NaCl) single crystal is shown in Figure 2.1, and highlights the typical form of an XP spectrum. The sharp, intense features labelled with the names of atomic orbitals (1s, 2s, 2p *etc.*) are known as *core lines*, because they arise from photoemission from core atomic orbitals. Auger–Meitner peaks (labelled Na KLL and Cl LMM, and historically called simply Auger peaks) arise from a decay process of the excited state that includes emission of another electron. The background (highlighted in red) arises mostly from inelastic emission processes. The contribution of photoemission from the valence orbitals, just below zero on the binding energy scale, is almost too small to see at this scale.

To understand and interpret the features of a spectrum, we need to have a physical model of photoemission, and so our first task is to build up such a physical model to the point where it can usefully be used for interpretation of spectra.

2.3 The Photoemission Process

2.3.1 The Single Particle Approximation of Photoemission

A suitable place to begin is with the *single particle approximation*. The single particle in question is the photoelectron: the electron ejected from an atom (known as the emitting atom) by absorption of the energy of one incoming photon. We can explain some basic features of an XP spectrum quite successfully by only considering the energy change that the photoelectron undergoes, and we will initially proceed as proposed by Einstein in his 1905 paper: '*Let us assume that when a light quantum is absorbed, its energy is transferred to a single electron*.'[1]

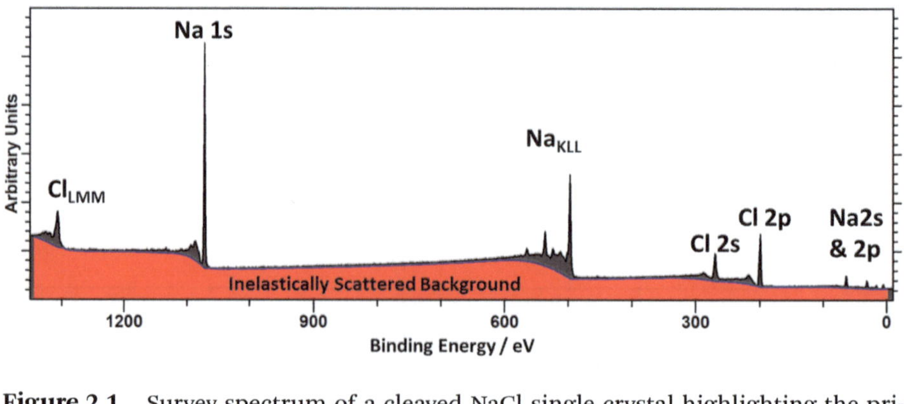

Figure 2.1 Survey spectrum of a cleaved NaCl single crystal highlighting the primary features of photoelectron spectra.

In this single particle approximation, for photoemission to occur the energy of the incoming photon must be sufficient to remove the electron from the atom to which it is bound, then any excess energy is used to give the now free electron (the photoelectron) velocity relative to the emitting atom, *i.e.* impart kinetic energy. We can write a simple equation to describe this:

$$h\nu = E_{\mathrm{K}} + E_{\mathrm{ion}} \tag{2.1}$$

In this equation, $h\nu$ is the photon energy, E_{K} is the kinetic energy of the photoelectron, and E_{ion} is the ionisation energy, the energy needed to move the electron from its bound state in the atom to a free state in the vacuum outside the surface of the sample. This equation is sufficient for photoemission from gas phase species, but for reasons we will discuss below, in analysis of condensed samples the ionisation energy is usefully divided into two parts. The binding energy (abbreviated BE, or written E_{B}) can be defined as the energy difference between the bound electron and the Fermi level of the material. The workfunction (abbreviated WF or given the symbol Φ) is the energy needed to raise the electron from the Fermi level to the vacuum level. Expressing E_{ion} as the sum of the binding energy and the workfunction is important for establishing the Fermi level as the reference point on the XPS energy scale (to be discussed shortly), and by dividing the ionisation energy into these two parts we arrive at the famous equation for XPS:

$$h\nu = E_{\mathrm{K}} + E_{\mathrm{B}} + \Phi_{\mathrm{spec}} \tag{2.2}$$

This equation is represented graphically in Figure 2.2, where the energy levels of the sample and the spectrometer are shown. Note that the common level between the sample and spectrometer is the Fermi level (E_{F}), which acts as the origin of the binding energy scale in XPS. The reason for this is that in experiment, the Fermi level can be aligned between the sample and spectrometer by placing them in electrical contact. Given sufficient electrical conductivity of the sample, the Fermi levels will align by flow of electrical charge, quickly reaching equilibrium. The implications of the electrical conductivity not being sufficiently high are discussed at length below, but for now we will assume the Fermi levels are aligned as shown in Figure 2.2.

A notable feature of the left-hand side of Figure 2.2 is that if the workfunctions of the sample and spectrometer are different, as they often are, then the vacuum levels are not equal between the spectrometer and sample. This means that a photoelectron leaving the sample with a workfunction larger (smaller) than the spectrometer will gain (lose) kinetic energy as it passes from the sample to spectrometer. In other words, there is an electric field present between the sample and spectrometer if the workfunctions are not equal. The result is that when the kinetic energy of the electron is measured *in the spectrometer*, it is influenced only by the workfunction of the spectrometer, not by that of the sample. The workfunction term in eqn (2.2) is therefore that of the spectrometer, which is a constant known for each instrument. This has the fortunate consequence that the WF of the sample does not need to be known in order to carry out XPS analysis.

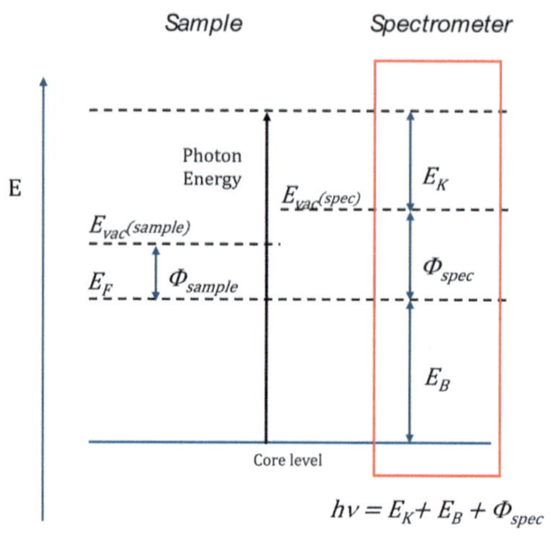

Figure 2.2 Two models of photoemission. Left: the single particle approximation. The electron is removed from the core level of a sample in electronic equilibrium with the spectrometer, signified by the Fermi levels (E_F) being equal. The measured kinetic energy (E_K) at the spectrometer can be related to the workfunction of the spectrometer (Φ_{spec}) and the binding energy (E_B) (red box) to give a graphical derivation of eqn (2.2). Right: the many body model. The energy of the initial N electron state (E_N^i) is separated from that of the final N electron state (E_N^f) by the photon energy. The ($N-1$) state refers to the sample excluding the photoelectron, and this can be calculated by subtracting E_K and Φ_{spec} from the photon energy as shown.

2.3.1.1 *Understanding Binding Energies*

When the workfunction is treated as a constant of the spectrometer, and the kinetic energy measured, we can obtain the binding energy (BE) of the photoelectron. We will now develop a theoretical framework to understand the binding energy and what factors affect it.

In the single particle approximation, the binding energy is a measure of the potential felt by the bound electron before photoemission, which must be overcome to remove the electron from the atom. By far the most important contributor to the potential of a bound core electron is the nuclear charge of its own atom. In multi-electron atoms, electrons screen each other and so each feels only a fraction of the full nuclear charge; this fraction is known as the effective nuclear charge (Z_{eff}). Considering only the nuclear charge of the emitting atom and the screening by the other electrons and their interaction through Coulomb's Law, the binding energies should then be proportional to the square of Z_{eff}. This is the trend found across the periodic table, as shown in Figure 2.3, where the binding energies of 1s, 2s and 3s orbitals are plotted against Z_{eff} calculated by Slater's method: all follow closely the quadratic trend expected.

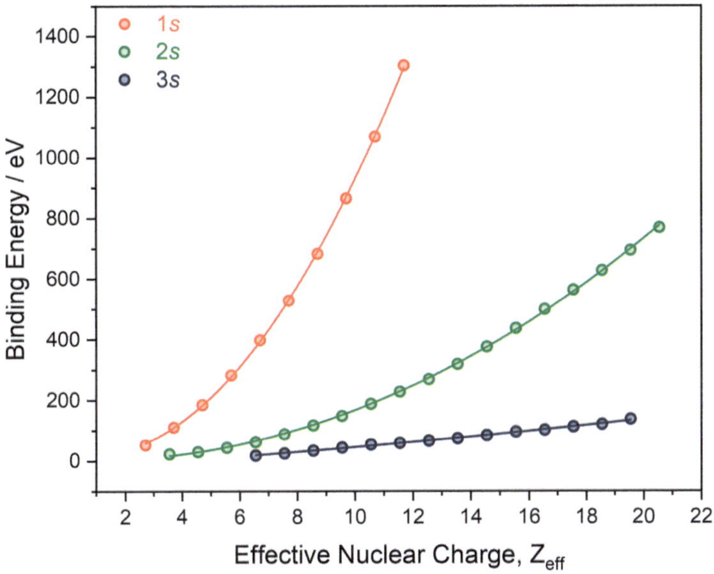

Figure 2.3 Variation in experimental binding energy of 1s, 2s, and 3s orbitals with effective nuclear charge, Z_{eff}, calculated using Slater's method. Lines are quadratic best fits that pass the origin.

2.3.1.2 Chemical Shifts

The effective nuclear charge can explain the broad trends in BE for orbitals across the periodic table, but it is not the only contributor to the potential felt by a bound electron. When core line BEs are examined more closely, it is found that they vary by as much as 12 eV for the same element in different compounds. For example, nitrogen might be found in a nitrate (NO_3^-) environment or a nitride (N^{3-}) environment. The former has a N 1s core line with BE around 408 eV, and the latter around 397 eV.[2,3] This 11 eV difference between nitride and nitrate means that XPS can be used to distinguish between nitrogen in these two forms. Other chemical environments of nitrogen, for example organic nitrogen, ammonium or nitrite groups, will have N 1s core lines that fall somewhere between these two extremes. This is the chemical shift, and leads to one of the most important capabilities of XPS: the ability to distinguish and quantify different chemical environments of the same element. Nitrogen has one of the larger ranges of chemical shifts, but most elements show at least small changes in core line binding energy as their chemical environment changes.

We have established that the binding energy is determined by electrostatic potential. Extending this idea, the XPS chemical shift should also be linked to some small changes in the electrostatic potential felt by core electrons. This seems to make sense when we consider the C 1s spectrum of polyethylene terephthalate (PET, Figure 2.4). We see the spectrum of the carbon core electrons is comprised of a series of three sharp peaks at distinct binding

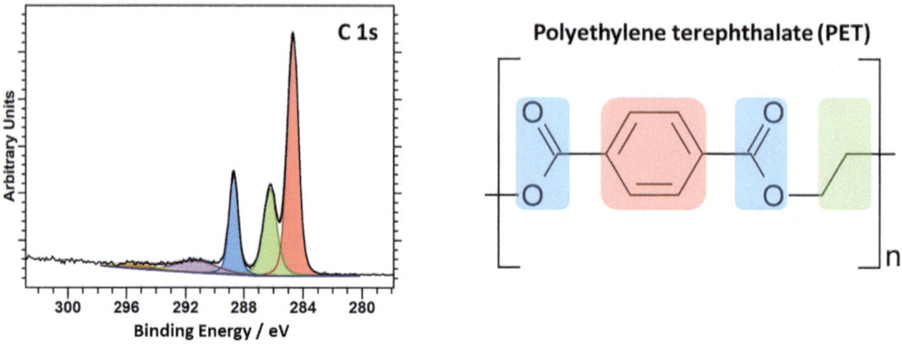

Figure 2.4 C 1s spectrum for the polymer polyethylene terephthalate (PET), with its structure shown on the right. The three distinct chemical environments in the polymer are colour coded to match the spectrum. The origin of the additional peaks in the spectrum is described in Section 2.3.2.3 and Figure 2.6.

energies, relating to the three different chemical environments shown. The observed peak positions can be related to the chemical functionality, which is in turn related to the electronegativity of the atoms to which the carbon is bound. Intuitively we can imagine that the increased electronegativity of the oxygen atoms polarises the C–O bond, resulting in a partial positive charge on the carbon and therefore to eject the C 1s electron more energy is required. We can also see the area of the three peaks looks very close to the expected atomic ratio of the chemical environments – quantification will be covered in later sections. Closer examination of the spectrum in Figure 2.4 reveals some additional broad, small features at higher binding energy (shaded purple) that don't obviously relate to a chemical environment. This is a clue to the need for a more advanced model that we will return to later.

The first attempt to turn intuition of chemical shift into a quantitative understanding was the *charge potential model*. This model is based on changes to the potential felt by core electrons as the atom gains or loses valence electrons as part of bonding. We can represent the valence charge, q, as distributed on a sphere of radius r, approximating the size of valence orbitals. If the core orbitals are completely inside this sphere, they will all experience an equal classical potential of q/r due to the valence charge. The value of q might be the formal oxidation state in ionic compounds,[4] or a partial charge based on the polarity of the bond in more covalent systems.

The potential created by the valence charge will shift the BE of the core electron by making it easier (for negative potential) or harder (for positive potential) to remove the electron from the atom. In the left-hand side of Figure 2.2, this amounts to slightly raising or lowering the energy of the core orbital relative to the Fermi level.

The charge potential model makes several additional predictions. Firstly, chemical shifts should depend linearly on the valence charge: positive valence charge should increase BE and *vice versa*. This prediction works well for some elements, for example nitrogen where the plot of BE against formal oxidation states from −3 to +5 is roughly linear. Even where the BE is not linear with formal oxidation state, in almost all cases the direction of the chemical shift is as expected, *i.e.* higher positive formal oxidation states show higher core line BEs. But in some elements in some compounds, we see an anomalous chemical shift – the more positive oxidation states actually have lower binding energy: examples such as Ag and Ba are explored later, but the existence of these anomalous chemical shifts indicates the limitations of the charge potential model as we currently state it.

The second prediction is that the chemical shift should be inversely proportional to the atom size, due to the $1/r$ dependence. A good illustration of this effect can be found with the halogens – the change in BE of the Cl 2p core line from NaCl (−1 formal charge on Cl) to $NaClO_4$ (+7 formal charge on Cl) is +10.5 eV. In contrast the same change between NaI and $NaIO_4$ is only +5.6 eV. We can understand the smaller chemical shift shown by iodine as being a result of its larger size. The group 2 oxides show a similar trend, this time in the cation chemical shift as its radius expands down the group.[5]

A final prediction from the charge potential model is that the size of the chemical shift should be the same for all core lines of a particular element, because the potential is constant inside the sphere of valence charge. Several elements show very close agreement, with less than 10% discrepancy, in the chemical shifts of their different core lines; some examples are Ge, As, Se and Br.[6] However, many elements show much larger deviations which cannot be explained by the charge potential model.

The mismatches with observation highlighted above point to some deficiencies of the model. Like any model challenged by disagreement with observation, we are able to add additional terms to rescue it. To build the model further, and gain better quantitative agreement with experiment, additional contributions to the electrostatic potential of the core electrons should be considered.

In a solid (or any condensed phase), we can also take into account charges on the surrounding atoms as well as the valence charge. The potential at a specific atomic position due to the surrounding charge in a solid is known as the Madelung potential. In a stable ionic compound, a cation, with its positive valence potential, will always be found surrounded by anions, so will experience a negative Madelung potential. Likewise an anion, surrounded by cations, will experience a positive Madelung potential. The same idea can equally apply to compounds that contain partial valence charges, rather than full ionic bonding. Elemental metals, on the other hand, do not have Madelung potential as all the atoms can be thought of as neutral. Thus, where present, the Madelung potential will always act in the opposite direction to the valence potential in shifting the BE of the core electron.[5] In almost all cases, the Madelung potential is smaller than the valence potential, and this means that BE values still shift to larger values with increasing positive valence charge, and to smaller values for increasing negative valence charge.

As mentioned previously, some rare examples of negative or anomalous chemical shift are known – this is where photoemission from a higher formal oxidation state in fact appears at lower binding energy. One example is in barium where in BaO the Ba 3d core line appears around 0.3 eV lower than Ba^0 in barium metal.[7] Other examples are found in Ag oxides and in some Cu compounds. A possible reason for this is already clear – if the Madelung potential is larger than the valence charge potential, then the normal direction of chemical shift is reversed. This is thought to be the explanation for the BaO example. Other causes of anomalous chemical shift will have to wait for introduction of the many body model, below.[8]

After including Madelung potential, valence potential, and the effective nuclear charge, we come to the end of what can be explained easily using the single particle approximation. While additional terms can be added to explain some of the features described in the next section, this becomes increasingly *ad hoc*. To move forwards from here we need to change our perspective on the photoemission process. But before doing so, we can pause and consider that, despite its simplicity, many of the features of XP spectra can be explained very well using the single particle approximation. The variation in BE of atomic orbitals across the periodic table based on effective nuclear charge, and chemical shifts based on valence charge, can be predicted, in some cases with very good quantitative agreement. The single particle approximation serves as a useful framework for understanding how factors such as changing bonding environments might affect core line energies, and this may be enough for a good proportion of the analytical problems in materials science that XPS is put towards solving. However, some important spectral features cannot be explained by the picture we have built so far, and these features are relevant for many important classes of materials. These will be tackled in the following sections.

2.3.2 The Many Body Model of Photoemission

Where the single particle approximation fails, we can attribute the failure to the assumption that all the photon's energy is transferred to a single electron. Making this assumption leaves us unable to say anything about the effect of photoionisation on the atom left behind and its surroundings. To move forwards, we need to think of the photoemission process in a different way. Instead of considering the energy change of the photoelectron alone, we now move to a *many body model* – so called because the energy of the photon can be distributed across many bodies (in practice, this almost always means many electrons). The foundation of this model is the idea that absorption of the photon causes a transition between an *initial state* of the system and a higher energy *final state*. Considering these two states allows us room to include the effects of photoemission not only on the emitted photoelectron itself, but on all other electrons in the sample as well. Hence, now we abandon Einstein's approximation, and doing so allows us to understand far better the fine detail of the features we find in XP spectra.

Let us define the initial state as the system of N electrons before photoemission. The energy of this N-electron initial state is represented as E_N^{i}. After absorbing a photon of energy $h\nu$, the system must naturally have increased its energy by this amount. The energy of the final N electron state energy is therefore just:

$$E_N^{\mathrm{f}} = E_N^{\mathrm{i}} + h\nu \tag{2.3}$$

Of course, the final N electron state includes the now free photoelectron as well as the remaining $(N - 1)$ electrons. Taking the photoelectron to be isolated from the $(N - 1)$ electron system, the portion of the N electron final state energy that is due to the photoelectron is the sum of its kinetic energy, E_{K}, and the workfunction, ϕ_{spec}, that had to be overcome for the electron to escape to the spectrometer. Therefore, the energy of the $(N - 1)$ final state is given by

$$E_{N-1}^{\mathrm{f}} = E_N^{\mathrm{f}} - E_{\mathrm{K}} - \phi_{\mathrm{spec}} \tag{2.4}$$

Substituting E_N^{f} from eqn (2.3) gives

$$E_{N-1}^{\mathrm{f}} = E_N^{\mathrm{i}} + h\nu - E_{\mathrm{K}} - \phi_{\mathrm{spec}} \tag{2.5}$$

Recalling the definition of binding energy from eqn (2.2):

$$E_{\mathrm{B}} = h\nu - E_{\mathrm{K}} - \phi_{\mathrm{spec}} \tag{2.6}$$

substituting this in gives:

$$E_{N-1}^{\mathrm{f}} = E_N^{\mathrm{i}} + E_{\mathrm{B}} \tag{2.7}$$

$$E_{\mathrm{B}} = E_{N-1}^{\mathrm{f}} - E_N^{\mathrm{i}} \tag{2.8}$$

Thus, the binding energy can be thought of as the energy difference between the N electron initial state and the $(N - 1)$ electron final state. This is illustrated graphically in Figure 2.2. Note that in some texts the signs are reversed *i.e.* $E_{\mathrm{B}} = E_N^{\mathrm{i}} - E_{N-1}^{\mathrm{f}}$, but here we prefer the E_{B} to be positive (as it is always plotted on XPS plots) and the energies of the initial and final states to be negative.

An important implication of eqn (2.8), illustrated in Figure 2.5, is that when reached from the same initial state, more stable final states lead to *lower binding* energy peaks, while less stable final states lead to *higher binding* energy peaks. Keeping this in mind is very helpful for understanding how the various final states we will go on to discuss appear in spectra. The diagrammatic spectrum in Figure 2.5 shows the result of a photoemission process with two possible final states. This gives us two peaks; in the example shown these are resolved (meaning separated sufficiently in binding energy to observe each of their maxima) but that isn't always the case in practice.

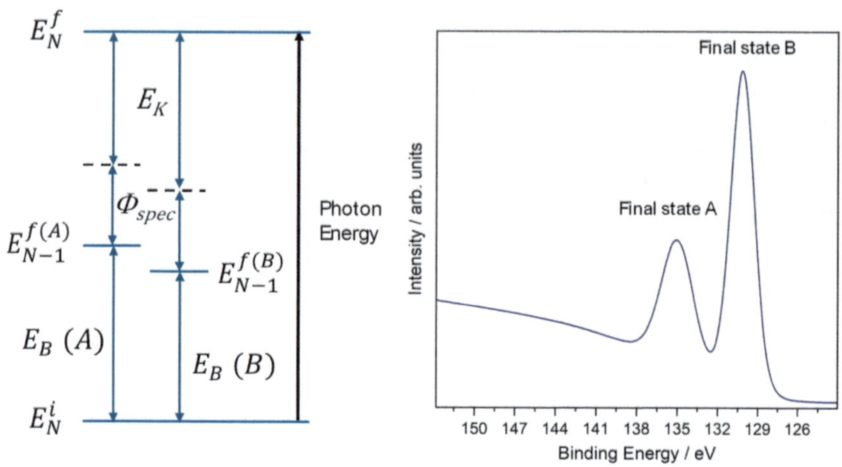

Figure 2.5 Left: energy level diagram to show the effect of different final state energies on the binding energy. Final state A, with energy $E_{N-1}^{f(A)}$, is less stable than final state B, which has energy $E_{N-1}^{f(B)}$, so the binding energy that results from transition to final state A is larger. Right: diagrammatic spectrum showing possible appearance of the two final states A and B.

The relative intensities of the peaks (meaning their area) correspond to the relative probability of the system arriving at each of the two final states, *i.e.* the transition probability from the initial state. Figure 2.5 also illustrates another important point, which is that photoemission can very often lead to multiple final states of different energy, which will be observed in the spectrum at different binding energies if they have sufficient intensity. This means that an observed high-binding-energy shoulder might not represent a distinct chemical environment as we might have assumed in the single particle approximation, but might instead be a less stable final state.

2.3.2.1 Choice of N-Electron System

An important question, which is often overlooked, is what exactly do we mean by the '*N-electron initial state*', or in other words, what is the *N*-electron system we are considering? Our choice is important, as the *N*-electron system we choose defines the size of the system on which we are able to model the effect of photoemission. Furthermore, to use eqn (2.8) quantitatively, we need a single value for the energy of both the initial and each final state, so their precise definition is important if we ever wish to calculate the binding energy in this way, for example if we wished to do so though *ab initio* calculations of the energy of our system. But more generally, our choice of *N*-electron system influences how we understand the causes of binding energy changes in XP spectra.

The phrases '*initial state effect*' and '*final state effect*' are commonly used to describe different influences on the observed binding energy. Initial state effects can be thought of as any effect that alters E_N^{i}, and final state effects are those that alter E_{N-1}^{f}. While this seems straightforward enough at first glance, whether a particular effect is 'initial state' or 'final state' is dependent entirely on the definition of the *N*-electron system. Disagreements on attribution of a particular spectral feature to initial or final states are usually in fact disagreements on the definition of the *N*-electron system. How then can we choose our *N*-electron system in a useful way?

We already encountered the minimal possible choice for the *N*-electron system – that is the single particle approximation. We could say this is the case where $N = 1$, *i.e.* no electrons other than the photoelectron are considered in understanding the binding energy. The initial state, E_1^{i}, is then then energy of that one electron in the core orbital, and once we exclude the photoelectron, the $(N - 1)$ final state actually contains no electrons, and so doesn't contribute at all to the binding energy. In this choice of system, discussed in Section 2.3.1, there are no final state effects, only initial state ones.

The maximal useful choice for the *N*-electron system is the entire sample under analysis. The initial state energy is then a single value encompassing the energy of all electrons in all the atoms of the sample. Using our example of the XP spectrum from the polymer PET (Figure 2.4), recall that we saw three different binding energies which previously we assigned to different chemical environments of the carbon atoms. By choosing the *N*-electron system as the whole sample, there is only one single value of E_N^{i} that includes all the electrons in all the atoms – there is no different initial state energy for the aromatic carbon and ester carbon. In this view the different binding energies we observe must therefore arise from differences in E_{N-1}^{f}. The argument works as follows: when we localise a hole in the C 1s orbital of an aromatic carbon, this is a more stable final state than if the C 1s hole were created on the more positively charged ester carbon. More stable final states appear at lower binding energies, so we see the aromatic carbon peak at *ca.* 285 eV and the ester carbon peak at *ca.* 288 eV binding energy. In this interpretation, the only cause of binding energy shifts is the different final states reached by localising the core hole on orbitals on different atoms, and the intensity of the peaks comes from the probability of localising the hole on the different carbons, which in turn comes from the ratio of the number of each type of carbon present. In other words, for this maximal choice of *N* electron system, there are no initial state effects (beyond the choice of sample itself), only final state effects.

Many XPS practitioners (especially experimentalists) actually prefer a kind of middle way between the maximal and minimal definitions of the *N*-electron system. While the minimal system is too simplistic, many find the maximal *N*-electron system unattractive: to many scientists it seems intuitive that the different carbon chemical environments in PET *must* exist before photoemission is carried out, after all, the stated aim of virtually all XPS analysis studies is to discover something about the atoms in the sample before photoemission – *i.e.* to study the initial state. Eliminating chemical environments from the initial state entirely therefore seems unnatural in a sense.

Although not normally articulated this way, in practice many analysts choose the N-electron system as the emitting atom itself. With this choice we can say that the initial state is different for the various carbon atoms in PET, and this allows us to talk about the chemical environment as an initial state effect, and factors which involve rearrangement of the other electrons in the emitting atom, or neighbouring atoms, (to be discussed below) as final state effects.

The choice of N-electron system will be mentioned again when discussing quantitative *ab initio* calculation methods, but for now it is sufficient to note that our choice really only affects how we categorise contributions to the binding energy as initial or final state effects, and while this is a diverting philosophical question, it isn't of any great importance to most analytical studies. We now move forwards to describe what many body effects exist, how they appear in spectra and their physical basis.

2.3.2.2 Many Body Effects

Thinking back to the single particle approximation, we can recognise that we had assumed that during photoemission, the remaining electrons and orbitals of the emitting atom and its neighbours were 'frozen' *i.e.* didn't change in energy or configuration in response to photoemission. The *frozen orbital approximation* is that the final state electrons are exactly the same as the initial state ones, only with one missing core electron.

The important perspective shift represented in eqn (2.8) allows us to see that the orbitals don't have to be frozen, and so the binding energy of core lines can be influenced by the $(N-1)$ final state left behind after photoemission. But we can use the hypothetical frozen orbital final state as a benchmark to categorise the changes that we see in real final states. Doing so we can identify three types of many body effects (often called final state effects, but see discussion in the previous Section 2.3.2.1):

Relaxation is any effect that produces a lower energy, more stable final state compared with the frozen orbital final state. Relaxation causes a shift in the XP peak to lower binding energy (see Figure 2.5).

Shake is any effect that produces a higher energy, less stable final state compared with the frozen orbital final state. Shake causes a shift in the XP peak to higher binding energy.

Splitting is any effect that splits the energy of the final state into two or more levels due to the presence of unpaired electrons. This is driven by either spin orbit coupling (between electron spin and angular momentum) or the exchange interaction (between two or more electron spins).

As already mentioned, photoemission does not always lead to only one final state – in many cases the ejection of the core electron will lead to significant populations of several different final states, and this distribution will lead to the appearance of different peaks in the photoemission spectrum.

2.3.2.3 Shake up and Shake off Processes

Shake, as defined above, is the name for the process where photoemission leads to a less stable final state compared with the frozen orbital final state. This normally happens where one or more of the valence electrons of the emitting atom has been excited to a higher level (known as *shake up*), but can also occur when a valence electron is ejected from the atom in addition to the core electron (known as *shake off*). We will look first at a shake up process that is often seen in aromatic compounds where an electron in the highest occupied molecular orbital (HOMO), which is a π orbital, can be 'shaken up' to the lowest occupied molecular orbital (LUMO), a π^* orbital, during photoemission from the C 1s core orbital.

Now there are two final states available for the photoemission of the C 1s electron. The $(N-1)$ final state with $1s^1\pi^{*1}$ configuration is less stable than the $(N-1)$ final state with $1s^1\pi^{*0}$ configuration. Relating this to Figure 2.5, final state A, having energy $E_{N-1}^{f(A)}$ is less stable, and for this example corresponds to the π^{*1} state. Final state B, having energy $E_{N-1}^{f(B)}$ is more stable and corresponds to the π^{*0} state. It is clear from Figure 2.5 that the binding energy of final state A is greater than final state B. Thus we should see two peaks in our XP spectrum, the higher BE one representing the less stable final state, in this case the one with the electron in the π^* orbital.

This is exactly what is observed, for example in the C 1s spectrum of polystyrene, shown in Figure 2.6a, where two peaks are seen with around 6.7 eV separation, while in comparison, the spectrum from non-aromatic polypropylene has no such additional feature. The energy difference between the two peaks in the spectrum represents the difference in energy between the π^{*0} and π^{*1} final states. It is worth pointing out that this does not necessarily imply that the HOMO–LUMO energy gap in the initial state is 6.7 eV (although this is an inference often made), as there could be other differences between the two final states – for example, the missing C 1s electron may affect the nuclear charge felt by π and π^* electrons differently, so their energy gap may not be the same as in the initial state. It is clear from the spectrum in Figure 2.6a that the intensity of the higher BE peak is much lower. This is because the probability of the transition from the initial state to the final state with the π^* electron is much smaller than to the final state without.

In fact, a shake up peak was present in the PET spectrum shown earlier; this is shaded purple in Figure 2.4 and arises from the aromatic ring present in the polymer, in the same fashion as discussed for polystyrene.

Where the shake up is between molecular orbitals with well-defined energy, the shake up peaks are distinct and sharp, as in the case of polystyrene. In materials with highly delocalised orbitals, these excitations are typically broader and less prominent due to the continuous distribution of

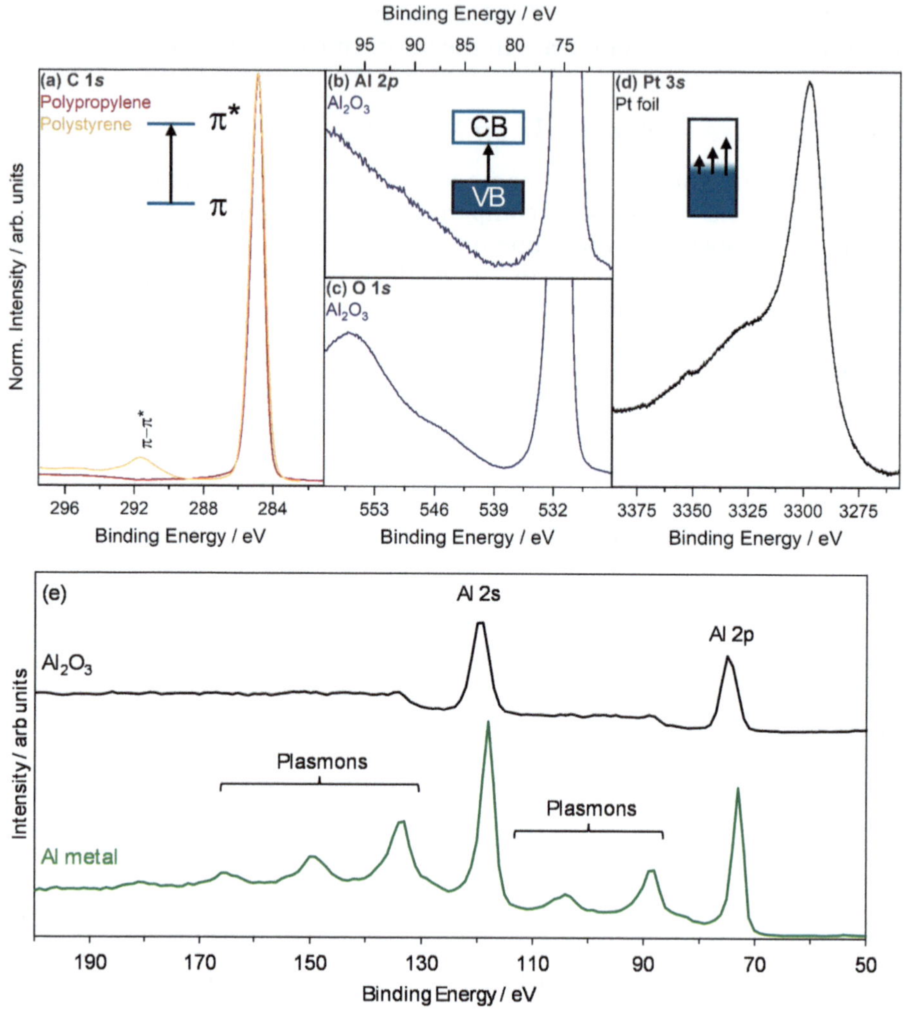

Figure 2.6 Shake up and plasmon processes in samples with different electronic structures. (a) C 1s spectrum of polystyrene (PS) and polypropylene (PP). PS contains an aromatic ring and exhibits a shake up feature between molecular orbitals around 6.7 eV higher binding energy than the main peak. PP has no aromatic groups and has no shake up feature. (b) and (c) Al 2p and O 1s spectra of a cleaned Al_2O_3 single crystal surface. Both core lines show shake up due to transitions between valence and conduction band, which have a minimum energy onset corresponding to the band gap. (d) Pt 3s spectrum from clean metallic Pt foil. As the conduction band is part filled, there is no onset energy to the shake up features, leading to an asymmetric peak shape, which is common for metals. (e) Wider scans of Al 2p and Al 2s core lines from Al_2O_3 and Al metal. The metal shows intense plasmon loss peaks up to 60 eV above the core lines.

available states. In semiconductors, shake-up processes involve interband transitions; an electron can be excited from the valence band into the conduction band during photoemission, creating final states with electron–hole pairs in the valence shells. The example of Al_2O_3, with a band gap around 7 eV, is shown in Figure 2.6b and c. In the absence of defects, the band gap imposes a minimum energy threshold for shake-up transitions in a semiconductor. The shake up final states appear as an onset on the high binding energy side of the main peak, the beginning of the onset corresponding to the minimum excitation energy required to promote an electron across the band gap. For narrow band gap semiconductors, the edge may not be resolved from the main peak.

In metals, the partially filled conduction band provides a continuum of states for valence electrons to be excited into, meaning that shake-up effects in metals are closely linked to the density of states just above the Fermi level. Instead of a distinct shake up peak or edge, the excitation of electrons to empty conduction band states often causes a tailing on the high binding energy side of a core line. An example is shown in Figure 2.6d.

2.3.2.4 Intrinsic Plasmon Loss

In metallic systems or materials with free-electron-like behaviour, plasmon excitations are an important many body effect. When a photoelectron is ejected, it can lose energy by exciting collective oscillations of the conduction electrons, known as plasmons. The final states therefore include one or more quanta of plasma excitation. Being higher energy, less stable than the frozen orbital final states, we classify these final states in our scheme in Section 2.3.2.2 as resulting from a type of *shake* process, although not all literature would classify them this way. The resulting peaks in the XP spectrum, called *plasmon loss peaks*, appear at fixed energy intervals above (higher binding energy than) the main photoemission peak. An example is shown for Al metal in Figure 2.6. The energy separation of these peaks from the main peak corresponds to the plasmon excitation energy of the material. By measuring this energy, properties such as the free electron density and dielectric function of the material can be obtained, if other materials properties are known. Intrinsic plasmon loss occurs as part of the photoemission process, and is contrasted with extrinsic plasmon loss, which occurs as the atom moves through the material and is discussed later.

2.3.2.5 Spin Orbit Splitting

The first, and most important, type of splitting we encounter is spin orbital splitting, which is observed in any core line from an orbital that has an angular momentum quantum number, l, greater than zero, *i.e.* every type of orbital except for an s orbital. The final state after photoionisation will have an unpaired core electron, which can couple with the angular momentum of the core orbital to give two different final states with different total angular momentum, J. Traditionally two approaches have been used to understand spin orbit contributions to determining the energy of an electron configuration: *L–S coupling* (also known as Russell–Saunders coupling) and *j–j coupling*. In L–S coupling, the electron–electron interaction is assumed to be stronger than the spin orbit coupling, so we calculate the total angular momentum and total spin of all the electrons, then couple these totals to get the spin orbit interactions. In j–j coupling, the situation is reversed, and we calculate a spin–orbit coupling for each electron individually, then add them together to get a total spin orbit value.

Many XPS texts recommend using one or the other scheme, but if we only are interested in the total angular momentum, J (which is the case for most XPS-based applications of this theory), it doesn't matter which we use, they both give the same answer. Because it facilitates later discussions on multiplet splitting, we will use L–S coupling terminology here.

2.3.2.5.1 Worked Example: Au 4f. We can take an example of emission from the 4f core orbital in Au. In the ground state, the 4f orbital, being a core orbital, is filled and all electrons are paired. For a filled subshell every allowed m_ℓ state is occupied with two opposite spins, so the contributions to the orbital angular momentum cancel. The total orbital angular momentum quantum number, L, is therefore zero, Likewise, the total spin, S, is the sum of all the spins of the 4f electrons, and for a filled orbital this too is always zero. With $L = S = 0$, the term symbol for these and any filled core states (ignoring valence electrons) can be written as ^{1}S.

When one electron is removed from the Au 4f orbital, we are left with a $4f^{13}$ core configuration. All the electron spins cancel out except one, so the total spin is $S = ½$. Since we have one missing electron from a full configuration ($4f^{13}$), we will have the same result as if we had only one electron in the orbital ($4f^1$) which greatly simplifies the calculation, and it can immediately be seen that $L = 3$ because an f electron has an orbital angular momentum quantum number $\ell = 3$; for a single f electron (or equivalently a single f hole) the total orbital angular momentum of the shell is therefore $L = \ell = 3$. Since an L value of 3 corresponds to an F term, we therefore have a ^{2}F final state term symbol for the core orbitals.

The ^{2}F term has two possible values of the spin orbit coupling, J. When L and S are coupled, the possible values of the total angular momentum quantum number J run from $L + S$ to $|L - S|$ in steps of 1. Recalling that $S = ½$, we arrive at the two possible values of J, which are $J = {}^7/_2, {}^5/_2$. For a more than half filled orbital (and $4f^{13}$ is more than half filled), the higher J value has lower energy, so the lower energy final state is $^2F_{7/2}$ and the higher energy final state is $^2F_{5/2}$. As noted above, a more energetically stable final state leads to a lower binding energy, therefore we see the $J = {}^7/_2$ state at lower binding energy than the $J = {}^5/_2$ state. These two final states are known as the two *spin branches*.

The relative intensities of the spin orbit doublet are given by the degeneracies of the two final states. The degeneracy of a spin state is $2J + 1$, therefore calculating the degeneracies for $J = {}^7/_2, {}^5/_2$ shows that two parts of the 4f core line should have the ratio 4 : 3. The two parts of the spectrum are conventionally labelled $4f_{7/2}$ and $4f_{5/2}$, note the lower-case f (Figure 2.7).

The separation between the two J states is the spin orbit splitting α, and is dependent on the nuclear charge, Z, the principal quantum number, n, and the angular momentum quantum number, l, according to eqn (2.9):[9]

$$a \propto \frac{Z^4}{n^3\left(l+\frac{1}{2}\right)(l+1)} \tag{2.9}$$

From eqn (2.9) we can see that orbitals with larger values of Z, lower values of n, and lower values of l, will lead to larger splitting. The observed splitting for some Au orbitals, in comparison with some Bi orbitals, is shown in Table 2.1. Some examples of core lines from the same orbital in different elements with very different splitting energies are also shown in Figure 2.7.

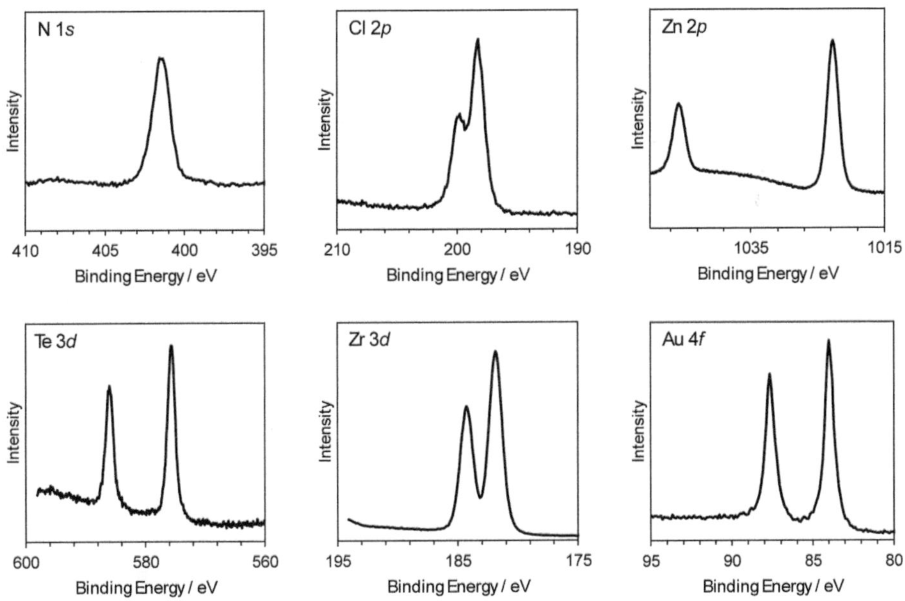

Figure 2.7 Examples of core lines showing different spin orbit splitting. N 1s shows no spin orbit splitting, and appears as a single peak. Core lines from p, d, f orbitals appear as doublets; their intensity ratio depends only on the angular momentum quantum number, but the splitting also depends on Z and n (eqn (2.9)). Cl 2p and Zn 2p doublets have a 2 : 1 intensity ratio. Te 3d and Zr 3d doublets have a 3 : 2 intensity ratio, and the Au 4f doublet has a 4 : 3 intensity ratio.

2.3.2.6 Spin Orbit Coupling Across the Periodic Table

In the example above, the *L–S* coupling scheme can explain the observed spin orbit coupling in Au 4f. In general, the same features are reproduced for all p, d, and f core orbitals. Core line photoemission will always lead to one missing electron in a core orbital, so following the arguments made for Au 4f, there will always be two final states with *J* values of $L+\frac{1}{2}$ and $L-\frac{1}{2}$. We call these pairs of final states *spin branches*. The values of *L* will just be the angular momentum values of the orbital in question (p = 1, d = 2, f = 3). The spectra from p, d, and f orbitals will therefore always consist of a high spin, higher intensity component at low binding energy, and a lower intensity, low spin component at higher binding energy. The ratio of the intensities of the spin branches, given by the ratio of $2J+1$ values for each is characteristic of the type of orbital, and the expected values are shown in Table 2.2. In practice, the observed intensity ratios are very close to the expected values from this table.

Because spin orbit final states arise from different core electron configurations, spin orbit splitting can be seen to manifest almost completely uniformly across different compounds of the same element. The splitting energy is virtually unaffected by the valence orbitals and hence by any bonding phenomena. The splitting in the Au 4f core line is the same in all compounds of Au, whether the elemental metal, $AuCl_3$, an atomic gas of Au or any other state, and this is true for all elements. Note that the separation between two spin branches may *appear* to vary with chemical environment in some spectra, but this is not because the underlying spin orbit spitting is

Table 2.1 Experimental spin orbit splitting constants for various Au and Bi orbitals.

	Spin orbit splitting/eV	
Orbital	Au ($Z=79$)	Bi ($Z=83$)
4p	96.5	126.6
4d	18.1	23.6
4f	3.7	5.3
5p	17.0	26.3

Table 2.2 Comparison of spin orbit splitting in core lines observed in XP spectra.

Core orbital	Angular momentum of singly ionised core orbital	Possible *J*-values	Intensity ratio
s	0	$^1/_2$	—
p	1	$^1/_2$ and $^3/_2$	1:2
d	2	$^3/_2$ and $^5/_2$	2:3
f	3	$^5/_2$ and $^7/_2$	3:4

changing, but rather that other effects are influencing the two spin branches differently, and these are discussed in the next section.

2.3.2.7 Multiplet Splitting

In the previous discussion of spin orbit coupling, we have dealt with the case where the only unpaired electron in the system is that created by the photoemission from the core orbital. We can now consider what would happen if the valence shell in the initial state also contains unpaired electrons. The final state in such a system would contain unpaired electrons in both the core and valence shell, and these can couple to produce several final states with differing energy. This is known as multiplet splitting and is most prevalent in the transition metals and lanthanides, where compounds often have unpaired valence electrons in their ground state.

2.3.2.7.1 Multiplet Splitting of s Orbitals. The simplest manifestation of multiplet splitting occurs in core lines from s orbitals. As already established, because the angular momentum quantum number of the s orbital is zero, core lines from s orbitals do not show spin orbit splitting. However, when the valence orbitals have unpaired electrons, the spin of the valence orbital couples with the spin of the remaining core electron in the s orbital, leading to splitting of the final state and the appearance of a multiplet.[10]

Multiplet splitting is caused by the exchange interaction between the unpaired electron in the core orbital and those in the valence orbital – this makes final states where the core electron is spinning parallel or antiparallel to the valence electrons non-degenerate, and the high spin final state will be more stable.

A good initial example is the compound MnF_2, where the electron configuration of the Mn^{2+} in the ground state is high spin with 5 unpaired electrons. This leads to a total spin, S, of $^5/_2$, and a ground state term symbol of ^6S. Removal of one of the 3s electrons leads to a core level with spin of ½ and a core level term symbol of ^2S. Combining these two term symbols allows us to calculate the possible states of the final state. The only resulting S terms (*i.e.* with zero total angular momentum) – are a high-spin state ^7S term and low spin ^5S term. These can be thought of as the core electron spinning with or against the valence spins. The expected ratio of the multiplicity of these two states is 7 : 5, hence we expect to see a doublet due to the two final states; the high spin final state being at lower binding energy as it is the more stable of the two. A second example is FeF_2, which has one extra valence electron making it d^6 valence d electrons, leading to a pairing of two electrons, a total spin $S = 2$, and ground state with total angular momentum of 1. Here the initial state term symbol is ^5D, and the final states with the same total angular momentum that result from coupling to the $3s^1$ unpaired electron are ^6D and ^4D. These will have multiplicities of 6 : 4, and as a result we expect a different ratio of intensities in the Fe 3s orbital in FeF_2 compared with the Mn 3s orbital in MnF_2. Contrast this to spin orbit splitting where the intensity ratio for a given orbital is always the same, and already we can see that multiplet splitting is more complex than spin orbit coupling. But we still have some way to go.

The separation in energy of the multiplet final states is a measure of the exchange interaction between the core and valence unpaired electrons; this will be proportional to the number of unpaired electrons in the valence shell.[11] Exchange interaction between orbitals in the same principal quantum shell will also be much greater than between those of different shells.[12] So the 3s–3d exchange interaction discussed here is large. For the first row transition metal compounds for example, the observed splitting of the 3s is around 6 eV, far larger than the energy resolution of XPS, so these doublets are easily resolved. Careful measurement of the size of the splitting may reveal subtle changes in the exchange energy across a series of compounds, and this has found use in the study of magnetic materials.[13] Splitting of deeper s core levels (2s, 1s) will be smaller, due to the lower exchange interaction between valence and inner core electrons; this will also apply to deeper p and d states discussed in the next section.

2.3.2.7.2 Multiplet Splitting of p, d, f Orbitals. When the core orbital being ionised is not an s orbital, we will have to consider the angular momentum of the core electron, and how this couples with the spin and angular momentum of the valence electrons.[14] This is important for photoemission from p, d or f orbitals in transition metal or lanthanide elements. As an example, we can consider the 2p photoemission from an Fe^{2+} compound, having a $3d^6$ initial state with a ground state term symbol ^5D. When combined with the ^2P core level term that exists after removal of one 2p electron, we can produce the terms ^6F, ^6D, ^6P, ^4F, ^4D and ^4P. Additional terms can be produced by coupling excited state terms from the initial state with the core hole term. Because in all cases the angular momentum is non-zero, we must also consider spin orbit coupling in these terms. In total for the $3d^6$ configuration of Fe^{2+}, there are 70 final states that exist when spin orbit coupling is accounted for. Note, that multiplet splitting is the only phenomenon that can affect the different spin branches of a spin orbit doublet differently, as the terms that arise from the different *J* values in the core orbitals combining with the terms of the valence orbital are different. Kotani *et al.* highlighted this effect for Ti 2p orbitals, where the $2p_{3/2}$ and $2p_{1/2}$ are affected differently by multiplet splitting.[15] As mentioned above different values of the separation between spin branches are measured for different chemical states of the same atom. Briggs and Gibson measured apparent splitting of Co compounds and found they varied by up to 0.7 eV.[16] They explain that this is not because of any dependence of the spin orbit splitting on the chemical state (eqn (2.9) shows no such dependence), but rather the multiplet splitting acts differently on the different spin branches, and this latter effect is chemical state dependent. It's important to note that every other many body effect discussed beside multiplet splitting must be present in each of the spin orbit branches identically.

The number and energy of the final multiplet states have been determined for each d-electron configuration for the first-row transition metals, and the spectra calculated by considering the transition probabilities for each. These are known as the Gupta–Sen spectra, and form one basis for understanding multiplet splitting in the first-row transition metals.[17] Gupta–Sen spectra are, however, missing important contributions. As they are free ion calculations, there are many effects of coordination environment of the atom that are not accounted for. They do not take into account crystal field splitting of the d-orbitals, which will not be constant for different compounds and will change the number of unpaired electrons in the valence shell. For example, the form of the Fe^{2+} multiplet spectrum is strongly influenced by the ligands attached to Fe, and whether the ion is high or low spin: low spin Fe^{2+} has no unpaired valence electrons so does not show any mutiplet splitting.[18] Additionally, covalency will affect the energies and intensity of different final state peaks: the formal oxidation state of the transition metal ion may not be adequate in describing the valence shell occupancy where covalency is significant. Once these considerations are added in, the calculation of the expected multiplet splitting for a given compound becomes a daunting task. A good example of such an undertaking is the work of Bagus and coworkers on the Fe^{3+} 2p multiplet.[19,20]

The magnitude of multiplet splitting decreases very significantly in the 2nd and 3rd transition series, due to lower exchange interaction between the valence and core orbitals as *n* increases. Usually, multiplet splitting is seen in these orbitals as a slight asymmetry, if detectable at all.

The lanthanides and actinides provide another chance to observe multiplet effects.[20] Having variable 4f or 5f occupancy, the lanthanides (Ln) and actinides (An) respectively can produce a large number of possible final states, leading to complex spectra if unpaired electrons are present. There is considerable multiplet splitting of the Ln 4d and An 5d core orbitals as these are in the same shell as the unpaired valence electrons. Experimental and calculated spectra are in good agreement for many of these.[20]

In contrast the magnitude of the multiplet splitting for Ln and An is small for the 3d and 4d core electrons respectively, as they are in the shell below the valence electrons.[21] The effect of the small multiplet splitting is to slightly broaden Ln 3d and An 4d spectra, although note these core lines do undergo additional screening processes that cause some complexity which will be discussed in the next section.

2.3.2.8 Relaxation: Core Hole Screening

We have already seen how in shake processes, the valence electrons on the emitting atom can adopt higher energy configurations (*e.g.* populating π^* orbitals in aromatic compounds). In relaxation processes, the remaining electrons in the sample respond to the creation of a core hole in ways that *lower* the energy of the final state compared with the frozen orbital approximation. Siegbahn identified two forms of relaxation: *contraction* and *flow*.[22] Contraction is the reduction in radius of the emitting atom when it loses a core electron due to the greater effective nuclear charge now felt by the remaining electrons. Contraction is a general stabilising effect on all remaining electrons in the emitting atom and can be thought of as being roughly constant for a given element, no matter what its charge state or bonding arrangement. It will influence every spectrum but is not normally separately accounted for. Siegbahn's 'flow' (which is now an outdated term, now replaced by the terms *screening* and *charge transfer*, discussed below) is more impactful on spectral shape, as it represents movement of one or a small number of valence electrons between the emitting atom and neighbouring atoms.

Creation of a core hole leads to the stabilisation of empty valence electrons of the emitting atom through contraction, and as a result these may now fall close to or below the Fermi level in the material, making it possible for electrons to fill them.[23] When one or more electrons from a neighbouring atom, or from the delocalised free electrons in a metal, is transferred to the localised valence shell of the emitting atom, this is called *screening* (also known as *non-local screening* or *charge transfer*, to emphasise that the screening electrons have moved from a different atom). In a system where a screened and unscreened final state are possible, these will have different energy and so appear at different BE values in the XP spectrum. In some cases, there are many possible screened final states, which represent different amounts of electron transfer from the ligands to the emitting atom – the transition metals are a good place to find such instances.

A good example of screening can be seen in the compounds of Cu. Found in group 11, and commonly adopting the +1 or +2 oxidation states, Cu can form open shell d^9 ions (Cu^{2+}) or closed shell d^{10} ions (Cu^{+}). In the case of a Cu^{2+} ion, photoemission from the 2p core orbital, while freezing the other electrons, would leave a $2p^5 3d^9$ configuration in the final state. Screening occurs by moving an electron from a neighbouring atom onto the Cu valence orbitals, doing so creates a $2p^5 3d^{10}\underline{L}$ final state, where $\underline{L}$ represents a hole in the valence orbital of a neighbouring atom (L stands for ligand). This final state with a Cu $3d^{10}$ valence configuration is known as the 'well screened' state. Importantly, the $3d^9$ unscreened state should show multiplet splitting (due to the unpaired valence electron), while the screened $3d^{10}$ state will not. This makes the Cu^{2+} 2p spectrum a sensitive probe of the degree of screening. Figure 2.8 shows $2p_{3/2}$ spectra taken from CuF_2, $CuCl_2$, and $CuBr_2$.[24–26] Only the $2p_{3/2}$ spin branch is shown for simplicity. Each spectrum is split into two main regions, a low BE peak that varies in intensity and position across the three samples, and a higher BE feature that is reasonably constant in both intensity and position.

The low BE component is the screened peak; its lower BE is because the screened final state is more stable than the unscreened one. The energy of the screened state depends on the difficulty of the charge transfer from

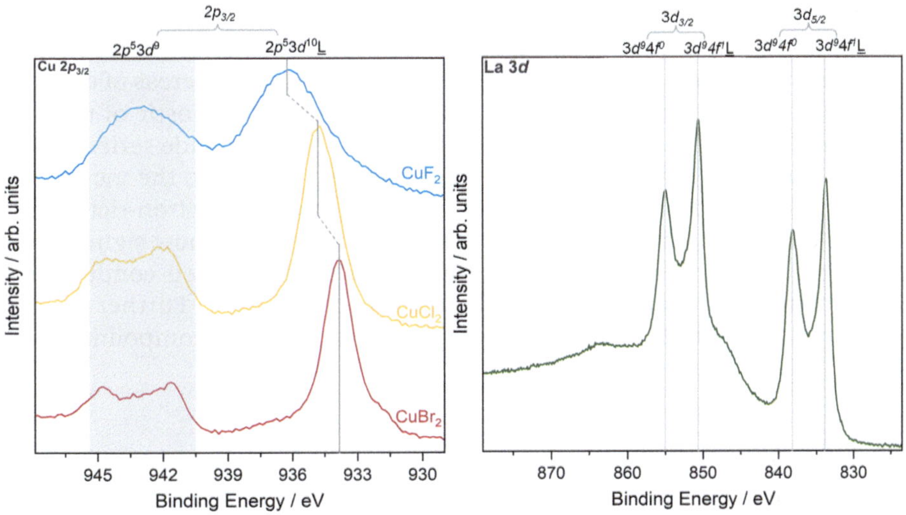

Figure 2.8 Left: Cu $2p_{3/2}$ spectra from the Cu^{2+} compounds CuF_2, $CuCl_2$, and $CuBr_2$ replotted from ref. 24–26. The low binding energy feature in all cases is the screened final state $2p^5 3d^{10}\underline{L}$. The screened state varies strongly with the identity of the halide due to the energy required to transfer an electron from the halide valence shell to the copper. It is also symmetrical as it does not suffer multiplet splitting, as the Cu has no unpaired valence electrons. The higher BE feature is the unscreened $2p^5 3d^9$ state, which only weakly depends on the halide, but is broadened by multiplet splitting. Right: La 3d spectrum from La_2O_3 showing the splitting of the La $3d_{5/2}$ and $3d_{3/2}$ peaks due to screened and unscreened final states.

halide to metal. The screened peak appears at 936.4 eV for CuF_2, compared with 934.9 eV for $CuCl_2$ and 934.1 eV for $CuBr_2$. The decreasing BE from F to Br, represents the increasing stability of the screened final state, due to easier charge transfer from halide to metal as the electronegativity decreases down the halogen group.

In contrast, the unscreened peaks, representing final states with Cu $3d^9$ valence configuration, are almost invariant in energy as the halide changes, showing the halide does not contribute significantly to the energy of these final states. They are also broad with complex shapes, characteristic of multiplet splitting expected for a $3d^9$ configuration.

Further important examples of core hole screening are found in the lanthanides. Having in some cases partially filled 4f orbitals, it is expected that the lanthanides will show multiplet splitting where unpaired valence electrons are present. Compounds of La^{3+}, with electron configuration $4f^0$, might be expected to show simple core line spectra, with only spin orbit splitting and no multiplet splitting. In fact, La compounds typically show a 'double doublet' in the 3d core line: one doublet produced by spin orbit splitting, with a large splitting of around 16.1 eV, and a second doublet which occurs due to an unscreened $3d^9 4f^0$ final state and a screened $3d^9 4f^1\underline{L}$ final state

(Figure 2.8). The screened final state is more stable and hence appears at lower binding energy; the separation between the screened and unscreened peak is around 4 eV but varies for different La compounds (whereas of course the spin orbit splitting is virtually constant), so can be diagnostic of the La compound present.[27] Similar effects occur across the lanthanide series, with final states consisting of electron transfer from the ligands to the metal for the early lanthanides, or the metal to the ligands for the electron-rich later lanthanides.[28] Metallic alloys containing Ln elements also show significant splitting due to screening, in this case caused by delocalised conduction band electrons that become localised on the emitting atom.[29] Further examples of relaxation and screening in specific transition metal compounds are given in Section 2.14.4.

2.3.2.9 Peak Width

The width of a core line peak, expressed as the full width at half maximum (FWHM), will depend on several contributing factors. These can be split into: lifetime broadening, vibrational broadening, instrumental broadening and differential charging. Lifetime broadening reflects the uncertainty in the energy of the final state that arises from the very short lifetime of the final state itself. As a highly energetic excited state, the core hole will decay rapidly, on the order of femtoseconds, either by emission of an X-ray, an electron, or some other process. The shorter the lifetime, the broader the XPS peak due to greater uncertainty. The lifetimes of many core holes have been tabulated by Campbell and Papp,[30] and are largely independent of chemical factors – the lifetime for a particular core line will be very similar in all compounds.

The timescale of photoemission is fast compared with atomic motion. The vibrational motion of atoms in the sample will lead to a range of different initial states with variable interatomic distances. This leads to broadening of the peak which is proportional to temperature.[31]

Instrumental broadening comes from the linewidth of the X-rays (the linewidth of monochromated Al Kα, the most commonly used X-ray source for lab XPS, is 0.25 eV)[32] and the resolution of the detector, which in modern instruments is less than 0.5 eV. This will be constant for all core lines measured on the spectrometer.

Lastly, differential charging can broaden peaks if the sample charges differently across the analysis area. This rarely occurs in metals but can be an issue for semiconductors and insulators. The topic of charging is discussed further below. When charging is an issue, the broadening will be sensitive to factors such as charge compensation (flood gun) settings, the sample roughness and the mounting of the sample.

The square of the total broadening is the sum of squares of each of the contributions described above and listed in Table 2.3. In modern instruments, the overall core line broadening is usually found somewhere in the region of 0.3 to 1.7 eV. The lower end of this range will be more likely found in clean

Table 2.3 Summary of peak broadening contributions in XPS. The shape refers to the distribution of binding energy, either L (Lorentzian) or G (Gaussian).

Broadening effect	Cause	Shape	Approximate magnitude
Lifetime	Finite lifetime of final state leads to uncertainty in energy	L	0.1–1 eV (depends on core-hole lifetime; smaller for lighter elements)
Vibrational	Motion of atoms leads to a range of initial and final states	G	0.01–0.3 eV (significant for samples with light elements)
Instrumental	Energy range of X-rays, pass energy, detector resolution	G	0.25–0.6 eV (with modern instruments)
Differential charging	Parts of sample charge differently	G	0–1 eV or more (depends on material and charging compensation)

metal foils, *e.g.* Au or Ag measured on synchrotron or high-performance laboratory spectrometers with settings optimised for high energy resolution. Values outside this range should be treated with some scepticism: a FWHM below 0.3 eV is usually not physically possible, and for very broad peaks consideration should be given that the broad peak feature may in fact consist of multiple chemical environments or overlapping peaks.

2.3.3 Summary of Intrinsic Photoemission

We have come to the end of the discussion of the energy changes that the photoelectron undergoes when it is removed from a core orbital. Collectively, these are known as intrinsic photoemission processes (or sometimes called primary photoemission processes), meaning they are fundamental to the physical process of ejection of an electron from an atom by a photon. The signal we detect from photoelectrons that reach the detector without undergoing any more energy loss processes is called the intrinsic signal. Other processes, known as *extrinsic processes*, can occur after the electron leaves the atom before it reaches the detector, but before discussing these it is worth taking stock of what has been established so far. Figure 2.9 shows a schematic of the interplay of some of the important intrinsic processes in photoemission, going from an initial state to the eventual spectrum.

We have so far viewed this process from bottom to top – envisaging the way the electrons of the sample respond to the incident photon. Of course, as analysts, all we ever see is the spectrum – the top line in Figure 2.9. It is not unusual that the physical cause of features in the spectrum is not yet fully understood at the time of data collection, hence there must be a language to describe spectral features without assigning a specific physical cause. Complex spectra, such as those arising from several many body effects combined, are often labelled with a 'main line', the most intense part of the spectrum,

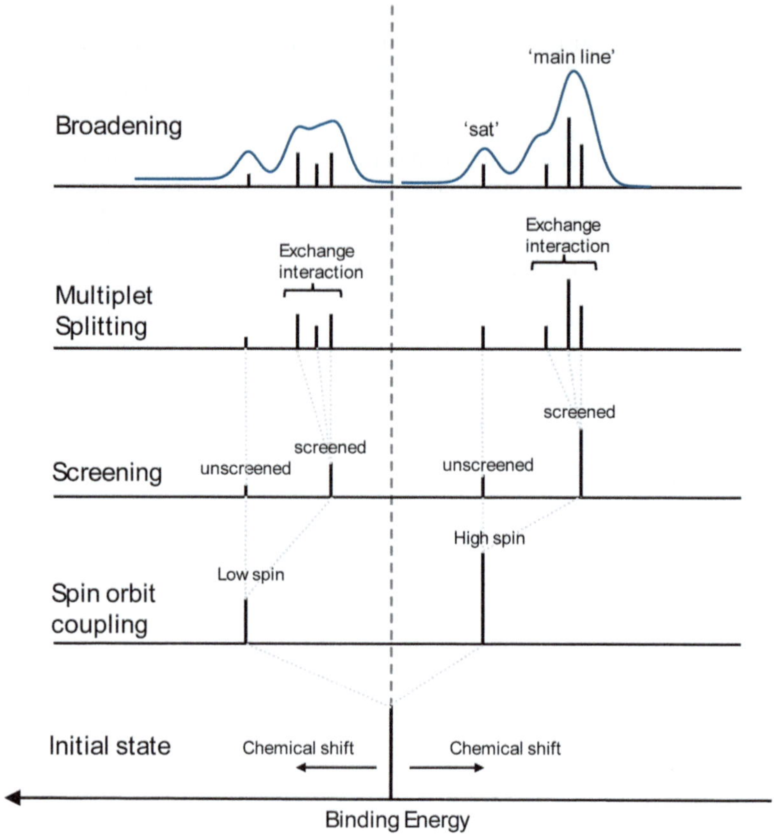

Figure 2.9 Schematic of possible intrinsic processes that can occur in photoemission. Not all processes will occur for every core line. Here we describe chemical shift as an initial state effect, but see discussion in Section 2.3.2.1. The order from bottom to top is for conceptual understanding, not to imply the processes happen in a sequence. In this example, a screened state undergoes multiplet splitting while the unscreened state does not – this is dependent on the valence electron configuration; here we show the possibility that differently screened states can undergo different multiplet splitting. The labelling 'main line' and 'sat' (for satellite) on the topmost spectrum indicates how peaks are often described in the literature. Not all processes are included in this example, for example, shake up is not included. The vertical dashed line can be thought of as the frozen orbital approximation.

and other resolved or semi-resolved features as 'satellites'. These terms are agnostic to the physical origin for the peaks (*e.g.* shake up, relaxation, splitting *etc.*), which might be discovered only later and after some effort.

A very common motif seen in XP spectra is that the feature at lowest binding energy, *i.e.* that associated with most stable final states, has the largest intensity. What this signifies is that the probability of transition from the initial state to the more stable final states is often higher. As can be seen in the schematic in Figure 2.9, it quite possible that the 'main line' or 'satellite' is in fact made up of several unresolved components, *e.g.* from multiplet splitting.

2.4 Extrinsic Processes

The photoelectron, once free of the atom, can be influenced by extrinsic processes as it makes its way to the detector. In the XPS of solids (or liquids), the vast majority of electrons that are emitted from their atoms will undergo one or more scattering events as they pass through the condensed phase (the sample). This scattering may be elastic, in which case the trajectory of the photoelectron changes but not its kinetic energy, or inelastic, in which case the photoelectron loses some kinetic energy.[33] Inelastically scattered electrons will have lower kinetic energy, and therefore higher binding energy, than those that are not scattered. Each photoemission peak in the spectrum will be accompanied by an intensity of extrinsic, inelastic emission to the higher binding energy side.

The intensity of this extrinsic portion relative to the intrinsic peak will depend on a great number of factors, including details of the larger scale sample structure; the spatial distribution of the emitting atoms is reflected in the ratio of elastic to inelastic emission. Separating the extrinsic from intrinsic signal is vital for correct identification of spectral features and quantification, which is discussed in the next section. Before that, we must introduce the different types of inelastic processes that may occur as the electron moves through the sample and methods that can help separate the intrinsic from extrinsic signal.

The most important interaction is inelastic collisions of the photoelectron with atoms.[34] Vicanek explored the frequency and energy of inelastic collision in a variety of homogeneous and inhomogeneous sample types.[35] Inelastic collisions tend to yield very broad loss features 10–100 eV wide or more. As well as inelastic collision with atoms, the photoelectron may lose energy in other ways. In a metal, it can excite plasmons either in the bulk or the surface. These are collective oscillations of free electrons, and the frequency depends on the electron density and dielectric constant.[36] These are different to the intrinsic plasmon loss processes described above. Another possibility is to excite interband transitions as the photoelectron passes through a semiconducting medium. These will have a minimum energy loss corresponding to the band gap energy of the semiconductor.

2.4.1 Reflectance Electron Energy Loss Spectroscopy (REELS)

Reflectance electron energy loss spectroscopy (REELS) is an effective tool for characterising some extrinsic processes. In this method, a beam of monoenergetic electrons, typically with a kinetic energy of 1 keV, impinges on the sample, and the kinetic energy distribution of the reflected beam is measured. REELS spectra consist of a large peak due to the elastically scattered electrons, and an inelastic tail consisting of electrons that have lost energy somehow.[37] The utility of the REELS spectrum is that it can be thought of as a version of the XP spectrum that only contains features due to inelastic scattering. REELS can be compared with the XP spectrum by aligning the elastic REELS peak and the core line energy, and considering features relative to them. If features appear in both the XP and REEL spectra, this is good evidence that they originate from extrinsic inelastic processes. Features in the XP spectrum that are not in REELS may well be intrinsic photoemission processes. More details are given in Chapter 3, As seen in Section 3.4.4.3.

2.5 Summary of Intrinsic and Extrinsic Processes in XPS

Table 2.4 shows a summary of the intrinsic and extrinsic processes that occur in XPS, affecting the initial state, final state and which can occur after photoemission before the photoelectron reaches the detector.

2.6 Peak Intensities and Quantification

In a general quantification procedure, we would like to calculate the atom or mass concentration of a species of interest, expressed as a percentage of all atoms, or all mass. In XPS we have no way to know the mass of the sample, or count all the atoms within the analysis volume. The compositional information we can obtain is through measuring the intensities of core lines from different elements, and comparing these with each other to yield relative atomic percentages.

Our ability to do this rests on the fact that the intensity of a core line is proportional to the concentration of the emitting atom in the sample. What's more, XPS can be seen with some justification as a 'standardless' quantification method, requiring no measurements of standard samples of known concentration as a prerequisite to analysis. This fact alone makes it very attractive as a method of chemical quantification. However, as we will see, there are many other factors aside from the concentration that contribute to the intensity of a core line in XPS. If we wish to reliably obtain the elemental composition from XPS peak intensities, it is important that we can account for all of the other contributions to intensity. We will begin with the dependence of intensity on depth, as this gives rise to another major feature of XPS: its surface sensitivity.

2.6.1 Information Depth and the Inelastic Background

XPS is surface sensitive, and this is because the measured intensity of the photoemitted electrons is much greater from emitting atoms on the surface than from bulk. Photoelectrons are strongly attenuated by passing through

Table 2.4 Summary of intrinsic and extrinsic processes in photoemission.

State	Process	Description	Manifestation	Principally observed
Initial state (but see discussion in 2.3.2.1)	Chemical shift	Initial state effect, changing the potential felt by core electrons due to valence electron charge. Counteracted by shift in Madelung potential.	Shift of core lines, usually increase in +ve charge causes shift to higher BE, but anomalous chemical shift is known in rare cases.	All core lines
Final state: splitting	Spin–orbit coupling	Spin of core electrons couple with angular momentum of core orbital to give two possible final state energies.	Peak split into doublet with well-defined separation and intensity ratios; doublet almost independent of chemical environment.	All p, d, f core line spectra
Final state: splitting	Multiplet splitting	Spin of core electrons couple with spin of valence electrons to give a large number of possible final states.	Complex peak shape, can be strongly dependent on chemical environment.	Any material with unpaired valence electrons. Strongest in 1st row transition metals, and in spectra of core orbitals of the same principal quantum shell as the valence
Final state: relaxation	Contraction	Core hole increases effective nuclear charge felt by all electrons in emitting atom.	Shift of all orbitals to lower BE	All atoms
Final state: relaxation	Non-local screening	Electron from neighbouring atom moves on to emitting atom, creating a more stable final state. Sometimes known as charge transfer.	Higher energy unscreened peak. Can combine with multiplet splitting *e.g.* in Cu^{2+}	Many elements, especially lanthanides and late transition metals.
Final state: shake	Shake up	Valence electron moves to higher orbital.	Higher BE shake up peak, usually low intensity.	Aromatic compounds, semiconductors, covalent materials.
Final state: shake	Shake off	Valence electron is emitted from atom.	Higher BE shake off peak, usually low intensity.	Can occur in any material but especially prevalent in molecules and adsorbates.[42]

(*Continued*)

Table 2.4 (*Continued*)

State	Process	Description	Manifestation	Principally observed
Intrinsic or extrinsic	Plasmon excitation	Excitation of free electrons leads to energy loss. Can be intrinsic or extrinsic.	Series of high binding energy satellites at regular intervals.	Materials with high concentration of free electrons.
Extrinsic	Inelastic scattering	Kinetic energy loss as photoelectron travels through the condensed phase.	Background at high binding energy side of each peak.	Ratio of intrinsic to extrinsic signal depends on depth distribution of emitting species.

condensed matter. Electrons emitted deep inside the sample have only a small chance of escaping the surface and reaching the detector, whereas those emitted on the surface have a much higher probability of doing so. The attenuation of the intrinsic signal is described by eqn (2.10), where the elastic signal intensity, $I(z)$, drops exponentially from the original value, I_0, as electrons travel a distance, z, through the condensed material.

$$I(z) = I_0 e^{-\frac{z}{\lambda \cos\theta}} \tag{2.10}$$

Here λ is the inelastic mean free path (IMFP) which is the average distance an electron travels between inelastic scattering events. The angle of emission, θ, is defined as zero for normal emission, meaning emission is perpendicular to the sample surface; this is the standard geometry for XP spectrometers, but some can record emission at different angles (Angle-Resolved XPS – ARXPS), as discussed below.

A common question from those studying XPS is, '*to what depth does XPS probe?*'. You may already see from eqn (2.10) that there is no simple answer to that question, as the signal intensity exponentially decreases with depth, with no natural cut-off. However, when a simple answer is nonetheless required, we can quote the *information depth*. The information depth is defined as '*the maximum depth, normal to the specimen surface, from which useful signal information is obtained*'.[33] 'Useful' information here is normally taken to mean the intrinsic signal (although that is not to say the extrinsic signal has no use), and a cut-off is the thickness of the surface layer from which 95% of the detected photoelectrons originate. The information depth – the depth above which 95% of the signal emerges – can be expressed using the terms in eqn (2.10), and is almost exactly $z = 3\lambda \cos\theta$. Emission at angles greater than zero (away from the surface normal) will have a lower information depth, and this forms the basis of the ARXPS technique.

An additional complication is that the IMFP of an electron is a function of the kinetic energy (KE) and so will be different for different core lines. The relationship between IMFP and KE is non-linear (Figure 2.10) which can be explained by changing photoelectron scattering mechanisms at different KE ranges.[38] At very low KE, the photoelectron doesn't have sufficient energy to trigger many scattering processes. For example, it will lack energy to induce plasmon oscillations in metals, or to promote bound electrons to higher energy states.

As the KE increases, the electron gains sufficient energy to efficiently excite plasmons and interband transitions (in semiconductors) and activate other scattering mechanisms. This increases the scattering rate and shortens the IMFP, reaching a minimum around a KE of 30–100 eV, making this the most surface-sensitive KE range. Above ~200 eV, the IMFP increases with increasing KE. In this regime, electrons behave increasingly like ballistic free particles, and the probability of inelastic scattering decreases because the cross-section for plasmon excitation and other inelastic processes becomes smaller. Additionally, at high kinetic energies (short electron wavelength),

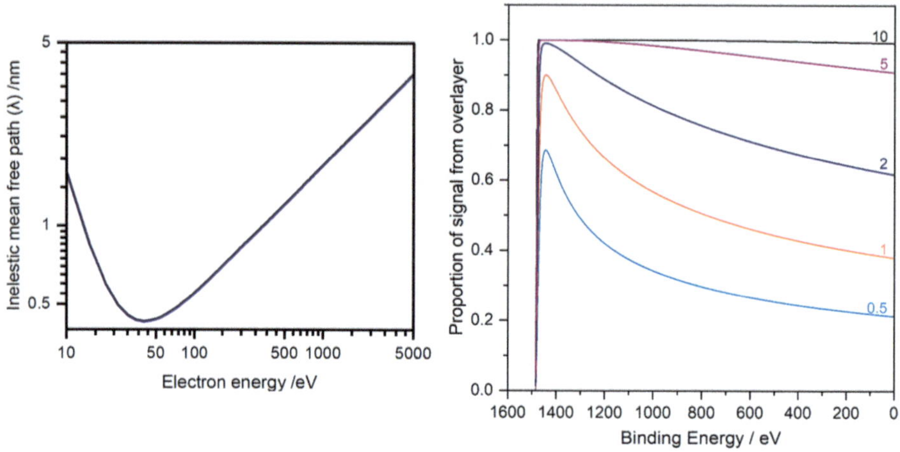

Figure 2.10 Left: the "Universal Curve", the inelastic mean free path of an electron in a solid empirically derived from fitting experimental observations on a large number of solids. Note the log–log scale. Right: the proportion of signal from an overlayer of the stated thickness in nm, as it varies with binding energy for an X-ray excitation energy of 1487 eV (Al Kα X-rays).

electrons are less sensitive to the fine details of the band structure, further reducing the overall scattering probability. The IMFP energy dependence curve (Figure 2.10) is often considered universal because, while IMFP values vary slightly between different materials, the overall shape of the curve remains consistent. This is because the primary scattering mechanisms at intermediate energies and reduced cross-sections at high energies act similarly in most solids.

Since core electrons from different elements have different binding energies, their corresponding photoelectrons will have different kinetic energies for the same X-ray energy. This means in an XPS experiment, core lines from different elements will have different information depths.

While the universal curve for IMFP is accurate enough for most applications, in fact the information depth does depend on the material type.[39] Differences in electronic structure and the corresponding dielectric response are the underlying cause, but taking into account the variation from material to material is rarely necessary for basic surface analysis, and in any case will be impossible if the dielectric properties of the sample are not known. Tougaard has written extensively on this topic and readers who wish to go deeper are directed to his works.[40]

Eqn (2.10) gives the dependence of intensity on the depth of the emitting atom, z. But the total intensity we measure for a core line will be proportional to the sum of the intensities from all depths where that element is present. To find this for a slab of defined thickness, we can take the definite integral of the right-hand side of eqn (2.10) with respect to depth, between limits that represent the depth range in which we find the element. Eqn (2.11) is an example of this approach, calculating the total intensity, I_A, for an element A that is present homogeneously at all depths in a semi-infinite sample (meaning that we integrate between zero and positive infinity). The solution in this case is that the intensity will simply be proportional to λ_A, the IMFP for photoelectrons with the kinetic energy corresponding to the core line.

$$I_A \propto \int_{\infty}^{0} e^{\left(-\frac{z}{\lambda_A \cos\theta}\right)} dz = \lambda_A \cos\theta \tag{2.11}$$

The practical result is that, even in a completely homogeneous sample, photoelectrons from orbitals with greater IMFP will be more likely to escape the sample from a given depth, and so the signal will be of higher intensity. For example, in a homogeneous Au:Cu alloy of 1:1 composition analysed by standard Al Kα radiation (E = 1467 eV), the signal from the Au 4f, with KE = 1403 eV, λ_{Au4f} = 3.5 nm, will be more intense than that of Cu 2p, with KE = 554 eV, λ_{Cu2p} = 1.6 nm. Therefore the intensity ratio will be $\lambda_{Au\,4f}$:λ_{Cu2p}, or approximately 68:32 in favour of Au due to the IMFP difference alone, without taking into account other factors that influence relative intensity. Luckily, this effect is easily included in the standard sensitivity factors, discussed in full in the next section, and so can be corrected for without issue for homogeneous samples.

However, we cannot so easily cancel out the effect of differing IMFP where the sample is not homogeneous. Figure 2.11 shows some example situations with the homogeneous CuAu alloy compared to samples where we have either a 0.5 nm Cu layer atop a Au substrate, or the reverse, 0.5 nm Au atop Cu.

To calculate the relative intensities from the overlayer and substrates, we can adapt eqn (2.11) by using limits representing the depth region each material is present in. The relative intensity from an overlayer of thickness t would be obtained by integrating between 0 and t, and that of the substrate would be obtained by integrating between t and infinity. This is represented graphically in Figure 2.11.

As can be seen in Figure 2.11, when we have a thin film of Cu on Au, the corrected intensity from the overlayer is 24% of the total signal, whereas in the reverse situation with an Au thin film, the corrected intensity from the overlayer is only 17%. The cause for this asymmetry can be seen graphically, the Au signal contribution persists at higher level, further into the sample, than the Cu signal does, because of its greater IMFP.

Figure 2.11 also illustrates the difficulties in assigning a 'composition' to a heterogeneous surface. For the homogeneous example in Figure 2.11, it is easy to assign 1:1 as the true Au:Cu ratio, but what is the 'true' composition of the heterogeneous, overlayer samples? Any answer must specify the spatial region that we are considering. We will return to this point in our general consideration of the accuracy of quantification in Section 2.6.4.

Au and Cu are chosen for this example because photoelectrons from their principal core lines (Au 4f and Cu 2p), are very different in KE, and hence have very different IMFPs, giving a large effect on the composition. A more

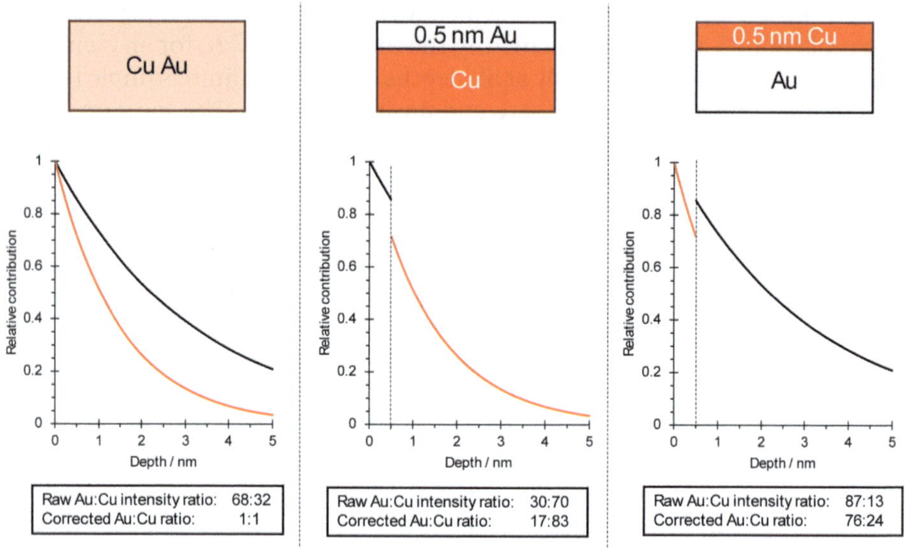

Figure 2.11 Three spatial arrangements of Cu and Au and the resulting contributions to the photoemission signal from different depths from Cu (orange) and Au (black). In the homogeneous arrangement (left), the ratio of intensities is corrected to the true value (1:1) by the relative sensitivity factors. For non-homogeneous samples (centre and right) the corrected intensity ratios differ despite the samples being structurally identical, with just the identity of the materials reversed.

practical example of overlayers which will be regularly encountered by the analyst is adventitious carbon, the organic material present on the surface of most samples exposed to atmosphere. A uniform layer of carbon, of 1 nm thickness, atop different metals, will give different intensity ratios due to differences in IMFP. For example, 1 nm of carbon on Cu will give a Cu : C intensity ratio of almost exactly 1 : 1 (when using C 1s and Cu 2p). While the same thickness of carbon on Au will give an almost 3 : 1 Au : C ratio (using C 1s and Au 4f). At first glance the Cu substrate may seem more dirty than the Au one, but in this example the difference is only due to IMFP differences.

As we have seen, even with a sample made of two planar slabs, the relative intensities can be non-trivial to interpret. Of course, in practice there are many other spatial arrangements that elements may take in our samples. In some cases, there is good prior information on the physical arrangement of different components of the sample, *e.g.* in metals there is likely a native oxide layer on top of a metallic phase, but very often we do not have good knowledge of what the near-surface arrangement is.

Tougaard has shown that many different arrangements of two elements can give rise to the same intensity of intrinsic signal.[34] One way to distinguish such arrangements is by examining the inelastic background, which will be discussed below. An alternative is through angle-resolved XPS (ARXPS) which involves varying the angle of emission, θ, relative to the sample surface that photoemission is measured at, thereby altering the effective information depth (eqn (2.10)). ARXPS helps in identifying non-uniform surface layers, such as thin films with a gradient composition or islands of one material on another. The approach is to build a proposed model of the spatial arrangement, and calculate the angular dependence of each element by adapting eqn (2.11). Different models can then be compared against experimentally observed angular dependence of concentration.

2.6.2 Photoionisation Cross Sections

The probability that a photon with a given energy, E, will be absorbed by a particular atomic orbital and lead to photoemission is described by the photoionisation cross section, $\sigma(E)$. These cross sections vary strongly with both photon energy and the electronic shell involved. For XPS at Al Kα energies, the cross sections for different orbitals can differ by orders of magnitude. This is one of the main reasons why certain core lines dominate the spectrum, and why some elements or particular orbitals, such as Li 1s, are much harder to detect under standard laboratory excitation.

The cross section for an orbital is relatively low for photon energies just above the binding energy of that orbital (representing the ionisation threshold for that orbital), and then increases sharply with increasing photon energy, reaching a maximum of 50–100 eV above the binding energy.[50] From this point, increasing photon energy further leads to a decrease in cross section. Examples are shown in Figure 2.12 for the 4f orbitals of Eu and Au, which have binding energies of around 2 eV and 84 eV respectively. It can be seen that the ratio of the cross sections of the two elements changes dramatically over the photon energy range, and this will be true for all core lines.

2.6.3 Relative Sensitivity Factors

We have already identified two main factors that contribute to the peak intensity: the IMFP and photoionisation cross section. There are several other factors to add to these: the X-ray flux, the angle over which the photoelectrons are collected, the transmission of the electrons from the sample to the detector, and the sample area over which electrons are collected.[43] Lastly, of course, the concentration of the element present will affect the core line intensity. Usually, this concentration is the quantity we wish to discover, and so we must account for the influence of all the other factors listed in this paragraph.

In practice, this is done by using Relative Sensitivity Factors (RSFs). RSFs collect all of the terms that contribute to the intensity, apart from the concentration, into one factor. To see how these are derived and used, we can first consider eqn (2.12), containing all of the factors listed in the previous paragraph that contribute to the intrinsic intensity of a core line:

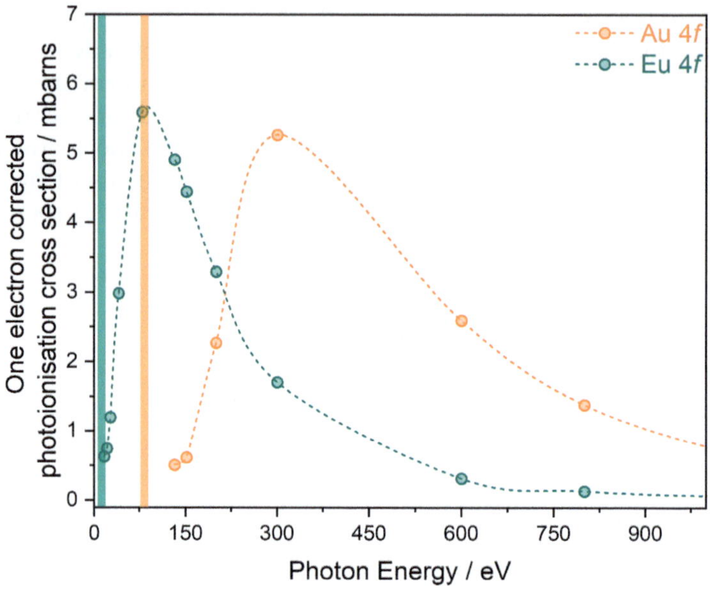

Figure 2.12 Variation in photoionisation cross sections of Au 4f and Eu 4f with photon energy. The vertical lines indicate the binding energies of the core lines (Eu 4f at 13 eV, Au 4f at 83 eV).

$$I = A \times f \times T \times n \times \varphi \times \sigma \times \lambda \tag{2.12}$$

where, I = integrated photoelectron peak area, f = flux of X-rays at the collection area (photons cm^{-2}), n = the concentration of the atom being investigated (atoms cm^{-3}), σ = the photo-ionisation cross section for the orbital of interest, A = collection area for photoelectrons (defined by the spectrometer lens system or the illuminated area), T = the transmission function of the spectrometer which describes the efficiency of which electrons with kinetic energy E are transmitted through the spectrometer detection system, φ = an angular distribution function which depends on the element's atomic number and both the angle between the X-ray source and analyser and the asymmetry parameter of the orbital from which the electron is ejected, λ = the inelastic mean free path (IMFP).

Some terms are instrument specific (A, f, T and φ), and some terms are intrinsic to a given element (n and σ), whereas we have to take λ as the value for homogeneous distribution of elements, which as discussed above is not always the case.

Considering the ratio of the concentrations of two atoms (n_A and n_B), the stoichiometry is given by eqn (2.13):

$$\frac{n_A}{n_B} = \frac{I_A}{\sigma_A \varphi_A T_A \lambda_A} \div \frac{I_B}{\sigma_B \varphi_B T_B \lambda_B} \tag{2.13}$$

In eqn (2.13) we have lost the dependence of A and f since these are assumed constant for a given spectrometer operating mode and so cancel. For the other terms in the denominators in eqn (2.13), we have already noted the that φ has a dependency on the orbital angular momentum quantum number, then if a user references peak intensities for s orbitals only, as the angular momenta are equal, the ratio of φ_A/φ_B reduces to unity. To avoid the need for angular corrections for non-s orbitals, some systems are designed to operate at a so-called 'magic angle' of 54.7° between source and analyser which reduces this ratio to unity for all orbitals.[44,45] Modern spectrometers have well-characterised transmission functions (T) and are generally corrected either by the data acquisition system or by analysis software, so require little or no user input, whilst as discussed, λ has a kinetic energy dependence.[46]

Applying these simplifications, we can collect the remaining terms into a relative sensitivity factor (RSF), s_A:

$$s_A = \sigma_A \lambda_A \tag{2.14}$$

$$\frac{n_A}{n_B} = \frac{I_A}{s_A} \div \frac{I_B}{s_A} \tag{2.15}$$

Note that RSFs are specific to each element and orbital. The concentration ratio of two elements can now be expressed in terms of the intensity of each core line divided by the relevant RSFs.

The RSFs as defined above technically refer to the entire intrinsic signal. As pointed out in our discussion of shake, splitting, and relaxation phenomena, the intrinsic signal can in some cases be distributed over a number of spectral features. The difficulties associated with this point are discussed below when covering background functions, but it can be pointed out here that for some common compounds, the satellite features can account for 40% of the total intrinsic intensity, and as seen in Figure 2.6, some intrinsic features such as plasmons can be separated from the main core line by 60 eV or more, making this a significant challenge in correct application of the RSFs.[47]

RSFs can be obtained in two main ways. Theoretical RSFs are derived from photo-ionisation cross-sections calculated for specific photon energies, with the most widely used set being the theoretical photo-ionisation cross-sections for aluminium Kα (1486.6 eV) and magnesium Kα (1253.6 eV) radiation reported by Scofield,[48] with the λ_A term estimated using a universal curve, and the values normalised to C 1s having a RSF of unity. Scofield RSFs further assume unpolarised X-rays and magic angle detection.

Empirical RSFs such as those reported by Wagner *et al.*,[49] are determined for a specific instrument, using a range of standard materials. As these empirical RSFs include factors such as analyser transmission characteristics, instrument geometry and the electron escape depth for the materials used to calculate the RSF value, they are not directly transferrable to another XPS system. A further complication of quantification using empirical RSFs, that applies to the Wagner values, is that it is unclear if the measured intensity is of the whole intrinsic peak, or if some or all satellites are omitted. Despite these challenges, some analysts have attempted to modify the Wagner RSFs by applying correction factors to utilise them on their own instruments; this practice has been criticised.[43] However, locally sourced, well documented empirical RSFs obtained from high quality standards – costly to obtain – can be much more accurate than theoretical RSFs. An example of this kind of work is by Shallenberger *et al.* who derived RSFs for silicate glasses.[47]

Because of the photon energy dependence of the cross sections, RSFs apply only to one photon energy, and cannot be used for other photon energies, as the ratio of cross sections of two orbitals will change significantly. Thus a challenge in moving away from the most popular photon energy (Al Kα) is that there are far fewer empirical RSF collections, so an analyst wishing to use different energies may find they have to produce their own set of RSFs, rely on theoretical RSFs only, or forgo quantitative analysis.

2.6.4 The Inelastic Background

The electrons that form the sharp core lines can be thought of as elastically scattered – they have reached the detector without losing energy to extrinsic processes, and form the primary or intrinsic signal. But most of the electrons that are ionised in an XPS experiment will undergo some energy loss process, by interacting with electrons in the atoms through which they pass. These inelastically scattered electrons may still reach the detector, and manifest in the spectrum as the *inelastic background*.[41] Inelastically scattered electrons from an atomic orbital will always appear at higher binding energy (lower kinetic energy) compared with the orbital's elastically scattered core line. For this reason the core line is sometimes known in this context as the *zero loss line*. The inelastic background is contributed to by electrons that have undergone different numbers of these loss processes, losing different amounts of energy, and thus the background represents a distribution of the probabilities of energy loss from a particular core line. The background is typically very broad, often extending 100s of eV or more on the high binding energy side of a core line, representing the large distribution of energy losses suffered by the electrons. An inelastic background typically rises immediately after the core line, reaches a maximum, then gradually decreases as binding energy increases. The maximum in the background represents the most likely energy loss value of electrons from that core line, and the position of this maximum relative to the zero loss line, and the shape of the background in general, can reveal useful information on the distribution of elements in the sample. Maxima 5–30 eV above the core line correspond to electrons that have undergone one or a small number of inelastic interactions. Coupled with a decreasing, 'concave' background shape at higher BE, this is typical of an element homogeneously distributed within the information depth. A background shape with an intense maximum much further from the core line, perhaps at 50–200 eV separation, is also known as a 'convex' background, and this is typical of a buried layer, originating from an element which is not present at the surface but is only found deeper inside the sample (deeper in this context is still relative to the information depth, so around 5–10 nm with Al Kα X-rays). Photoelectrons emitted from such buried elements are likely to undergo multiple energy loss processes, and so the most probable energy loss, and hence the maximum in the background, is much higher than if that element were homogeneously distributed. Lastly, a core line with no background, or a small maximum around 5–30 eV above the core line but with no extended tail, is characteristic of an element that only appears at the very surface of the sample. Photoelectrons from such samples are most likely to undergo zero energy loss, and so the background itself is very small compared to the core line. Of those photoelectrons that do lose energy, the vast majority undergo only one loss event, and so the inelastic background is a discrete peak. Some examples of such spectra are shown in Figure 2.13. Figure 2.13a shows the C 1s spectrum from a polymer film which had residual silicone contamination on the surface. This could be removed with gentle etching, and doing so strongly changes the intensity and shape of the inelastic background of the C 1s peak. Figure 2.13c shows a spectrum taken from a sample of Fe metal. The sample has a native oxide layer, and carbon contamination. Examination of the inelastic backgrounds of the C 1s, O 1s and Fe 2p core lines reveals important differences. The C 1s core line especially has almost no inelastic background, and this is consistent with it being present only at the very surface. Fe 2p and O 1s peaks in contrast have much larger inelastic backgrounds and so can be concluded to be deeper within the information depth.

As we have seen, the inelastic background can be used to gives clues on the spatial arrangement of elements within the sample. This can be supplemented with angle-resolved measurements as described in the previous section to yield significant, non-destructive depth information, at least up to the information depth of the measurement.

2.6.5 Background Subtraction

As we have seen, spectral intensity is composed of the intrinsic signal, made up of photoelectrons that have lost no energy on their journey to the detector, and extrinsic signal, made up of photoelectrons that have lost energy through the processes described in Section 2.6.4. An important task is in spectral analysis is to separate the spectral intensity into intrinsic and extrinsic parts. This is vital if we wish to quantify elemental concentration by applying RSFs only to the intrinsic part; inclusion of significant extrinsic intensity, or exclusion of intrinsic intensity in the measured peak areas, will cause inaccuracies in quantification.

In almost all practical cases, the extrinsic spectrum is not calculated from *ab initio* physical principals. Instead, a heuristic background function is applied to divide the spectrum into intrinsic and extrinsic parts.[51] The choice

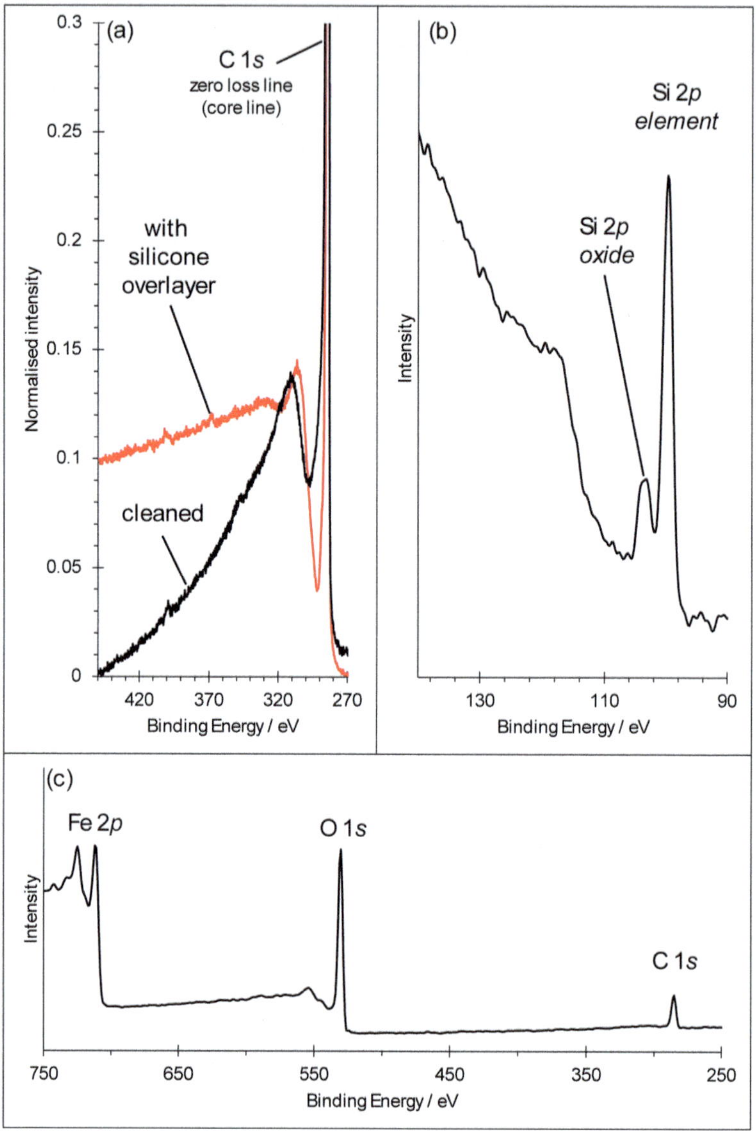

Figure 2.13 Some examples of the effect of element depth on background. (a) Spectra from the C 1s peak of a polymer thin film, with and without surface contamination of silicone, normalised to the zero loss peak (the core line). The overlayer of silicone causes the inelastic background of the C 1s to increase significantly, as the C 1s electrons are scattered passing through the overlayer. (b) Si 2p spectrum taken from a silicon substrate with a thin coating of organic polymer. The Si oxide and element environments can be seen, along with a large inelastic background at high binding energy, indicating the buried Si substrate. (c) A survey spectrum of an iron sheet, showing C 1s, O 1s and Fe 2p core lines. The C 1s peak has almost no inelastic background, indicating it is on the surface. The O 1s and Fe 2p peaks both have significant backgrounds, suggesting they are beneath the carbon layer.

of background function and its limits must be made with care; poorly chosen or implemented background functions can be one of the greatest sources of error in XPS quantification.[52]

The background around a particular core line comes from two main sources. Firstly, inelastically scattered electrons from features with lower binding energy than the core line of interest, and secondly inelastically scattered electrons from the core line itself. This can lead to a variety of background shapes around a core line which is a complex function of the depth distribution of the element of interest, and all those elements with core lines at lower binding energy.

The commonly encountered background functions are known as linear, Shirley and Tougaard,[40] and are shown overlaid in Figure 2.14. All three involve placing a function between two limits, either side of the feature of interest (typically the core line) and each has their own advantages and disadvantages. We will discuss the Shirley and linear first as these are the most widely used and easiest to apply.

The linear background is simply a straight line between the two limits. Such a background is suitable where there is little or no increase in the background signal on the high binding energy side of the peak, such as wide band gap materials (for example polymers or some closed-shell oxides), or where a peak is superimposed on a *decreasing* background from a lower binding energy feature. But it does not deal well with peaks that have a significantly different background level on either side.

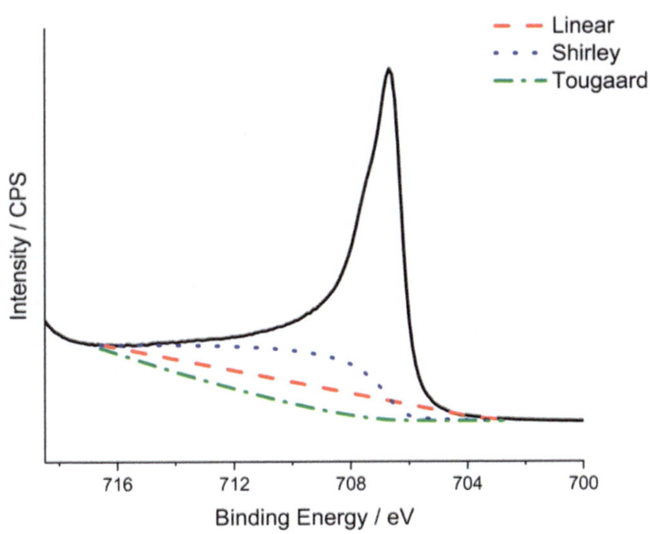

Figure 2.14 Overlay of linear, Shirley and Tougaard backgrounds on the Fe $2p_{3/2}$ peak. Note how the background intersects the peak differently resulting in different integrated areas for the photoelectron signal.

The Shirley background is the most commonly used in XPS quantification since it is simple to apply and, in some cases, gives a good approximation to the true background.[41,53] The Shirley function is S-shaped and attempts to account for the way in which the cumulative peak intensity above the background contributes to the inelastically scattered electrons. However, as the background was developed for the valence band of gold[54] it assumes the inflection point is at the peak centre, as would be the case for metallic elements. For insulators and semiconductors, the point of inflection may be offset to the high binding energy side of the peak maximum by the value of the bandgap, hence developments have been made to modify the original Shirley background to account for these.[55,56]

The positioning of the limits between which the background is applied can also have a strong effect on the results of quantification. The Shirley background is mostly insensitive to the high binding energy endpoint, giving virtually constant areas (often less than 1% difference) with the upper limit placed across a wide range of values. The low binding energy endpoint however, is more important. The linear background has the opposite sensitivity: the calculated area usually shows greater variation with the high binding energy endpoint than it does with the low BE endpoint. The exact situation will depend on the shape of the feature under analysis, but it is important to carefully choose the endpoints and be aware that a 5–10% change in measured area can result from different functions and endpoints.

The Tougaard background is derived from a description of the physics of the electron scattering as it moves through the sample.[40] Unlike the linear and Shirley backgrounds, the Tougaard background has parameters which can be altered to fit it to the spectrum, and to account for the scattering behaviour of different materials. Fitting of Tougaard backgrounds is an iterative process but allows more advanced understanding of background shapes.

2.6.6 Accuracy of Quantification

While it can be readily seen that the precision of XPS quantitative analysis is high – one only has to repeatedly measure the same sample and see the minimal variation in measured composition, quantifying the accuracy of XPS compositional measurements has been a long-standing challenge. Several approaches have been taken, including Bayesian analysis,[57] but unlike many analytical methods, XPS has no standardised solution for quantifying uncertainties in composition, even after 70 years of development.

One major issue affecting XPS accuracy is the depth dependence of signal intensity, which was introduced in Section 2.6.1. As discussed there, the recorded intensity of a photoelectron signal is exponentially dependent on the depth of the emitting species (eqn (2.10)). This means that variations in the composition at the nanometre scale can significantly influence the measured intensity. In practice, the assumption that the surface composition is homogeneous within the outermost few nanometres often does not hold true. This lack of knowledge regarding the depth composition of a sample adds significant uncertainty to the results. A second issue already mentioned is the difficulty in determining the intensity of the intrinsic signal, separate from the extrinsic signal.

A third issue compounding these problems is the lack of a reliable alternative method for determining the 'true' surface composition of a sample, meaning there is no easy way to evaluate the deviation of XPS measurements from actual values. While the XPS composition may be compared with bulk chemical analysis, or with the expected formula from chemical laws, surfaces can very often differ from bulk stoichiometry, and issues of surface contamination or oxidation, which may not be apparent in bulk composition analysis, can completely dominate the surface composition. Even if we had perfect knowledge of the atomic arrangement of the surface, it is still hard to define what we mean by the 'true' surface composition. Unless the sample is completely compositionally homogeneous, any composition value we quote must refer to a specific volume of the sample, and how is that volume to be defined? Thus, the accuracy of XPS compositional analysis is fundamentally limited by surface inhomogeneity at the nanoscale, poorly defined background functions, and the difficulty of determining (or even defining) the true value of the surface composition.

Despite these challenges, quantitative studies are still possible using XPS for compositional analysis. By taking care in the mode of calibration, analyses accurate to 5% are achievable. A more often quoted accuracy level is ±10% for compositions. While these estimates are based on accumulated experience rather than a robust physical model or statistical treatment, they have proven to be useful guidelines for XPS practitioners.

2.6.7 Detection Limits

Detection limits, *i.e.* the minimum amount of an element that can be detected, are dependent on the interplay between the background intensity at the expected peak position and the signal strength of the element being measured. As outlined in Shard's comprehensive analysis of this topic,[58] the detection threshold for a given element is not a fixed value but varies significantly depending on the matrix in which it is embedded. In many cases, elements can be detected at concentrations between 1 at% and 0.1 at%, but these limits can depend on the atomic number of the trace element and the matrix. Heavy elements in a light matrix generally yield lower detection limits, often below 0.01 at%, due to the relatively weak background signal in such matrices. Conversely, light elements in a heavy matrix tend to be far more difficult to detect, with detection limits exceeding 10 at% in some cases. XPS detection limits can be estimated for specific instruments using a reference measurement, such as the Ag $3d_{5/2}$ peak intensity, allowing scaling of the results to different experimental conditions. This provides a valuable framework for analysts seeking to determine whether an element is present below a detectable threshold and, if so, how experimental parameters must be modified to improve sensitivity. The case-by-case nature of detection limit calculations underscores the importance of accounting for both matrix effects and signal-to-noise considerations in quantitative XPS analysis.

2.7 Auger–Meitner Effect

In addition to photoelectron peaks, XP spectra may include contributions from one of the mechanisms by which the high energy core hole created by photoemission decays to a lower energy state. This happens through the transition of an electron from an outer level to fill the core hole. There are then two possible routes for the release of the energy the decay produces. For elements with atomic numbers greater than ~50, the dominant decay route is the emission of an X-ray photon in a process known as X-ray fluorescence, but for elements with lower atomic numbers the excess energy more likely results in the emission of a second electron, and these show up in the photoelectron spectrum. Known as *Auger–Meitner* electrons after their discoverers,[59] the resulting peaks have been used as a separate form of spectroscopy for many years, especially since the same effect can be stimulated with an electron beam and could therefore be measured with the low energy electron diffraction equipment that was commonly available to surface scientists from the 60s onwards. Because the Auger–Meitner peaks arise from a process involving three different energy levels, all of which are affected differently by the presence of the core hole, direct information on the chemical environment of the atoms concerned is difficult to obtain, but the Auger–Meitner features have a great advantage over photoemission peaks in that they are independent of the exciting energy. Peak positions for these electrons are well documented (typically in kinetic energy) and peak shapes have been studied for many years so that a considerable knowledge base is now available that can provide many insights into the chemistry of a system.

2.8 Sample Charging and the Absolute Binding Energy Scale

Measuring the binding energy of photoelectrons is the central premise of XPS. Despite this, obtaining reliable values of the binding energy can be challenging in many sample types due to charging. In this section we will explore the physical basis for sample charging, and describe what methods are available to obtain reliable BE values. Although identified as an issue in the early days of XPS, a totally satisfactory solution has not been found to sample charging, and currently established methods of using adventitious carbon for calibration of the energy scale are under renewed criticism.

2.8.1 The Fermi Level as the Binding Energy Origin

Like any energy scale, we must define a zero point, or origin, if we wish to state absolute binding energy values. Since we are concerned with moving electrons from the sample to the vacuum, it might be expected that the vacuum level (E_{vac}) – the energy at which an electron is removed from the electrostatic influence of the sample – would be chosen for the zero point. However, this is not the case; in XPS, the convention is to set the Fermi level as the origin of the binding energy scale. A simple reason can be seen in Figure 2.2. Objects that are in electrical contact with each other will transfer electrons by conduction and equilibrate with the same Fermi level. Figure 2.2 shows the energy level diagram of an XPS experiment where a sample is in electrical contact with the spectrometer, and so their Fermi levels are the same. In contrast, the vacuum levels of sample and spectrometer are offset due to differences in workfunction.

The Fermi level (E_F) in a metal is the highest occupied electron energy level at absolute zero temperature. At $T > 0$, electrons can occupy states above the Fermi level due to thermal excitation according to the Fermi–Dirac distribution, and the E_F is then defined as the energy at which there is a 50% probability of finding the state filled.

For bulk metallic samples, such as plates, foil, and ingots, which can easily be mounted to be in electrical contact with the spectrometer, we have a convenient and reproducible method of defining the BE scale. Metals in electrical contact with the spectrometer should always show their core lines at the expected energies. This fact is used to calibrate the linearity of the spectrometer energy scale by measuring several pure metal samples, usually copper, silver and gold, and ensuring the core levels appear at the correct energies.

With the Fermi level as the BE origin, the binding energies of the filled states in a metal are then defined by their energy below the Fermi level. While metals have a band of electron states filled only up to the Fermi level, ideal semiconductors have completely filled and completely empty bands separated by a band gap, a forbidden energy range where in the perfect compound no electron energy levels exist. The definition of Fermi level in metals cannot be applied so easily to semiconductors. In the perfect, defect-free semiconductor, it may be that no half-filled energy level exists; within the band gap the probability of finding an electron is zero, and at the top of the valence band and bottom of the conduction band the probability is much high and lower than 50%, respectively. Instead, viewing the Fermi level as a measure of chemical potential is more appropriate for semiconductors, and its position is midway between the highest filled and lowest unfilled states. In a perfect semiconductor, the Fermi level would be exactly between the Valence Band Maximum (VBM) and Conduction Band Minimum (CBM), *i.e.* at the centre of the band gap. Adding filled states within the band gap (n-type doping) would move the Fermi level up towards the CBM, while adding empty states (p-type doping) would move the Fermi level down, towards the VBM. Such doping can occur through intentional or unintentional addition of impurities, or simply by formation of native defects, which have low enough formation energy to be present in all practical cases.

Even once the effects of defects and dopants are taken into account, there are still more complications. The position of the Fermi level relative to the band edges can vary considerably between the bulk and the surface. This effect, known as band bending, can occur due to a difference in the defects present at the surface compared with the bulk.

The Fermi level still acts as the reference point for our BE scale even for semiconductors, and therefore it can be seen that changes in E_F in a semiconductor due to doping, defects, surface states or any other reason, will lead to a change in the measured BE of all spectral features from that semiconductor. N-type doping leads to the Fermi level rising, and a consequent increase in all measured BEs from that material, while for p-type doping the same effect leads to a decrease in measured BE.

A demonstration of these effects was carried out by Suzer and coworkers,[60] who studied XPS of n- and p-type silicon wafers. They used *in situ* laser irradiation to counteract surface band bending and found the Si 2p peak appeared around 0.5 eV higher BE in n-type silicon compared with p-type; this difference was not due to chemical shift in the normal meaning, but due to the difference in Fermi level between the differently doped samples.

2.8.2 Charging

As is familiar by now, X-ray irradiation of a sample causes emission of electrons, by both photoemission and Auger–Meitner emission, and will leave behind a positive charge. A charge on the sample will produce a potential difference between sample and spectrometer, and so as discussed previously, will alter the measured binding energies of any emitted electrons. Therefore, identifying and counteracting charging is an important concern in modern photoemission practice. Here we will discuss basic ideas of the causes of charging, how it affects different sample types, and what measures can be taken to counteract it.

2.8.2.1 Physical Model of Charging in Photoelectron Spectroscopy

The removal of electrons under X-ray irradiation leaves a positive charge on the sample. The rate of electron removal by photons (the *photoemission yield*) will depend on the intensity and photon energy of the irradiating light, the electronic structure of the sample and the potential of the sample relative to the spectrometer.

We define the total positive charge on the sample as Q. For a sample bigger than the X-ray area of X-ray illumination, we assume the sample itself is sufficiently conducting for the charge to be able to spread out and become evenly distributed over the sample. For samples where that assumption is valid, the surface potential V_S is given by eqn (2.16):

$$V_S = \frac{Q}{C} \tag{2.16}$$

where C is the capacitance of the sample. As discussed previously, V_S directly affects the measured binding energy of photoemission features, shifting the observed binding energy by 1 eV V^{-1}.

If a sample were irradiated by X-ray while perfectly insulated from the environment, allowing no replacement of the electron removed by photoemission, Q would increase until V_S became so large that it could not be overcome by the photon energy, at which point the photoemission yield would drop to zero.

In a real experiment, the sample cannot be perfectly insulated, and the build-up of charge, Q, is regulated by migration of electrons into the sample from the surroundings. Electrons may migrate into the sample from the ground by a number of routes:

- Ohmic contact with the spectrometer, causing direct current to flow to the sample.
- From the gas phase, by ionisation of gas molecules at the sample surface.
- From secondary electrons or photoelectrons produced by X-ray irradiation of the sample holder or surroundings.
- From electrons supplied by a flood gun – a component of the spectrometer designed to reduce charging.

The current due to ohmic contact will be directly proportional to Q. The rate of electrons arriving from the gas phase or from secondary electrons will have more complex relationships, but will have some positive dependence on the sample charge as well. The flood gun current, if used, is largely independent of Q.

Under continuous X-ray irradiation, the size of the positive charge will grow until the rate of electron emission from the sample, r_1, is equal to the total rate of electron migration into the sample, r_2, at which point an equilibrium is reached where Q is constant.

If we assume r_1 is independent of Q, while r_2 is linearly dependent on Q, then the ratio of these two rates determines the equilibrium concentration of positive charge on the sample, Q_{eq}:

$$\frac{r_1}{r_2} = Q_{eq} \tag{2.17}$$

Neither of these assumptions is strictly accurate, but eqn (2.17) can be useful for understanding trends in charging behaviour, which is sufficient for dealing with charging in many practical situations. For a more detailed treatment of the different contributions to charging, the work of Cazauz is recommended.[61,62]

2.8.2.2 Charge Compensation Methods

In order to combat these issues, charge compensation methods have been developed ranging from sample preparation methods to electron flood guns.

For conducting samples, it may be sufficient to ground the samples to the spectrometer, ensuring that electrons from the ground can compensate the positive charge. In this case there should be little to no BE shift, and no peak distortion. In practice, it has been found that samples with conductivity greater than 10^{-10} S m^{-1} can typically be measured without any significant build-up of charge, although this will depend significantly on the experimental conditions.[63] This means materials with significantly lower conductivity than is usually thought of as a 'metallic', like silicon and graphite, might be measured without charge compensation, as long as they are electrically grounded.

Samples that are not sufficiently conducting can be mounted to reduce the amount of differential charging as much as possible. Methods such as using a metallic mask, which is a thin sheet of metal foil with a small hole exposing the sample for analysis, can help improve electrical contact between the sample and the ground. This can be especially useful if the sample is thick, for example, a coated sample of glass, and grounding by contact from the stage below is not feasible. A similar effect can be obtained by placing metal contacts on the sample surface, although use of metallic paint (such as silver DAG) is usually avoided due to risk of contamination of the surface.

An electron flood gun is included on almost all modern XPS instruments (although not necessarily synchrotron beamlines). The flood gun works to supply low energy electrons during measurement, providing another source of electron migration to the sample, increasing r_2, and decreasing Q_{eq}. Flood guns can be effective tools for reducing charging in insulating samples. Typically, the emission of the flood gun must be controlled to optimise the charge neutralisation – exposure of the sample to the electron beam can in fact cause a negative surface potential to form, and this leads to a shift of spectral features to low BE, just as the positive charge leads to a shift to higher BE. If this negative charge becomes too high, then similar unwanted effects such as differential charging and large BE shifts can occur. A usual procedure is to monitor the position, intensity and FWHM of a photoelectron peak while altering the emission characteristics (current and voltage) of the flood gun to obtain the correct spectral shape. This of course necessitates that the 'correct' spectrum is known or can be guessed.

An advanced type of flood gun is known as dual-beam, and provides a stream of low energy Ar^+ ions simultaneously with an electron beam. The purpose of both negative and positive charge compensation is to prevent build-up of either positive or negative charge on the sample. This can avoid the need for optimisation of the flood gun parameters for each sample, greatly saving time when setting up experiments. The dual-beam flood gun can have disadvantages however, for example, the low energy Ar^+ ions can sputter the surface, gradually changing surface chemistry over time.

Exposing the sample to a low pressure of gas can also help reduce charging, as gas molecules can donate electrons to the surface if it is at a sufficient potential. Techniques such as near-ambient pressure XPS exploit this phenomenon to prevent charging to a large degree.

2.8.2.3 *Types of Sample and Their Response to Charging*

The sample's positive charge, Q, results in a surface potential, V_S relative to the analyser. This causes all spectral features to be shifted to higher BE. The practical effect of charging on the photoemission spectrum depends on the nature of the sample and its electrical contact with the spectrometer. Some examples can illustrate this.

For the case of an electrically conducting sample that is grounded (in electrical contact with the spectrometer), direct current can easily flow to the sample from the ground: r_2 is large compared with r_1. Therefore, Q_{eq} would be small; metallic samples in good contact with the spectrometer typically do not experience measurable charging effects in their photoelectron spectra. This makes metals, especially noble metals which do not form a native oxide layer, good samples to use to calibrate spectrometer energy scales, as their core line BEs can be known with a high degree of accuracy.

A second example is a conducting sample that is electrically isolated from the ground, for example by being placed on an insulating substrate, or if the sample stage is floating – *i.e.* not in electrical contact with the spectrometer. Now the rate of migration of electrons into the sample will be much lower, because the direct current to the sample from the spectrometer has no low-resistance pathway. The lower value of r_2 will lead to a larger value of Q_{eq} than in the previous example. Because the sample itself is conducting, the equilibrium positive charge will be spread evenly over the whole sample, and so the surface potential at every point will be proportional to Q_{eq}. Photoemission peaks will be shifted to higher BE, but otherwise undergo no other changes, in shape or intensity, because the surface potential is uniform across the sample.

Figure 2.15 shows spectra taken from clean Pt foil that is grounded, floating, and floating with flood gun on. The grounded Pt foil shows the Pt $4f_{7/2}$ peak at BE of 71.1 eV, closely matching the established value. When electrical contact with the spectrometer is removed, the sample is said to be floating, and a large positive shift to the BE occurs; the Pt $4f_{7/2}$ peak is measured at 90.0 eV, a shift of +18.9 eV compared with the grounded sample. This means the surface potential on the Pt foil at the analysis position is +18.9 V. Finally, when the flood gun is used, the Pt $4f_{7/2}$ peak from the floating Pt foil is measured at 70.3 eV, a shift of −0.8 eV compared with the grounded measurement. This shows the flood gun is causing a negative surface potential on the Pt surface by adding an excess of electrons, shifting the core line to lower BE.

Samples that are not good conductors themselves, present additional problems. Electron migration from the ground, even if the sample is mounted

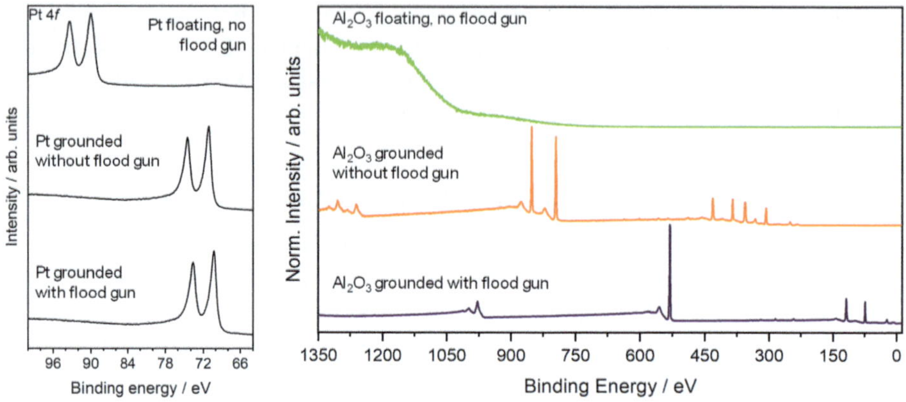

Figure 2.15 Response of metals and insulators to grounding and charge compensation. Left: Pt 4f spectra taken from Pt metal foil. Right: survey (wide) spectrum from an Al_2O_3 single crystal. See text for discussion.

on a grounded conductor, may be slow due to the insulating nature of the sample. A low value of r_2 will mean that a large Q_{eq} will be present, but worse, when the sample itself is insulating, this charge may not be evenly distributed. Different parts of the sample experience different X-ray fluxes, perhaps due to morphology of the sample or non-uniformity of the X-ray beam, and different rates of electron migration due to their position relative to the ground contact. The result is that different parts of the sample experience different potentials. This is known as *differential charging*, and can have serious negative effects on the recorded spectra. A small degree of differential charging will cause peak broadening, because electrons emitted from atoms in different parts of the analysis volume will experience slightly different potentials. The range of potentials manifest as a broader peak in the spectrum. This can be demonstrated with variable temperature measurements on semiconductors. As the temperature is decreased, the conductivity of a semiconductor falls. With decreasing temperature, the width of photoemission peaks from semiconductors tends to increase, due to the increase in differential charging. As the amount of differential charging increases, the broadening becomes worse, and for some very insulating samples, can cause such distortion in peaks shapes as to make analysis almost impossible. In the worst cases, the rate of migration of electrons into the sample is so low that very large charges build up, and these are so unevenly distributed that equilibrium is never established during the experiment. In this case the potential continually changes and the peaks move in binding energy during measurement. Such measurements cannot yield useful information, beyond confirming that the sample is an insulator.

These issues are illustrated in Figure 2.15, which shows spectra from a single crystal of Al_2O_3, an insulator, recorded under the same conditions as the Pt foil discussed earlier. With the flood gun on, core line peaks appear very close to the expected positions. But when the flood gun is switched off, peaks shift by hundreds of eV to higher BE, and worse, the spectra appear doubled, with two versions of each core line appearing. This is due to differential charging across the analysis area. Such spectra are of little use to the analyst. Lastly, when floating, the charging becomes so extreme that no core lines are detected at all, only a featureless emission at very high binding energies, likely from secondary electrons. This is characteristic of high and uncontrolled charging and makes any analysis impossible.

As difficult as insulating samples may be, they can normally be dealt with using the built-in charge compensation available in most instruments. Perhaps the most problematic class of samples (but one very commonly encountered in practice) is heterogeneous samples where different parts of the sample have different conductivities. A semiconductor coated with metal, a metal with a native oxide layer, a conducting substrate with an insulating film on top, or a sample with heavy surface contamination, might all be all examples of this class. In heterogeneous samples, different parts of the sample may acquire different surface potentials through a variety of mechanisms. For example, the conducting portion may be well grounded to the spectrometer, but the insulating part may not be. In such cases, a great deal of care is needed in interpreting observed binding energy shifts, as they might potentially be caused by charging. Practically, such samples can be hard to measure, but strategies such as comparing floating *vs.* grounded spectra can help identify whether one part of the sample is more strongly affected by charging than another.

2.8.3 Charge Correction of the Binding Energy Scale

The charging described in the previous section raises important questions over how to measure and report BEs of photoemission peaks. The measured BE of a peak will change if charge (positive or negative) is built up on the sample. But as mentioned in the previous section, the equilibrium charge is dependent on many factors such as the X-ray flux, the sample grounding and conductivity, the morphology of the sample *etc.* For poor conductors, these factors and their effect on the equilibrium sample charge cannot be quantitatively known in any practical case, and so the equilibrium charge, and the surface potential that results from it, cannot practically be determined. For metallic conductors that are well grounded, this does not cause much issue, as the equilibrium potential is immeasurably small. But even samples that are expected to be metallic and well-grounded may in fact not be – an oxide layer on a metal may prevent proper grounding or be thick and resistive enough to cause potential build up.

How then, are we to know the true BE of photoemission peaks if they have been shifted by a surface potential of unknown size? A popular answer is charge referencing: the process of correcting the measured BE scale to compensate for the effects of charging by use of a feature in the spectrum with known binding energy.

2.8.3.1 Charge Correction to Adventitious Carbon

Among the first things noticed by an XPS analyst is that almost every survey spectrum contains a carbon C 1s peak, even when no carbon is expected in the nominal composition of the sample. So-called *adventitious carbon*, or AdC, is found to be present on almost every sample unless special measures are taken to prevent or remove it. Early ideas that AdC came from pump oil from old style diffusion pumps were largely abandoned when the diffusion pumps were replaced with oil-free alternatives, yet the AdC remained. AdC is now believed to come from contamination of the sample surface by organic compounds from the environment; the extreme surface sensitivity of XPS means that only a very thin layer of contamination is needed to be detectable.

The use of AdC in charge referencing dates from the early days of XPS. To charge correct using AdC, the whole spectrum was shifted such that the C 1s peak moved to a specified BE, usually 285.0 eV, although other values have been used. For example, if the measured C 1s peak was at 286.0 eV, then a shift of −1.0 eV would bring the C 1s peak to the desired position. The charge correction process would then involve applying this same rigid shift to all spectra recorded from that sample.

Charge correction with AdC has several advantages which have made it very popular:[64]

- Since carbon is present in the spectrum of almost every sample it is easy to apply.
- In most cases it can account for any large changes in the BE scale caused by significant charging.[65]

However, there are several disadvantages to the method:

- Because the chemical nature of AdC is unknown, and in fact is certainly different in different samples,[66] we cannot have a high level of confidence that 285.0 eV (or any other value) represents the true BE value of AdC C 1s in any particular case.[67]
- Samples with a composition that includes carbon (virtually all organic compounds and many inorganic ones) will have their own C 1s peak. In this case, how is the AdC peak to be identified accurately?
- The assumption that the binding energy of C 1s in AdC shifts by the same amount as the sample under analysis may not be true in all circumstances. AdC is likely reasonably insulating, and so when used to charge correct more conducting samples can give incorrect results, due to different surface potentials on the two phases.[68]

Although widely used, charge correction with AdC has always been recognised as problematic for the reasons stated. There are several alternatives, but none have reached nearly the ubiquity of the AdC method.

Other peaks apart from C 1s might be used as internal standards, although of course whatever element is chosen must be present in the analysed sample. One possibility is to use a noble metal, either coated in a very thin layer onto the sample or added in the form of nanoparticles.[69] The advantage is that the chemical form of the noble metal is reasonably certain, and this gives confidence that the true value of the BE is known. However, the concern as to whether the noble metal has the same charging behaviour as the sample under analysis remains. Another possible method is to use implanted noble gas atoms as an internal ref. 70. Even more so than Au, Ar is chemically inert, but despite this the variation in BE of the Ar 2p peak has been shown to be rather large depending on which substrate Ar is implanted in ref. 71. Despite some advocates, referencing with Au, Ar or other added internal standards has not seen wide adoption by the XPS community, leaving AdC as the *de facto* standard method.

2.8.4 Abandoning Absolute Binding Energies

None of the methods of charge correction described above are completely satisfactory. While metals, which experience little charging, can be said to give reliable BE values, semiconducting and insulating samples rely on some known internal standard to account for charging. All the forms of internal standard have issues which makes us less than certain about their reliability in all circumstances. A solution is to avoid reliance on absolute binding energies, and place more emphasis on binding energy differences, ΔBE, between two features in the same spectrum. For example, the ΔBE between O 1s and Al 2p in Al_2O_3 might be expected to be constant even if the material itself is charging, as long as the charging is not too high. In physical mixtures consisting of multiple phases, the ΔBE of core lines that originate from the individual phases can be constant even if the phases themselves shift relative to each other due to different surface potentials. Such ideas are explored as an analytical tool in the section on correlation analysis (Section 2.13.1).

2.9 The Effect of Different Photon Sources

The X-rays for XPS are in almost every case generated either by an X-ray tube or a synchrotron. XPS using X-ray tubes is sometimes called '*lab XPS*' (short for '*home laboratory XPS*' with the presumption being that no one has a synchrotron in their '*home*' lab).

For X-ray tubes, the X-ray energy produced is fixed and is determined by the anode material. Today, the most widely used X-ray anode for XPS is Al, producing Kα X-rays with a photon energy of 1486.6 eV, which are considered *soft X-rays*. Other anodes that are sometimes used are Mg, Ag, Cr, Ga or Cu.

Carrying out XPS with different photon energies can have several important effects on the spectrum:

- Information depth: as the inelastic mean free path is dependent on kinetic energy, according to the universal curve (Figure 2.10), increasing photon energy will make most core lines have greater information depth, making the spectra less surface sensitive.
- Access to core lines: the photon energy must be greater than the ionisation energy to be able to excite a core electron to the vacuum. The higher the photon energy, the larger the number of core lines that are available to measure, although there will be an upper limit on the kinetic energy that is measurable by the spectrometer.
- Cross sections: the photoionisation cross sections of atomic orbitals increases to a maximum 50–100 eV above their binding energy, then falls with increasing photon energy (Figure 2.12). If the photon energy is much higher than the binding energy of a core line, the cross section is usually very small, meaning longer acquisition times are needed to achieve a certain signal-to-noise ratio.
- Auger–Meitner lines: as photon energy changes, Auger–Meitner lines appear at constant kinetic energy, whereas core lines appear at constant binding energy. Therefore, Auger–Meitner lines move relative to core lines when the photon energy is changed. This can be used to shift overlapping features.
- X-ray damage: higher energy photon sources can inflict greater damage on the sample, although the relations between X-ray damage and photon energy are complex and a topic of current research.

XPS carried out with high energy photons is called hard XPS, often abbreviated HAXPES (Hard X-ray Photo Emission Spectroscopy).[72] HAXPES can be carried out on lab instruments with anodes such as Cr or Ga, or at synchrotrons. In 2020, there were 25 HAXPES synchrotron beamlines worldwide, with high concentrations in Japan, USA and the EU.[73] HAXPES typically yields much greater information depth and access to higher binding energy core lines, which can make it extremely useful for characterising buried interfaces,[74] and distinguishing between surface and bulk effects in energy loss features.[75] It does however suffer from lower cross sections for many of the common core lines typically measured by Al Kα, and potentially greater beam damage.[76]

While X-ray tubes offer only fixed photon energies, synchrotron sources can allow the photon energy to be tuned to any value within the operational range. This might be used to progressively probe deeper into the sample, for example, work by Bagus *et al.* identified the contribution of surface chemical environments to the Ca 2p spectrum by incrementally increasing the photon energy to change the information depth.[77] Methodologies such as resonant photoemission are also possible,[78] where the photon energy is tuned to coincide with a specific electronic transition, useful for understanding valence band spectra in particular.

Synchrotron beamlines can also offer higher intensity X-rays with narrower energy range, control of X-ray polarisation and angular resolution, and detection of electron spin. These abilities open up a wide range of possible photoemission-based experiments, which are beyond the scope of this chapter.

2.10 *Ab Initio* Calculation of Core Line Binding Energies

There is strong motivation to calculate relative binding energy shifts between two chemical environments, or even the absolute binding energies of a particular chemical species, from *ab initio* theory. Reliable calculations would allow identification of ambiguous experimental features, better interpretation of spectra, and deeper understanding of physical processes in electron spectroscopy. Hartree–Fock (HF) theory and Kohn–Sham Density Functional Theory (DFT) are two frameworks used to calculate core line binding energies. Both methods are firmly established in quantum chemistry; here we discuss their application in XPS core line calculations.

The simplest theoretical approach follows the single-particle approximation, taking the binding energy of a core level as the ionisation energy of the core orbital in the ground state. This is known as *Koopmans' theorem*. By neglecting all final state effects, Koopmans' theorem can only provide qualitative trends and works best when final-state effects are negligible. Koopmans' theorem is best applied with orbital energies calculated from HF theory. It is generally not compatible with DFT, because Kohn–Sham eigenvalues do not directly correspond to ionisation energies of individual core orbitals.

More advanced treatments account for the effect of photoionisation on the final state. The most popular method is to calculate the energies of both the initial and final states: the difference between the two is the binding energy (eqn (2.8)). This is the delta self-consistent field (ΔSCF) method,[79,80] and it can be implemented using either HF or DFT to calculate the required energies. To calculate the initial and final state energies we need a precise definition of the N-electron system (see Section 2.3.2.1): for a periodic system a suitable supercell is chosen, for a molecular system it may be a single molecule.

For some lighter elements, both HF and DFT give almost equivalent results.[81,82] HF is usually more computationally expensive, and neglects electron correlation, meaning that screened final states are not properly calculated, and as a result, calculated BEs tend to be too high in systems where screening is important. In contrast DFT ΔSCF includes electron correlation and approximates the exchange interaction, so multiplet splitting can be accounted for.[83] But because the Kohn–Sham orbitals used in DFT are not interpretable as core atomic orbitals, DFT is not strictly able to inform about the contributions to the binding energy in such a physically interpretable way as HF.[81] For example, HF calculations can be used to explain the complex shape of the Fe 2p spectrum from Fe_2O_3 in terms of shake up and screening phenomena involving specific orbitals on Fe and O.[84]

While both HF and DFT have their advantages and proponents, perhaps because of the lower computational cost, DFT is more widely applied in ΔSCF calculations. It has been very successful in modelling molecular species containing lighter elements. Relative binding energies can be calculated accurately in many cases, where differences as low as 0.1 eV are found between theory and experiment, for simple 1s core lines in molecules.[85,86] Using this method, successful calculation of C 1s and O 1s BEs was achieved for a series of amino acid molecules.[87] There are far fewer DFT ΔSCF studies involving heavier elements, and where these do appear the accuracy tends to be lower. As the nuclear charge increases, so does the error associated with finite basis sets, meaning heavier elements require larger basis sets and more expensive calculations.[88] Kahk and Lischner surveyed the first row transition metals, and found a mean average error of 0.7 eV between predicted and experimental 2p BEs for molecular species, even after element-specific corrections had been applied.[89] There is considerable variation between the accuracy of the BEs calculated by different DFT functionals.[90]

Moving to periodic systems, adaptations need to be made to the ΔSCF approach, as any attempt to calculate the energy of the of a final state containing a core hole runs into the issue of interaction between the core holes in neighbouring supercells. Solutions to these problems exist but introduce extra computational expense (*e.g.* large supercells),[91] or additional approximations (*e.g.* pseudopotentials to neutralise the charge over the supercell). For metallic compounds, an older idea is to use an 'excited atom model', *i.e.* calculate the BE of an isolated neutral metal atom and assume the BE in the metallic compound is the same due to very short-range screening. This can work surprisingly well giving 1–2 eV errors,[80] but is not suitable for semiconductors.

There are alternatives to ΔSCF methods which don't involve calculation of the final state energy. In the GW method, the effect of removal of the core electron is calculated as a perturbation to the initial state wavefunction.[92] Traditional GW methods used plane wave basis sets which cannot model localised core holes but are more suitable for valence electrons. Use of localised basis sets needed for GW to predict core line energies is a relatively recent innovation,[93] and considerable improvements in accuracy have been achieved since their introduction.[94] Likewise time-dependent DFT (TDDFT) can be used to determine XPS binding energies, although more commonly used in X-ray Absorption Spectroscopy (XAS) where transitions are between bound states.[95] TDDFT can predict 2nd period XPS binding energies with some accuracy, but as with ΔSCF, accuracy decreases moving to heavier elements.[96]

A challenge in any theoretical calculation of absolute BEs is defining the zero point. In experimental XPS, the origin of the BE scale is the Fermi level, which is well-defined in a metal but ill-defined in a semiconductor, and able to vary significantly due to defects (Section 2.8.1). For DFT calculations, the BE is usually referenced to the valence band maximum in a semiconductor, *i.e.* the highest filled level at $T = 0$ K assuming no defects present in the material. This can introduce some significant inaccuracies in predicted absolute binding energies, but of course does not affect relative binding energies.

Overall, *ab initio* calculations of core line XPS binding energies are still at a formative stage. Good accuracy can be achieved on light element core lines in molecular species (which may be gas phase, solid state or adsorbates). Moving to heavier elements, periodic solids, or open shell systems, accuracy drops and computational cost increases significantly. As discussed above, for some systems such as open shell transition metals, a large number of final states are possible, and so to simulate spectra accurately not just one but many excited states must be calculated.[84] Quantitative prediction of solid state core lines is in its infancy, and is an active area of research that is expected to yield important advances in XPS as it develops. Machine learning methods may play an important part in this progress.[97]

2.11 Valence Band Spectra

As is clear from the discussion so far, much of XPS analysis is concerned with signals from core atomic orbitals, which give information on the chemical composition and environments present at the surface. Valence orbitals, which are the outermost orbitals of a material, also give rise to photoemission spectra, and can yield several important pieces of information distinct from that obtained from the core lines. Valence bands give more direct information on chemical bonding, the type of dopants present in a material and their effect on electronic structure; they can even be used to quantify parameters usually not considered to be within the purview of XPS, for example crystallographic phase composition.

A note on nomenclature: photoemission spectra arising from the valence orbitals are referred to as *valence band spectra* no matter what type of sample is under investigation. This can be confusing, as for a metal, the outermost electrons are usually described as being in the *conduction band*, and a molecular or organic polymeric species would not normally be thought of as having a valence band at all, but rather molecular orbitals. Yet the name *valence band spectrum* is usually used for all samples: metals and molecules, as well as for semiconductors, and we follow that convention here.

Interpretation of valence band spectra can be challenging, and the main reason for this is that the valence orbitals are not isolated atomic orbitals as we find in the core but are at least somewhat delocalised – spread over several atoms, and may be hybridised, combining contributions from different orbitals and different elements. This chemical interaction is of course what binds the atoms together, and so study of the valence band can reveal important facets of chemistry, as well as physical properties of the sample.

2.11.1 Approaches to Understanding Valence Band Spectra

Photoionisation of an electron from a valence orbital typically causes less severe perturbation in the electronic structure of a material compared with removal of a core electron, as the valence electrons are much less tightly held and can be more easily replaced due to the more delocalised nature of the valence band. Whereas core electron emission can lead to a range of final states which can have strong influence on the observed spectra, for valence band photoemission, final states vary much less in energy and are consequently less important in determining the shape of the spectrum. Instead, the valence band spectrum is principally influenced by the initial state electronic structure, *i.e.* the density of filled states.

Despite not suffering from the influence of final state effects to such a degree, in practice other factors make valence spectra challenging to interpret. The first of these is that the valence electronic structure itself is more complex than the core electronic structure. Allowed energy states are usually dispersed over a much wider energy range due to orbital overlap, and furthermore the valence orbitals will usually contain contributions from different orbitals and, in compounds, from different elements. An important consequence arises from the fact that the different orbitals that make up the valence band may have significantly different photoionisation cross sections. In metal oxides excited with the typical lab photon energy, Al Kα (1486.6 eV), the cross section of the metal orbitals that contribute to the valence band is often much larger than the cross section of O 2p. Thus, the contribution to the valence band spectrum of the metal density of states may be much larger than their actual contribution to the electronic structure. Furthermore, changing the photon energy will change the ratio of cross sections for the orbitals that make up the valence band, so often the shape of valence band spectra is strongly photon energy dependent. This can be helpful as a way to identify contributions to the valence band from different orbitals.

A second factor is that, like core photoemission spectra, valence spectra are affected by broadening due to instrumental and lifetime effects. A feature specific to valence spectra is that lifetime of the valence hole may vary considerably with binding energy, *i.e.* a hole created at the top of the valence band, near the valence band maximum or Fermi level in a metal, may have a lifetime significantly different to one created several electronvolts lower in energy at the bottom of the valence band. This means the lifetime broadening can change across the width of the valence band spectrum.[98] Theoretical approaches to understanding this broadening, such as Fermi liquid theory, have been attempted but none have been found to be universally applicable across all sample types – understanding quantitatively the broadening of valence band spectra due to lifetime remains a challenge to be overcome.

Thirdly, the intensity of valence spectra recorded with X-rays is low, because the X-rays have far higher energy than the binding energy of the valence electrons. This means the cross sections are usually much lower than some core level from the constituent elements (see Figure 2.12). The low intensity means that long collection times are often needed.

Lastly, while core lines of different elements usually do not overlap, every material has valence electrons, so any surface contamination or impurity will contribute some signal in the valence region. Thus, the observed valence band spectrum will be a combination of the VB spectrum of the sample under study, and that of any surface contamination present. Luckily, adventitious carbon (AdC) contamination usually has a rather featureless valence band, so will not significantly change the shape of the VB spectrum of a sample where it is present. However, even AdC contamination can complicate attempts to quantitatively interpret valence bands, which is discussed further below.

As a consequence of all the factors discussed above, spectra arising from the valence orbitals are usually broad and irregularly shaped, and while core lines can sometimes be successfully interpreted by fitting with Gaussian–Lorentzian components that represent individual chemical environments, this approach is not applicable for valence band spectra.

2.11.2 Basic Features of a Valence Band Spectrum

The valence band represents the highest filled orbitals of a material, and because of the strong link between density of filled states, the shape of the valence band spectrum is characteristic of the type of material under investigation (Figure 2.16). A semiconductor has a filled valence band and an empty conduction band, separated by a band gap, which may be several electronvolts wide, and in which (in the ideal crystalline material) there are no allowed energy levels. The highest energy point on the valence band, the valence band maximum (VBM), can be determined by observing the onset of the XPS valence band spectrum, but note this is always relative to the Fermi level as the zero point of the binding energy scale. Figure 2.16 shows the example of the valence band spectrum of TiO_2, a wide band gap semiconductor. In this spectrum, a gap can be seen between zero binding energy and the onset of the valence band intensity. Determining the VBM energy (E_{VBM}) relative to the Fermi level may be done by fitting an edge function to the low binding energy edge of the valence band spectrum. Various different edge functions have been applied, such as a complementary error function and Heaviside step function. Although these are convenient and useful for most applications, no simple function can be said to be physically accurate; the shape of the valence band maximum of a semiconductor is determined by the electronic structure and is different for each material, not necessarily conforming to a simple step function (for metals the situation is different as discussed below). In a semiconductor, the Fermi level is usually within the band gap, its exact position determined by the defects present in the semiconductor. Defects which add extra electrons (n-type defects) will move the Fermi level towards the conduction band, whilst those that remove electrons from the metal (p-type defects) will do the opposite. In this way the Fermi

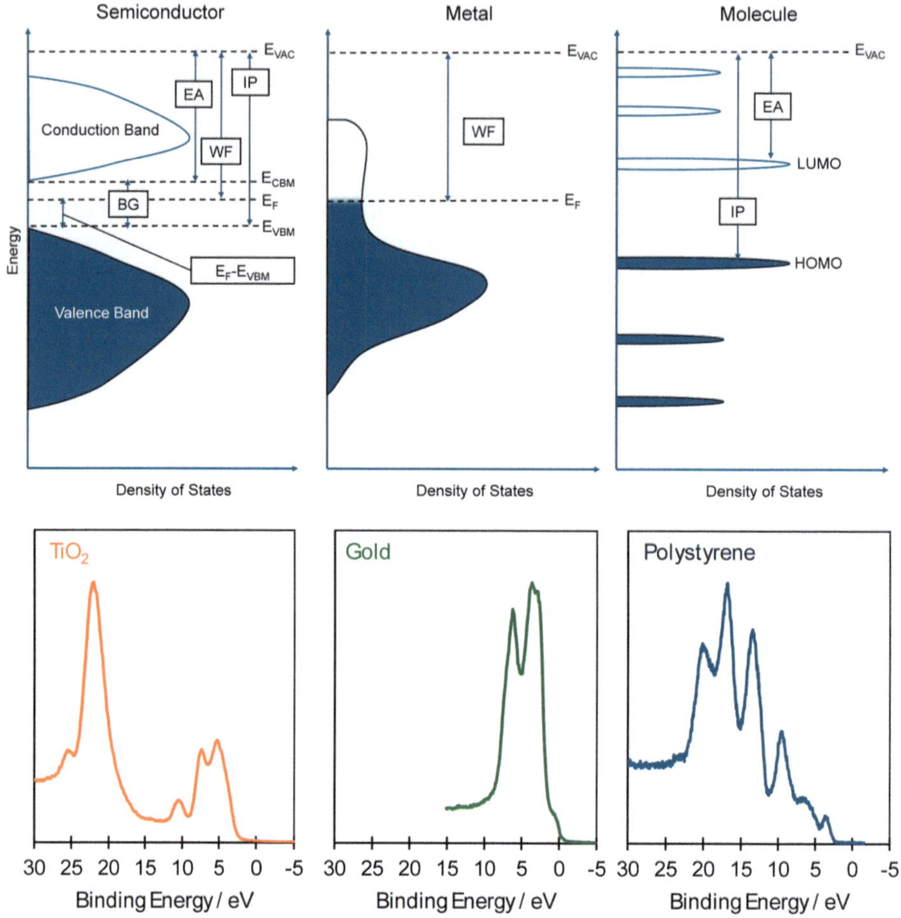

Figure 2.16 Top row: density of states for different materials classes: semiconductor, metal, molecule, and how these relate to observed energy differences in XPS: BG = bandgap, WF = workfunction, IP = ionisation potential, EA = electron affinity, E_F = Fermi level. Bottom row: examples of VB spectra taken from each class of compound (with a polymer standing in for molecular samples).

level can move sometimes by an electronvolt or more relative to the VBM through different defect or dopant chemistry. Since the Fermi level is the zero point on the binding energy scale, if E_F moves due to changes in defect concentrations, this will appear in the valence band spectrum as a shift in binding energy of the VBM (and all other spectral features including all core lines). If comparing several samples of the same semiconductor with different dopants, and if the Fermi level can be reliably determined (which can be difficult, see the section on charging), then the sample with the smallest E_{VBM} is likely to be the most p type. The workfunction, ionisation potential, and electron affinity are other characteristic energies of the semiconductor but cannot be extracted from a standard XP spectrum of the valence band. The workfunction can be obtained by electrically biasing the sample; this is a specialist measurement and is discussed further in Chapter 6.

A metal has its Fermi level within a band of allowed electronic states. Electrons fill the band up to the Fermi level, and at finite temperature some filled states exist above the Fermi level according to the Fermi–Dirac distribution. The VB spectrum from a metal (illustrated with a gold VB XP spectrum in Figure 2.16) has a measurable intensity at and just below zero binding energy, representing the Fermi–Dirac distribution convoluted with the instrumental resolution. Lastly, the electronic structure of molecules often consists of molecular orbitals, more localised, and therefore narrow in energy distribution, than in a semiconductor or metal. In Figure 2.16 this is illustrated with a polymer, polystyrene, and the VB spectrum consists of a series of sharp peaks representing the discrete molecular orbitals.

2.11.3 Qualitative Interpretation of Valence Band Spectra

This approach relies on simplified ideas of an inorganic material's electronic structure but can be surprisingly powerful as a heuristic to understanding valence band spectra. The orbitals that make up the valence region can be identified by consideration of the electronic configuration of the constituent elements, and the nature of the bonding. Valence band spectra, or more usefully, changes in spectra over a series of compounds, can then be interpreted in terms of changes to the density of the filled orbitals. Two examples will illustrate this idea.

Some valence band spectra from the well-studied oxide TiO_2 are shown in Figure 2.17. In this oxide we can consider as a first approximation that the bonding is purely ionic. This means that the titanium exists as Ti^{4+} ions and the oxygen as O^{2-} anions. The valence electronic configuration O^{2-} is $2s^22p^6$, *i.e.* it has filled 2s and 2p levels and a complete valence shell, the same overall configuration as the noble gas Ne. Ti^{4+} on the other hand has empty valence orbitals, which might be written $4s^03d^0$. In our purely ionic picture of the electronic configuration, it is clear that the highest energy filled states should be entirely made up of O 2p, possibly mixed with O 2s, while the unfilled states forming the conduction band would be comprised of Ti 3d,4s. Reality is more complex. DFT calculations by Lee *et al.* show that the TiO_2 valence band has non-negligible contributions from Ti 3d and other metal states (Figure 2.17c).[99]

We can now consider the valence band spectra from TiO_2, as shown in Figure 2.17a. The spectrum consists of a broad region with a width of approximately 9 eV. Also shown in Figure 2.17a are valence band spectra from Ir-doped TiO_2 with increasing concentration. One effect of doping open shell transition metals such as Ir into closed-shell metal oxides such as TiO_2, is that the partially filled d-orbitals of the dopant will be introduced into the electronic structure of the host oxide. As shown in Figure 2.17, introduction

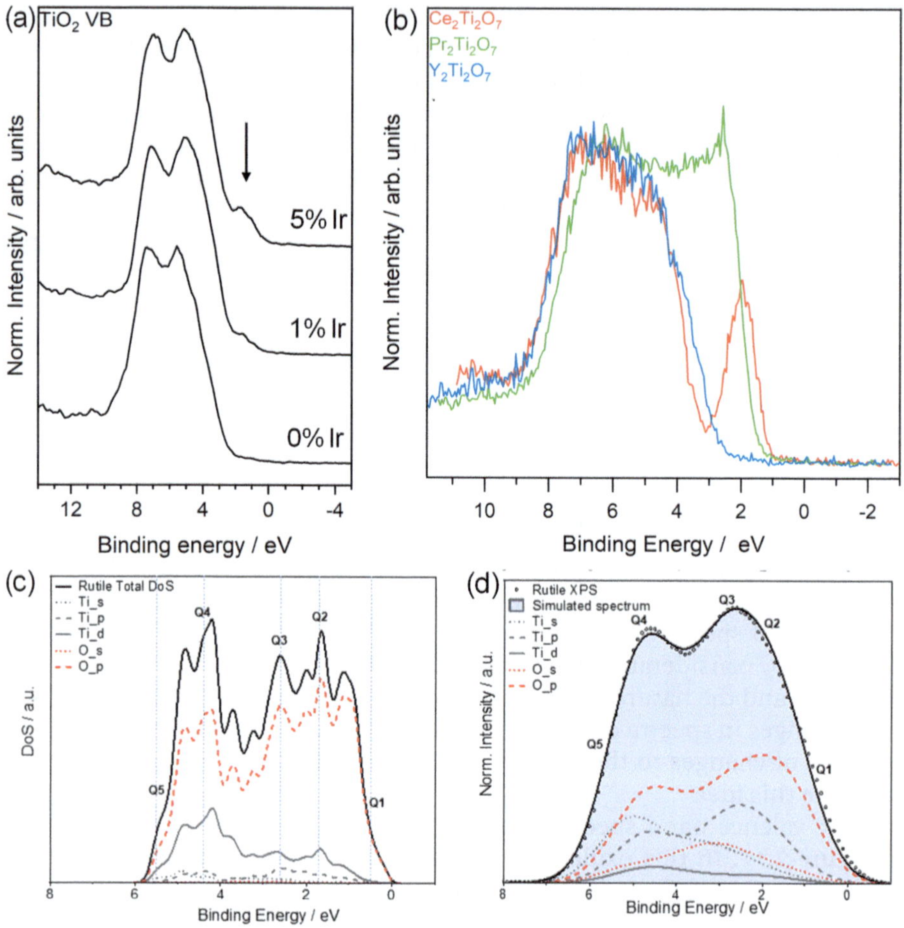

Figure 2.17 (a) XPS valence band spectra of undoped TiO_2 (bottom) 1% Ir-doped TiO_2 (middle) and 5% Ir-doped TiO_2 (top). The increasing intensity of the feature at the valence band maximum (around 1.5 eV binding energy) can be assigned as the Ir 5d electrons. (b) Valence band spectra of three pyrochlore oxides, $RE_2Ti_2O_7$. The spectra are normalised to the O 2s shallow core level. By comparing spectra in this series, the RE 4f electron contribution can be identified. (c) Calculated density of states of TiO_2 from DFT. (d) Simulated valence spectrum of rutile XPS obtained from (c) by applying sensitivity factors and broadening, compared with the experimental spectrum. (c) and (d) Reproduced from ref. 99, https://doi.org/10.1021/acsami.3c06638, under the terms of the CC BY 4.0 license, https://creativecommons.org/licenses/by/4.0/

of the Ir dopant leads to a new feature in the valence band spectrum (marked with an arrow) above the valence band maximum of undoped TiO_2, which scales with Ir concentration. This feature can be assigned as filled Ir 5d states. From this we can see that Ir 5d states are located close to or just above the TiO_2 valence band maximum, and this can be important for understanding the impact of this dopant on the electronic and optical properties of the material.

A second example involves a series of isostructural rare earth pyrochlore oxides, with formula $RE_2Ti_2O_7$, RE = Y^{3+}, Ce^{3+}, Pr^{3+}. The three RE cations have valence electron configurations of $4f^0$, $4f^1$, $4f^2$ respectively. The valence band spectra are shown in Figure 2.17b. For the RE = Y compound, we expect no 4f contribution to the filled states, so this spectrum represents closest to the 'pure' O 2p valence band. When RE = Ce, we expect to see some filled 4f states, and indeed a new feature appears, a relatively sharp peak around a binding energy of 2 eV. The position of this peak suggests the 4f states in $Ce_2Ti_2O_7$ are positioned above the valence band maximum of the O 2p valence band. When RE = Pr, we expect a roughly twice as intense 4f signal (due to the increase in 4f occupancy). Examining the spectrum of $Pr_2Ti_2O_7$ shows additional intensity in the region 1.5–4.0 eV, compared with the RE = Y analogue. As above, we can assign this intensity as arising from filled 4f states from the RE site. Finally, we can comment on the distribution of the 4f states in the RE = Ce and RE = Pr compounds. As mentioned, the RE = Ce case shows localised 4f states that appear separated from the O 2p dominated valence band. For RE = Pr however, the 4f states are now higher in binding energy, which is understandable as Pr is more electronegative than Ce, so we would expect the Pr 4f states to be more stable. Furthermore, the 4f states from Pr appear less localised, and overlapped more with the O 2p valence band.

The above discussion shows the potential for interpreting valence band spectra qualitatively by invoking knowledge of the electronic structure of the material. Care must be taken however, especially as XPS does not give definitive information on the identity of the compounds present. If such interpretations as these are to be made, there should be a high degree of certainty on the chemical identity of the material under investigation, for example with X-ray diffraction or similar methods where possible, and even where this is in place, an appreciation that the surface may differ from the bulk.

2.11.4 Calculation of Valence Band Spectra

While the qualitative approaches described above are easy to implement, for true quantitative analysis of the valence band spectrum, a calculation of the electronic structure is necessary. A good deal of success has been achieved in comparing experimental XPS valence band spectra to the results of calculations using Density Functional Theory. For valence band photoemission, creation of the valence hole has much less impact on the remaining electrons; this is partially because the valence electrons are much higher energy than core electrons, so do not partake significantly in shielding. It is also because the valence hole is often more delocalised than a core hole, spread over several atoms due to the delocalised nature of the valence orbitals themselves.

As a result, as far as the $(N-1)$ electrons are concerned, the final state is very similar to the initial state in valence band photoemission and so final state effects play a much smaller part. It becomes a reasonable approximation when dealing with valence spectra simply to ignore the final state entirely, go back to a single particle approximation, and calculate the spectrum using the initial state alone. Contrast this with the much more complex ΔSCF or alternative approaches needed to calculate core line BEs, and it can be seen why calculation of VB spectra from DFT is now a much more routine procedure.

Therefore, DFT partial densities of states can be scaled with photoionisation cross sections, and broadened with Gaussian and Lorentzian functions to simulate the experimental spectrum. Figure 2.17c and d show an example of this with rutile TiO_2. The density of states in Figure 2.17c shows the valence band is made up mostly of O 2p, with some Ti 3d and a small amount of Ti 4p mixture.[99] Once photoionisation cross sections are applied, however, it is seen in Figure 2.17d that the Ti 4p and 4s orbitals make a larger contribution to the spectrum than the Ti 3d orbitals do, on account of their much larger cross sections. Overall the match to the experimental spectrum is very good.

2.12 Peak Fitting

In the first part of this chapter, we examined the physical processes that occur when a photon ejects a core electron from an atom. The approach developed in Section 2.3 can be thought of as 'bottom up', in that it identifies different physical processes: relaxation, shake, splitting, chemical shift, inelastic scattering and so on, and assesses how these might appear in a spectrum. The main job of the experimental analyst however is in the opposite direction. We obtain spectra and need to interpret them in a 'top down' fashion, breaking spectra down into component parts that we can understand, ideally by invoking the physical processes just mentioned, all with the hope of answering our research questions about the sample. This is usually done with a process known as 'peak fitting' in which we match as closely as possible a mathematical model to our experimental spectrum. In fact, not only the peaks are fitted, as a background function has to be applied too (see Section 2.6.4), but the name has stuck.

Peak fitting in XPS lies at the intersection of physics, chemistry, and data analysis, and draws on each of these foundations. It can be a fraught process, with many perils for the unwary. As is general for modelling experimental observations, there are an infinite number of models that can be built that fit the data arbitrarily well. We can easily make a model that reproduces the data as closely as we wish by means of adding more and more parameters to the model, but despite looking superficially attractive, this approach fails to answer any relevant questions that we may have. The obvious consequence is that while a good fit to the data may be a necessary criterion for a good model, it is certainly not a sufficient one. Likewise, when deciding which of two contending models is better, comparing only their goodness of fit to the experimental data is a poor criterion.

So what does define a 'good' model? A model not only fits the data well, but also conforms with the known physical theory, laid out already in this chapter. As well as that, a good model conforms to chemical theory. And most importantly, it will help answer the research question of the investigator.

2.12.1 Synthetic Peak Shapes

A peak model is composed of synthetic peaks, known as *components*. These are mathematical functions that represent the shape of an experimental spectral feature. The synthetic peak shapes used in XPS fitting aim to model the underlying physical principles of peak broadening. As discussed in Section 2.3.2.9, peak broadening in XPS arises from both intrinsic and extrinsic factors. Intrinsic broadening is primarily due to the natural lifetime of the excited electronic states. Extrinsic broadening, on the other hand, results from instrumental factors such as the resolution of the X-ray source and the electron energy analyser.

The basic synthetic peak functions used in XPS fitting are Gaussian, which represents extrinsic broadening and vibrational broadening, and Lorentzian which represents lifetime broadening. Voigt peaks, resulting from the convolution of Gaussian and Lorentzian functions, provide a more realistic representation by incorporating both instrumental and intrinsic broadening effects. Pseudo-Voigt functions are computationally efficient approximations of Voigt profiles, combining Gaussian and Lorentzian components using a weighted sum. The proportion of Gaussian and Lorentzian contributions to the peak shape depends on the material and the resolution of the instrument, and proper selection is an important step in peak fitting. Here we refer to any peak shape made primarily from Gaussian and Lorentzian components as a *GL peak*, no matter how the components are combined mathematically.

Simple GL peak shapes can be effective at modelling some intrinsic photoemission features and combined to model more complex ones. Spectra from different core lines may be influenced by some or all of the intrinsic effects shown in Table 2.4, and so proper modelling of the intrinsic structure ranges from straightforward to extremely challenging.

2.12.2 Fitting Simple 1s Spectra

The simplest possible example of a core line would be a 1s spectrum of a light element (C, N, O, F *etc.*) that undergoes no multiplet splitting, has no differently screened final states or shake up satellites of measurable intensity, and arises from elements in only one chemical environment in an insulator which is not charging appreciably. In this simplest case, the core line can be fitted well with a simple GL peak shape. If the complexity is increased only by introducing additional chemical environments, then these environments can be fitted with one GL function each. This approach, sometimes called 'chemistry fitting',[100] is based on attributing each GL function to a different chemical environment. It can be a valid approach for simple spectra with no multiplet, screening or shake up effects, but can fail badly for anything that does suffer these complications, as we will see. Even in simple cases with multiple chemical environments, here we may encounter difficulty; it is possible that a spectrum could be well fitted by models consisting of different numbers or arrangements of GL peaks. How should such candidate models be assessed, and the best one chosen? An effective tool is to check the FWHM values of the GL components. A rule of thumb for s orbital core lines is that different chemical environments of the same element will be represented by components of equal FWHM. Like most rules of thumb there are exceptions. Some small variations will occur due to differences in vibrational broadening. For example, in polymers, carbon atoms bonded to hydrogen have broader C 1s peaks than quaternary carbons, due to the higher vibrational frequency of C–H bonds.[101] Another exception is commonly seen in the oxides, where the O^{2-} peak from a metal oxide is considerably narrower than those peaks from oxygen in organic environments. Nonetheless, in the absence of any outside information about a sample, when fitting a 1s spectrum it is advisable to use identical FWHM values between different components, as this will avoid the worst errors and lead to a reasonably useful model in most cases.

2.12.3 Fitting Spectra with Spin Orbit Coupling

Where no multiplet splitting is expected, spin orbit doublets should each be fitted with two GL components. These should be constrained to have the correct intensity ratio according to the degeneracy of the spin orbit states (Table 2.2) – note that the lower binding energy member of the doublet is always the high spin branch, so has greater intensity. The doublet separations for all common core lines are known and can be looked up in suitable references, and the component separation constrained to this value. In many cases the FWHM of the two spin orbit components is the same. Exceptions are where the Coster–Konig transition makes the lifetime of the low spin peak shorter, and therefore the FWHM larger, or if multiplet splitting affects the two spin branches differently. If multiple chemical environments are present, then one doublet is needed for each chemical environment, and in this case the separation between the spin orbit peaks in each doublet should certainly be constrained to ensure accurate fitting.

2.12.4 Fitting Spectra with Multiplet Splitting and Relaxation

Where multiplet splitting and relaxation is expected, fitting becomes much more difficult. Multiplet patterns are difficult to calculate and require detailed knowledge of the compound under analysis, which is often not available especially for new compounds or complex real-world samples. Simpler models may be based on Gupta–Sen free ion multiplets (which neglect screening or crystal field effects),[14,17] and these can work well in some circumstances. Going beyond this, full calculation of initial and final states and the probability of transitions between them is possible but not a straightforward task by any means: the approaches of Bagus and coworkers are a good starting point for those looking to undertake this.[84,102] Many XPS analysts use the outputs of such methods without performing calculations themselves. An alternative is to measure standard compounds and use their empirical spectra to fit complex core lines. Good examples are to be found in the work of Biesinger and coworkers.[100]

An approach that works very badly is just to extend the 'chemistry fitting' that is useful with simple 1s orbitals to more complex spectra. Attempting to fit open-shell transition metal 2p orbitals with two or more symmetrical line shapes, each assigned to a different charge state of the metal, is unfortunately not a rare occurrence in the literature, but is hardly ever appropriate when multiplet splitting is significant.

2.12.5 Examples of Good and Bad Fitting

Figure 2.18 summarises some good and bad peak fits for the types of spectra considered so far: spectra for core lines with and without spin orbit coupling (SOC), multiplet splitting (MS), and multiple chemical environments are shown.

In panel (a), one chemical environment is present with no SOC or MS. A poor fit might have the wrong peak shape, as shown here, where the ratio of Gaussian to Lorentzian is not optimised. On the improved fit, the GL ratio has been corrected.

In panel (b), we again have no SOC or MS, but now have multiple chemical environments. Here the poor fit uses three components with very different FWHM. While it is possible that some difference in FWHM can be physically realistic, in the absence of specific knowledge that this is the case, it is better to fit the spectrum with components of equal FWHM, as in the improved fit, where now four components fit the spectrum well.

In panel (c), we now have a spin orbit doublet, in this case a p orbital with small splitting that is unresolved. The poor fitting does not take into account the physics of spin orbit coupling: the components have the incorrect ratio and the smaller component at lower binding energy. The improved fit uses the correct 2 : 1 ratio between the spin orbit components, and the high spin branch is at lower binding energy.

In panel (d) we now have two doublets overlapped – for example this could be the 2p core line of an element present in two chemical environments. The poor fitting again doesn't take into account the physics, with only three components none of which are constrained by the appropriate intensity ratios. The improved fit has two doublets (the pairings indicated by similar shading). Both doublets have constrained intensity ratios, FWHM and separation. Each doublet is permitted a different FWHM, as the different chemical environments may have different vibrational broadening.

Panel (e) shows the high spin branch of the spin orbit doublet of a multiplet spectrum, in this case a $2p_{3/2}$ orbital from an open-shell transition metal. The poor fitting uses two symmetrical peaks and assigns these to different

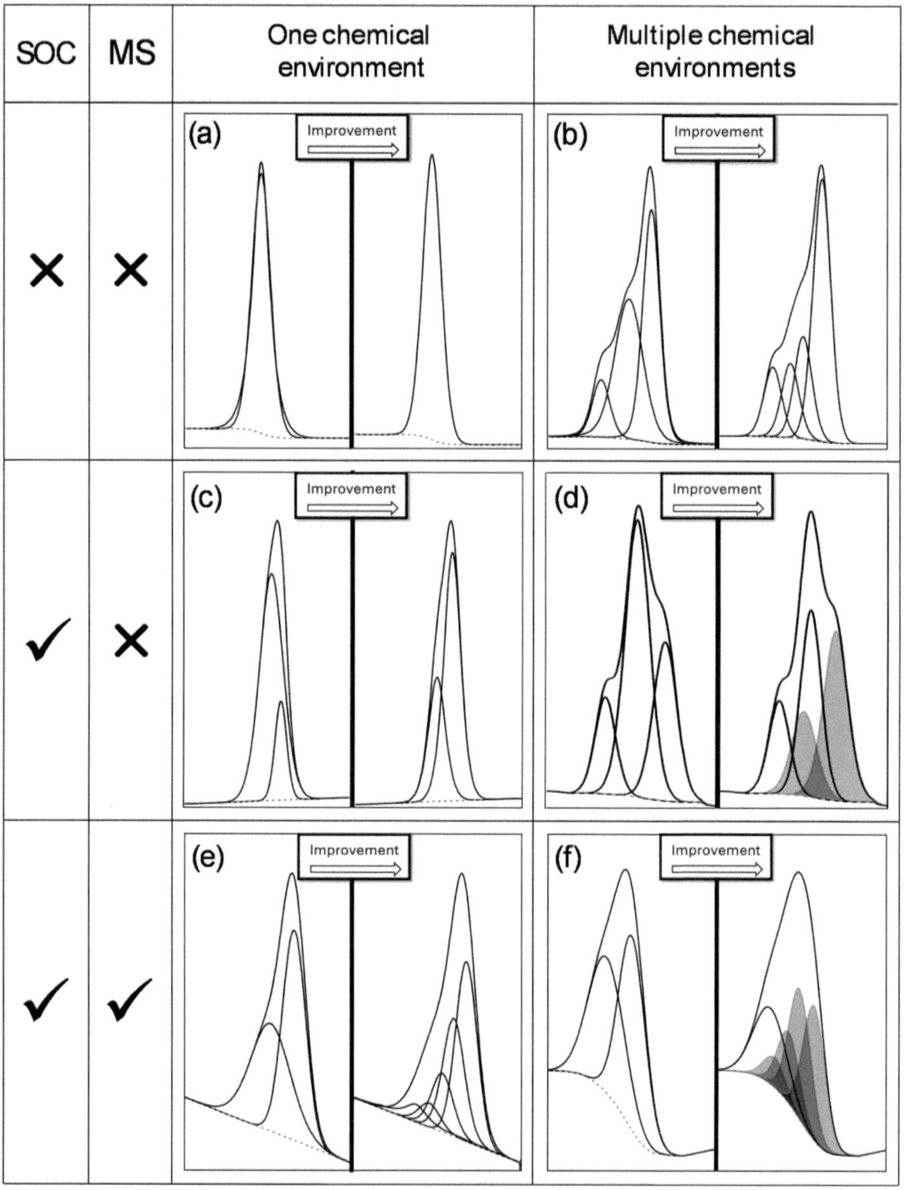

Figure 2.18 Examples of poor and improved fitting for core lines with and without spin orbit coupling (SOC), multiplet splitting (MS) and multiple chemical environments. Each example is discussed in the text. Binding energy increases from right to left as is usual for XPS plots. (a) Optimisation of the Gaussian/Lorentzian ratio; (b) imposing consistent FWHM; (c) using the correct area ratios for known spin orbit splittings; (d) and (f) including the expected spin orbit splitting of peaks; (e) using appropriate models to account for multiplet splitting.

oxidation states of the metal. In fact, this is one oxidation state, and can be more realistically fitted using the Gupta–Sen free ion multiplet components, constrained in intensity and separation.

Panel (f) again shows the high spin branch of a transition metal 2p orbital that has multiplet splitting. Here there are two chemical environments. The poor fitting uses two components with no regard for the multiplet. The improved fitting uses components from a metal oxide multiplet (shaded) and a broad peak (unshaded) to represent a hydroxide environment. A fit to such a challenging spectrum as (f) can be supported by using additional information on the chemical nature of the sample, *e.g.* here an analysis of the O 1s peak might support assignment of the oxide and hydroxide peaks.

2.12.6 Fitting Using Asymmetric Lineshapes

Asymmetric peak shapes in XP spectra arise because of closely spaced final states, where a main line has several closely spaced final states at higher binding energy.[103] This is commonly seen in metals, where shake up to the continuum of empty states in the conduction band above the Fermi level creates a characteristic tail in the spectrum at higher binding energies. Graphitic carbon materials also show asymmetry, for similar reasons. Multiplet splitting often leads to closely spaced unresolved final states, that appear as asymmetric peaks, although unlike shake up processes in metals, there is no requirement for the lowest binding energy peak to be the most intense.

From a peak fitting point of view, asymmetry can be addressed either by using asymmetric peak shapes or by fitting multiple symmetric peaks. Asymmetric peak shapes, such as the Doniach–Sunjic or asymmetric Lorentzian profiles, can be well-suited for metals and materials where asymmetry reflects intrinsic electronic properties. Alternatively, fitting multiple symmetrical peaks, typically GL profiles, can approximate asymmetry but may lead to over-parameterisation and misinterpretation, such as assigning peaks to non-existent chemical states. However the asymmetry is dealt with, a good model will be grounded in a clear understanding of the material's chemistry and electronic structure.

2.12.7 Fitting Overlapping Peaks from Different Elements

For peak fitting, the use of reference compounds is an excellent strategy and allows the analysts to capture satellite structure, peak asymmetry or be able to note overlapping chemical states. Figure 2.19 shows the fitting obtained for a Pd on ZrO_2 catalyst. The Pd 3d signal overlaps with the Zr 3p peaks, which are further complicated by Zr satellite structure which much be accounted for in any successful peak fitting. Using the spectrum from the Pd-free ZrO_2 material this can easily be modelled.[104]

Applying the peaks from the oxide, fitted with their known spin–orbit splitting and 2 : 1 intensity ratio and constrained widths and accounting for the satellite structure, it was found that both Pd^0 and Pd^{2+} states were required to model the data, and importantly the presence of both Pd oxidation states accounted for observed catalytic activity.

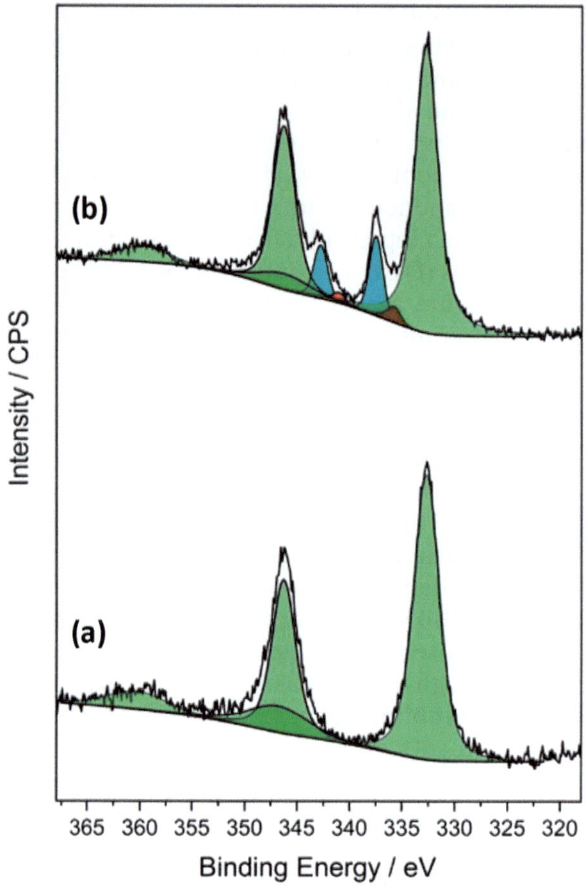

Figure 2.19 (a) Zr 3p peaks and satellite structure for a ZrO_2 support and (b) Zr 3p peaks fitted according to parameters in (a) together with peaks for Pd^0 (red) and Pd^{2+} (cyan).

As for chemical models, where functional groups or compounds are identified, the proportion of each constituent element is in the expected ratio. For example, if the sample is thought to contain PbS, then the Pb and S should be in the expected chemical environments, and also in 1 : 1 proportion. Likewise for organic functional groups: if the C–OH group is modelled, then there should be components of the C 1s and O 1s spectrum that correspond to this model in the correct BE regions and in the correct proportions.

2.12.8 Dealing with Complex Spectra

Having progressed this far, it is evident that for certain compounds, several final state effects may contribute significantly to a single core line spectrum, leading to complex peak shapes even for ideal, pure samples. Real-world analysis problems might often involve much less well-defined samples. How is the analyst to deal with complex spectra that can arise through the various processes described above?

The primary rule for peak fitting is that it must be physically meaningful. Many models fail this simple point for reasons such as a poorly described background, erroneous peak assignments, misinterpretation of satellite structure, incorrect spin–orbit splitting ratios *etc.* Analysts may also be swayed by statistics of a fit – whilst a peak model may have a good chi-squared value or residual spectrum, this is just a measure of how the model fits the raw data, not how accurate the interpretation is.

A common mistake is to fit all spectra irrespective of whether the research question to be answered requires it. Considering the lower two Ti 2p spectra in Figure 2.20, (a) and (b) are the spectra for TiO_2 and the titanosilicate TS-1 respectively. For the Ti 2p spectrum of TiO_2, no fitting is required as this is a single chemical state, so a suitable background can be applied and the concentration of Ti readily quantified. However, for the TS-1 spectra, whether to fit or not can be decided by defining what information we want from the analysis. If the object of the analysis is to ascertain the total Ti content in the near surface region, then addition of a background is all that is required. However, should we wish to know the amounts of Ti in octahedral and tetrahedral sites, then fitting would be required as shown in Figure 2.20c.

We can offer some general tips for dealing with complex spectra:

- Define your research question: do you need to do peak fitting to answer your question, or will a simple quantification suffice that can be done by taking the total area of the relevant core lines?
- Look for reliable assignments: many papers in the literature make assignments that are uninformed by the underlying theory. Peaks labelled 'shake up', 'multiplet' or other terms may in fact be something quite different. The literature has evolved and even in careful work, many older assignments have subsequently been reconsidered.
- Measure standards: understanding the spectral appearance of standard compounds, with as high purity as you can achieve, will always help. It also helps to record these on the same spectrometer you use for your research samples.
- Trend analysis: looking for trends in data is always preferable to seeking to make an assignment from one sample alone. If, for example, a series of samples have been produced with different Fe^{3+} concentration, when considered together the samples might be placed in order of the concentration of Fe^{3+} far more easily than the absolute Fe^{3+} concentration can be determined.
- Assess the confidence level of your model: for every set of experimental data, there will be multiple models that fit the data well. The first model produced isn't necessarily the best one, and the analyst should consider carefully if there are alternative candidate models that are equally good at matching the data whilst also being physically meaningful. How different are the conclusions that flow from these candidate models? If the conclusions are very different, but the models equally good, the confidence in the answers to your research questions may be low.

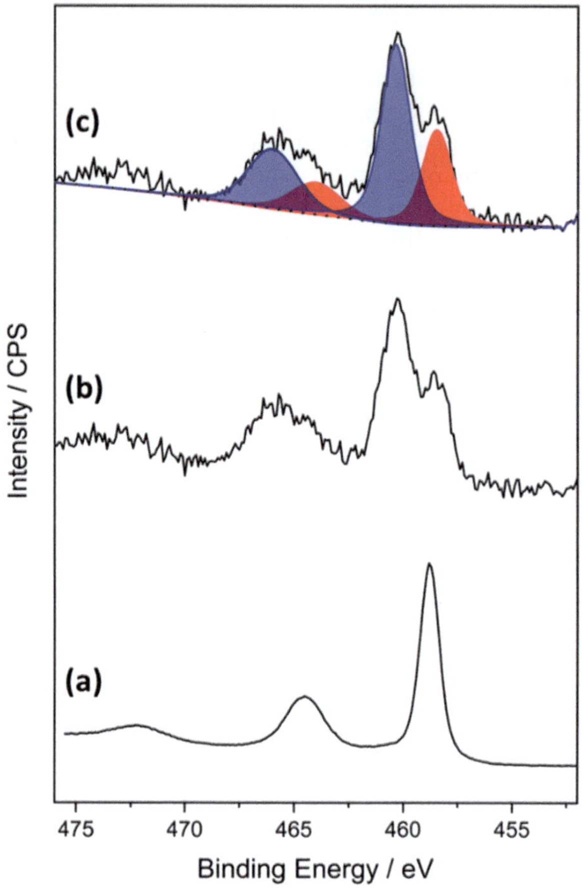

Figure 2.20 Ti 2p core-level spectra for (a) TiO_2 and (b) TS-1 titanosilicate material and (c) fitted version of (b) to allow the determination of the percentage of Ti in tetrahedral (lilac) and octahedral (red) sites.

2.13 Chemical Approaches to Peak Fitting

While the physical theory of photoemission presented above allows us to understand in great detail the spectra of a particular core line from a chemically homogeneous sample, XPS is increasingly used for non-ideal samples that have not been specifically produced for XPS, but the analysis of which is necessary as part of a wider programme of research. These samples may be non-homogeneous, multi-phase, and likely have unknown surface contaminants or reaction by-products. Such samples may exhibit complex spectra, for which many interpretations are possible. While several models may offer comparable statistical goodness of fit, and agreement with physical theory, they might lead to distinctly different conclusions about the sample's chemical composition and phases. For example, a spectrum fitted with a simple two-peak model might indicate two distinct phases, whereas a more complex model with overlapping peaks could suggest a gradient in chemical environments.

To distinguish between these competing models, additional information beyond the spectral data itself is often necessary. One approach is to incorporate chemical knowledge, either general chemical principles or specific details of the sample of interest. By conforming our XPS models to the expected laws of chemistry, as well as those of physics, we can gain more confidence in the models we build.

We have already encountered simple examples of this. If a carbonate (C=O) group is present, we expect to find a C 1s environment near to 288 eV, and an O 1s environment near to 532 eV, and the atomic % should be in roughly 1:3 ratio, taking into account the uncertainty in XPS-derived compositions. If a peak attributed to a $-CF_3$ group appears in the C 1s spectrum, the corresponding F 1s peak should also appear, and their atomic percentages should align with the expected 1:3 ratio. If the ratios are found to be far away from the expected values, then this indicates there may be another material present, in addition to or instead of that originally proposed. We can use this approach for chemical formula as well as functional groups. The perovskites have general formula ABX_3, so should give a ratio of the elements A:B:X of 1:1:3.

In addition to analysing composition, we can also investigate the binding energy separation between core lines. Although determining absolute binding energies in non-metals is often complicated by charging effects, binding energy differences between core lines of atoms within the same phase are far more constant. Core lines from elements within the same phase often exhibit consistent binding energy shifts due to uniform changes in the electrostatic potential. This means that in multi-phase samples, elements that belong to the same phase will typically exhibit uniform shifts in binding energy due to charging. Even if two phases charge by different amounts, we can use the idea of constant ΔBE to help build peak models. This type of analysis can be particularly effective in distinguishing phases with overlapping core lines or in validating proposed phase models against observed data.

By enforcing chemical consistency, both in atomic concentration ratios and in ΔBE, we can reduce ambiguity in interpretation of data. Deviations from expected ratios can indicate additional phases, or contamination. These deviations prompt further investigation using complementary methods or refined models.

2.13.1 Large Data Sets: Chemometric Approaches

Modern spectrometers are capable carrying out data acquisition with increasing speed, and can produce large volumes of data. This presents an opportunity to adopt data handling techniques that can extract information from such large datasets. Tools such as principal component analysis (PCA) and multivariate curve resolution (MCR) can help the analyst extract information from these large datasets.[105,106] Of these, MCR is the most straightforward for an XPS analyst to understand. It can be applied where a large number of spectra are recorded from the same core line, where there are different chemical environments present in different proportions across the dataset. MCR takes the idea that a set of spectra are made up of components, each representing a chemical environment, that can be summed together in different proportions to yield each individual spectrum.[107] The aim of MCR is to identify these component spectra: the analyst specifies how many components the divide the spectra into, and then must judge what number leads the algorithm to give the most suitable results. PCA is based on a similar idea, but now the components must be mathematically orthogonal to each other, and can also have negative intensity. These two points mean the components that result from PCA are sometimes impossible to interpret directly as XPS chemical environments.[108]

2.13.2 Correlation Analysis

The concept of correlation analysis is an extension of the chemical model approach. It is simpler to apply than PCA or MCR, but uses some of the same ideas.[109] As with MCR it uses large datasets consisting of many instances of one or more core line spectra taken from a sample system – this could be a depth profile or large area map make up of many analysis points. The difference from MCR is that correlation analysis builds chemical phase models by investigating relationships in XPS features across multiple spectra, to identify which belong to the same phase. The hypothesis behind correlation analysis is that the atomic concentration ratio and ΔBE will be fixed for elements in the same chemical phase. Therefore, if linear relationships are seen in in these relationships between core lines across a large dataset of multiple spectra from the sample type, this can indicate that the core lines originate from the same chemical phase. Likewise for atomic concentration, if two elements belong to the same phase, their concentrations should correlate linearly, with the slope determined by the stoichiometry of the phase.

2.13.2.1 Example: Native Oxide on Aluminium Foil

Consider the case of aluminium foil with a native oxide layer. We can assume the system consists of two chemical phases within the information depth of XPS: metallic aluminium (Al) and aluminium oxide (Al_2O_3).

During depth profiling, where the oxide layer is progressively removed by ion etching, the total concentrations of Al and O will vary. The Al signal may be decomposed into an Al(metal) and an Al(oxide) component, since these two chemical environments are discernible by XPS. Because of the Al_2O_3 stoichiometry, and because no other phase in our phase model contains either Al(oxide) or oxygen, the ratio of Al(oxide):O should be 2:3, whatever the proportion of the metal and oxide phase. Therefore if we plot the Al(oxide) concentration against the O concentration over all spectra in the depth profile, we expect a linear correlation with a gradient of 1.5. This known chemical correlation can be used to test different proposed peak fitting models for the Al and O core lines. Some examples are shown in Figure 2.21: three different models (A, B, and C) are shown to fit the Al 2p peak containing Al(oxide) and

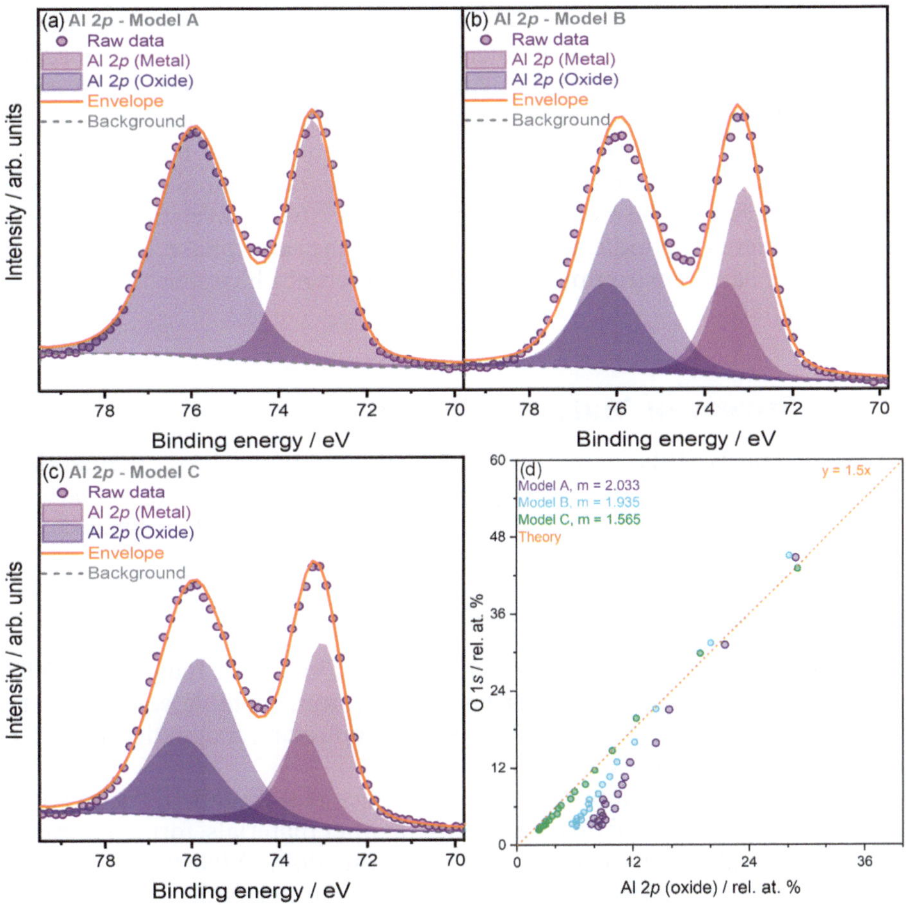

Figure 2.21 (a)–(c) Three different possible models for fitting the Al 2p peak of Al foil with a native oxide layer. Model A has a single peak for each environment, ignoring the presence of spin orbit coupling. Model B includes spin orbit coupling, but all components are symmetric. Model C include asymmetry of the Al metal peak. The correlation graph over the course of the depth profile shown in (d) shows the correlations of the Al(oxide) and O concentrations derived from each of the three models fitted to all the levels of a depth profile. The correlation reveals Model C has the best fit over the entire dataset to the expected composition. Reproduced from ref. 109, https://doi.org/10.1016/j.apsusc.2024.160808, under the terms of the CC BY 4.0 license, https://creativecommons.org/licenses/by/4.0/.

Al(metal) components. If the Al(oxide) and O components show constant ΔBE and composition ratio of 2:3, we can say our peak fitting is consistent with the chemical model. If a particular model predicts an incorrect ratio or fails to produce a linear relationship consistent with the expected stoichiometry, it can be discarded or refined. Model A is poor because it doesn't use a spin orbit doublet for the Al 2p. Model B is better, using a spin orbit doublet but using symmetric peaks for Al(metal), where we would expect asymmetry due to the shake up process. Model C incorporates both asymmetry of the metal peaks, and spin orbit doublets. The correlation plot in Figure 2.21d shows that models A and B fit the expected composition correlation less well than model C. In this way, consideration of the chemistry of the sample by means of correlation plots can help judge between different peak models.

For more complex systems with multiple phases and shared elements, correlation analysis can be more challenging. While linear correlations between two components can indicate their presence in the same phase, care must be taken if the same environments appear in different phases, as this complicates the analysis.

2.14 Survey of Material Classes

In the final part of this chapter, we survey some common classes of materials which very frequently are the subject of XPS analysis, pointing out special features or points of caution to keep in mind when undertaking analysis.

2.14.1 Oxides

Metal oxides represent the best studied class of inorganic materials by XPS (and maybe overall), with properties ranging from metallic conductors to high k dielectrics, high temperature superconductors, magnetic materials, catalysts, polar materials and so on.

Oxides also play an important role as surface passivation layers on some metals but are also found on a wide range of other materials, formed through reaction with atmospheric oxygen. Thus anyone doing XPS studies on inorganic materials regularly will sooner or later encounter oxides.

The O 1s peak from the oxide anion falls into a narrow range of binding energies, with most examples found in the interval 528–531 eV. The variation of O 1s BE in this range does follow some periodic trends, for example the group 2 oxides see a decrease in O 1s BE down the group: this is best explained by the influence of the Madelung potential rather than a difference in covalency.[110] However, inferring much about the bonding from the position of the oxide O 1s core line alone is usually impossible. Other chemical environments of oxygen do show significant chemical shift compared with the oxide anion however.

In an as-presented sample that has come from atmospheric conditions, there are often higher BE O 1s environments that can be assigned to species such as hydroxide, organic oxygen, carbonate, and silicones. Although many of these appear at very similar BE values, assignment of the other oxygen environments can be aided by cross-checking with other core lines expected, *e.g.* a suspected carbonate should display both an O 1s peak around 532 eV, and a C 1s peak at 288 eV, and it is good practice to check both before making the assignment. It is worth noting here that a mistake found in the literature is assignment of some part of the O 1s signal to 'oxygen vacancies'. Despite photoemission from the core orbitals of a vacancy being impossible (on account of it not having any orbitals, core or otherwise), this assignment has become increasingly common in the literature. Most examples of 'oxygen vacancy' components are better assigned as one of the higher BE oxygen environments mentioned above and listed in Table 2.5.

2.14.2 Halides

Inorganic halides are important for optoelectronic properties, ionic conductivity and many emerging applications. Care must be taken with sample suitability for XPS. Some molecular metal halides are volatile (*e.g.* SnI_4) and many decompose under X-ray irradiation, so testing for sample stability should be a first step in analysis. A special concern are iodides – since iodide is the most easily oxidised halide, metal iodides may contain elemental iodine from the synthesis or may decompose to form elemental iodide under vacuum or X-ray conditions. Elemental iodide will quickly vaporise in vacuum and may contaminate other samples and the spectrometer itself. Bromides also suffer this problem to a lesser extent. Fluorine contamination is less likely, but can have worse consequences, as being so light and forming such strong bonds, F is difficult to remove from a spectrometer by usual strategies such as baking.

XPS analysis of halides should include the relevant core line for the anion, F 1s, Cl 2p, Br 3d and I 3d are the best choices when using most conventional photon energies. Halide surface reactivity varies considerably. Compounds with electropositive metals such as NaF and CaF_2 suffer little oxidation on

Table 2.5 Some common oxygen chemical environments. Reproduced from ref. 111 with permission from Elsevier, Copyright 2021.

Common O 1s environment	BE range/eV	Notes
Metal oxide	528–531	Small vibrational broadening
Metal hydroxide	530–531	Large vibrational broadening
Organic oxygen	531–533	Associated C 1s environments 286–288 eV
Metal carbonate	531–532	Associated with C 1s environment at 288–289 eV
Silica/silicone	533	Associated with a Si 2p signal 102–104 eV

air exposure. Other compounds, for example the Li halides, react quickly forming surface oxide, carbonate and hydroxide species, and others, such as bismuth halides will form oxyhalides. Collection of O 1s is recommended to understand any surface oxidation.

2.14.3 Sulfides

Metal sulfides are important as solar cell materials and thermoelectrics, and form a class of 2D materials, the transition metal disulfides. As with halides, they may suffer oxidation at the surface due to reaction with atmospheric gases. The S 2p peak has a small spin orbit splitting and a wide range of chemical shifts, from sulfates at 169 eV to organic thiols at 165 eV and metal sulfides at around 162 eV. Sulfates have an associated O 1s peak around 533 eV.

2.14.4 Transition Metal Compounds

Transition metals are very commonly found in functional inorganic materials. One of their features is that they can adopt variable oxidation states, leading to important roles in technologies such as intercalation batteries, catalysts, magnetic materials and so on.

XPS analysis of transition metals compounds is often motivated by the need to qualitatively identify or quantitatively measure the oxidation state of the metals present. Depending on the metal, this can be a challenging task. The occurrence of multiplet splitting, due to the interaction of the core hole with unpaired spins in the valence shell, can produce complex spectra that depend not only on the oxidation state, but on the coordination environment of the metal. Where a mixture of metal oxidation states is present, it is often not possible to quantify an individual spectrum with any reliability.

Multiplet splitting, relaxation and shake up effects are important in the spectra of the first row transition metals (Ti to Cu). Multiplet splitting tends to be more significant here compared with the 2nd and 3rd row transition metals, as there is stronger exchange interaction between open 3d shells and core orbitals, compared with open 4d or 5d shells. The multiplet splitting in most open-shell 1st row transition metal 2p spectra is not resolved, and so its effect is to broaden and distort the 2p peak shape, giving broad asymmetric peaks even when only one chemical environment is present. Covalency mixes the metal 3d and ligand orbitals, and so decreases the exchange interaction between the metal core and valence: increasing covalency reduces the importance of multiplet splitting, and this can be significant even in metal oxides normally thought of as highly ionic.[112] For transition metal oxides, the 'main line' (ML) peak, being the most intense peak in the 2p spectrum, in virtually all cases represents a non-locally screened state, with electron transfer from the oxide ligands to the metal having occurred, resulting in a final state with higher 3d occupancy than expected from the formal electron configuration.

For example, the Ti 2p spectrum in TiO_2 consists of a main line made of several final states with an average d orbital occupancy of 1.8 electrons, despite the formal $3d^0$ configuration of Ti^{4+}.[113] A higher binding energy unscreened final state, with more $3d^0$ character, is seen around 13.5 eV away from the main line. For metals near to the middle of the series, a great number of possible charge transfer and multiplet states is possible; Bagus *et al.* commented that '*literally hundreds of final states contribute to the Fe 2p XPS of solid* Fe_2O_3'.[84]

Attempts to fit strongly multiplet and charge transfer affected transition metal 2p peaks using simple ideas of the chemical shift from the single particle approximation, for example fitting symmetrical components to represent different oxidation states as might be done for 1s orbitals of C, N, O or F, are doomed to failure (although sadly not uncommon in published literature). The complex natures of the initial and final states for transition metal spectra defy any useful application of the single particle approximation. Considerable effort has been made by many researchers to understand the TM core level spectra in solids, a task that is still ongoing.

A word on nomenclature: what we have referred to as screened or unscreened peaks are variously called charge transfer peaks, shake (up) peaks, or the general term, satellite; the latter is non-descriptive in terms of the mechanism. This may represent different words used for the same phenomena or maybe indicate genuinely different understanding of the physical process that led to particular features. In Table 2.6 we use the general term 'main line' to describe the lowest binding energy intense feature, and 'satellite' to describe a higher energy feature of significant intensity.

The 2nd and 3rd row transition metals tend to show far simpler spectra. Figure 2.22 shows the 2p spectrum from NiO,[120] and the 3d spectrum from PdO.[121] Moving down group 10 from Ni to Pd causes a dramatic change in the shape of the spectrum, from the highly complex Ni 2p, to the relatively simple looking Pd 3d. This is because the exchange interaction between the 2p and valence 3d is greater than that between the 3d and valence 4d orbitals. Screening features due to charge transfer are also considerably less prominent in PdO.

2.14.5 Carbon-based Materials

Carbon is ubiquitous in any sample exposed to atmosphere. Where carbon forms part of the material under study, special care must be taken to be sure the carbon of the sample is differentiated from the adventitious carbon present from surface contamination.

This principal core line is of course C 1s with a binding energy close to 285 eV. Graphite (or sp^2 carbon) and diamond (or sp^3 carbon) are distinguishable by their peak shape as well as BE: graphite having a population of free electrons has an asymmetric shape due to shake up to unfilled levels just above the Fermi level. Table 2.7 shows common carbon environments and their binding energies.

Table 2.6 Properties of the 2p core lines for the first row transition metals from Ti to Zn. ML = main line, sat = satellite, MS = multiplet splitting.

Core line	Spin orbit splitting (eV)	Multiplet splitting (MS)	Satellite (sat.)	Other features
Ti 2p	6.0	Very small multiplet splitting. Ti^{3+} and Ti^{2+} can be modelled as asymmetric peaks, and have considerable chemical shift from Ti^{4+}.	TiO_2: sat. 4 eV (weak), 13.5 eV (strong) above ML.[114] Strong sat. is around 25% of the ML.	Coster–Konig transition makes the $2p_{1/2}$ significantly broader than the $2p_{3/2}$.
V 2p	7.5	V^{3+}, V^{4+} show clear asymmetry.	V_2O_5: very weak sat. VO_2 and V_2O_3 have increasingly intense sat.	Overlaps with O 1s.
Cr 2p	9.3	All oxidation states show strong multiplet splitting.	Cr_2O_3: Sat. 12 eV above ML, 32% intensity.[114]	Hydroxide often forms on contact of Cr oxides with air.
Mn 2p	11.2	All oxidation states show strong multiplet splitting.[115]	MnO: sat. affects $2p_{1/2}$ more strongly.[116]	Distinguishing oxidation states is very hard.
Fe 2p	13.1	All oxidation states show strong multiplet splitting except low spin octahedral Fe^{2+} (which has $S = 0$).	Fe_2O_3: Sat. 7 and 14 eV above ML 85% intensity. [114]	Full peak shape not fully understood even for Fe_2O_3.[117] Distinguishing oxidation states is very hard
Co 2p	15	All oxidation states show strong multiplet splitting.	Cr_2O_3: no resolved satellites.	Cr^{3+} 2p peaks can be explained with MS and crystal field theory.[118]
Ni 2p	17.3	All oxidation states show strong multiplet splitting.	NiO: Sat. 7 eV above ML.	NiO spectral assignment still under debate.[102]
Cu 2p	19.8	Cu^+ is closed shell. Cu^{2+} shows strong MS dependent on coordination environment.	CuO: ML is $3d^{10}\underline{L}$, sat. peak around 6.5 eV above ML.[119]	Cu^+, Cu^{2+} easily distinguishable by MS and CT satellite structure.
Zn 2p	23	None, Zn^{2+} closed shell.	ZnO: sat. 13.8 eV above ML, 5% intensity.[117]	

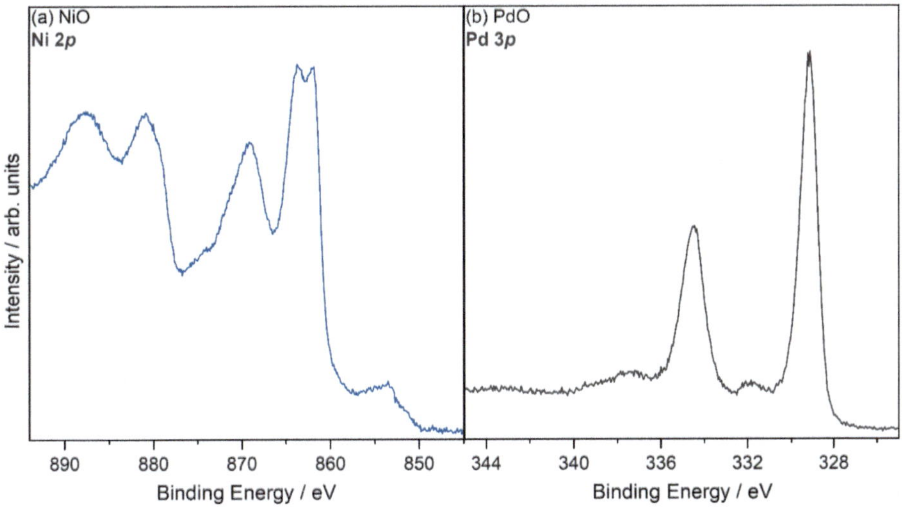

Figure 2.22 (a) NiO Ni 2p spectrum. (b) PdO Pd 3p spectrum.

Table 2.7 Common carbon environments observed in XPS.[a]

Common C 1s environment	BE range/eV	Notes
Metal carbides[122]	282–283	A rare carbon environment with lower BE than graphitic carbon
Graphite/graphene[123]	284.5	Asymmetric, and shake up satellites
Diamond	285	Symmetric peak
'Adventitious carbon'	284.5–285	From environmental contamination
Metal carbonate	288–289	Associated O 1s environments 531–532 eV
Alcohol C–OH	286–287	Associated with O 1s environment at 533 eV
Carboxylate C=O	288	Associated with O 1s peak at 532 eV
Fluorinated carbon (CF_2, CF_3)	290–292	Associated with F 1s peak at 688–689 eV
Aromatic shake up satellite	292–293	Energy loss associated with π–π* transition

[a]Graphene appears almost identical to clean graphite in XPS. Graphene oxide however contains significant high BE environments corresponding to C–O, C=O or other carbon-oxygen environments.[124] The ratio of these will depend on the degree of oxidation.

2.14.6 Polymers

Organic polymers are in some ways ideal samples for XPS analysis, and extensive work has been carried out on them. They have several favourable properties for XPS analysis. Polymers are often made of light elements, and so display simple to interpret 1s core lines. They are usually very involatile, meaning they are stable in vacuum unlike many organic molecular species. The structure and bonding are usually well known, meaning assignments of chemical environments can be assisted by knowledge of the monomer structure. A great deal of work was done on polymer XPS by Beamson and Briggs, and a comprehensive database of spectra from common polymers is available.[125]

2.14.7 Metals

XPS of metals and alloys plays a crucial role in understanding catalysis, corrosion, and electronic properties. Alloys often exhibit surface segregation; the most well-known example is stainless steel, where chromium segregates to the surface to form a protective Cr_2O_3 layer. XPS is extremely well suited to understanding surface composition in alloys.

Due to the high electrical conductivity of metals, as long as samples are grounded, then charging effects during XPS measurements are generally minimal. The delocalised electrons of the metallic state mean that multiplet splitting is minimal, so transition elements in their metallic state show much simpler peak shapes compared with their compounds where the valence electrons are localised.

2.14.8 Native Oxides

Most metals and alloys, when exposed to air, develop a thin native oxide layer. The thickness and composition of these layers can vary, affecting the observed XP spectra.

In cases where the metal oxide layer is thin enough, photoelectrons from the underlying metal can penetrate through and it is possible to observe both the metal and oxide chemical environments at once. For example, in the case of aluminium, one might observe Al 2p peaks for both metallic aluminium (at ~72 eV) and Al_2O_3 (at ~74 eV). The thickness of native oxide layers can be determined measuring the attenuation of the metallic core line intensity due to the overlying oxide layer. By comparing the intensity of the metallic signal to that of the oxide signal, the oxide layer thickness can be estimated using eqn (2.18):[126]

$$d = \lambda \sin\theta \ln\left[\frac{N_m I_O}{N_O I_m} + 1\right] \tag{2.18}$$

In this equation, λ is the mean free paths of the photoelectrons (assuming this to be the same for the metal and oxide peak), θ is the photoelectron take off angle, (which is usually 90°), I_O and I_m are the intensities of the oxide and metal peak respectively, N_m and N_O are the volume density of the metal atoms in the metal and oxide respectively.

2.14.8.1 *Plasmon Energy Loss*

Plasmon energy loss features arise due to the excitation of plasmons – collective oscillations of free electron density – when the emitted photoelectrons interact with the electron cloud of the metal. Plasmon peaks appear at higher binding energy than the main peak; the energy difference between them corresponds to the plasmon energy, which depends on the metal's electron density. For example, in aluminium, the bulk plasmon energy is around 15 eV. Al core line spectra from the metal show a series of plasmon loss peaks, appearing at multiples of the plasmon energy. Due to the large separation from the main core line position, for metals with large plasmon energies like Al, these loss peaks can be easy to misinterpret as core lines from other elements.

Metals with much smaller plasmon energies such as silver (3.8 eV) show instead plasmon peaks as shoulders on the high BE side of the core line, making the peaks asymmetric.

2.14.9 Ionic Liquids

Ionic liquids (ILs) are ionic compounds with a melting point below room temperature. They are a fascinating class of materials which have exceptionally low vapour pressures, and so are often stable enough in vacuum to be analysed in a standard XPS instrument. ILs generally have good conductivity (arising from ionic rather than electronic conductivity) and so do not usually charge significantly if grounded. Unlike solids, droplets of ionic liquids have a dynamic surface, and so are less susceptible to X-ray damage and contamination. Their spectra can be well fitted by reference to the molecular ion structure. Important work has been contributed by Lovelock and Licence in this field.[127,128] Conducting analysis on samples in the liquid state allows for unconventional XPS experiments, such as *in situ* electrochemistry.[129]

2.14.10 Alloys and Intermetallics

In metallic alloys or intermetallics, the binding energy of core lines can shift depending on the local chemical environment and electron density. For example, when alloyed with Al, Ni core lines show a 0.6 eV chemical shift compared to the pure metal.[130] In the Cu–Au alloy system, the chemical shift of both components was found to be sensitive to the alloy composition, and the temperature-driven order–disorder transition.[131] Similar results are seen in the Cu–Al alloy.[132]

2.15 Conclusions

In this chapter we have outlined the basic physical processes needed to understand and interpret XP spectra. We have attempted to take an accessible approach suitable for those with a wide range of backgrounds in physical sciences. For those readers who require more detail on the physics of XPS, we can recommend the texts listed in the Further Reading section.

Acknowledgements

Unless otherwise stated, spectra presented in this chapter were recorded at HarwellXPS, the UK National XPS Service or the National Centre for Electron Spectroscopy and Surface Analysis (NCESS), Daresbury. The authors are grateful to Dr Aisha Mumtaz for providing polymer samples.

Further Reading

F. De Groot and A. Kotani, *Core Level Spectroscopy of Solids*, Taylor & Francis Group, 2008, ISBN 9780849390715.
S. Hüfner, *Photoelectron Spectroscopy: Principles and Applications*, Springer Science & Business Media, 2003, ISBN 9783642075209.
C. R. Brundle and A. D. Baker, *Electron Spectroscopy: Theory, Techniques, and Applications, Volume 2*, Academic Press, 1997, ISBN 9780121378028.

References

1. A. Einstein, *Ann. Phys.*, 1905, **17**, 132.
2. Y.-C. Lin, Y.-Y. Chen, B.-Y. Yu, W.-C. Lin, C.-H. Kuo and J.-J. Shyue, *Analyst*, 2009, **134**, 945.
3. E. Kabachkov, E. Kurkin, N. Vershinin, I. Balikhin, V. Berestenko, A. Michtchenko and Y. Shulga, *J. Adv. Chem. Eng.*, 2021, **6**, 131.
4. A. Walsh, A. A. Sokol, J. Buckeridge, D. O. Scanlon and C. R. A. Catlow, *Nat. Mater.*, 2018, **17**, 958.
5. P. S. Bagus, G. Pacchioni, C. Sousa, T. Minerva and F. Parmigiani, *Chem. Phys. Lett.*, 1992, **196**, 641.
6. J. C. Rivière, J. A. A. Crossley and G. Moretti, *Surf. Interface Anal.*, 1989, **14**, 257.
7. W. V. Lampert, K. D. Rachocki, B. C. Lamartine and T. W. Haas, *J. Electron Spectrosc. Relat. Phenom.*, 1982, **26**, 133.
8. A. Grzelak, T. Jaroń, Z. Mazej, T. Michałowski, P. Szarek and W. Grochala, *J. Electron Spectrosc. Relat. Phenom.*, 2015, **202**, 38.
9. D. J. Griffiths and D. F. Schroeter, *Introduction to Quantum Mechanics*, Cambridge University Press, Cambridge, 3rd edn, 2018.
10. B. W. Veal and A. P. Paulikas, *Phys. Rev. Lett.*, 1983, **51**, 1995.
11. V. Kinsinger, I. Sinder, P. Steiner, R. Zimmermann and S. Hüfner, *Sol. State Commun.*, 1990, **73**, 527.
12. J. C. Slater, *Quantum Theory of Atomic Structure*, McGraw-Hill, New York, 1960, vol. 2.
13. V. Tsurkan, M. Demeter, B. Schneider, D. Hartmann and M. Neumann, *Sol. State Commun.*, 2000, **114**, 149.
14. R. P. Gupta and S. K. Sen, *Phys. Rev. B*, 1974, **10**, 71.
15. A. Kotani, T. Uozumi, K. Okada and J. C. Parlebas, Synchrotron radiation research: advances in surface and interface science, ed. R. Z. Bachrach, in *Handbook on Synchrotron Radiation*, Springer-Verlag, Berlin, Heidelberg, 1995, vol. 1B.

16. D. Briggs and V. A. Gibson, *Chem. Phys. Lett.*, 1974, **25**, 493.
17. R. P. Gupta and S. K. Sen, *Phys. Rev. B*, 1975, **12**, 15.
18. A. P. Grosvenor, B. A. Kobe, M. C. Biesinger and N. S. McIntyre, *Surf. Interface Analy*, 2004, **36**, 1564.
19. P. S. Bagus, C. J. Nelin, C. R. Brundle, N. Lahiri, E. S. Ilton and K. M. Rosso, *J. Chem. Phys.*, 2020, **152**(1), 014704.
20. P. S. Bagus, C. J. Nelin, Y. Al-Salik, E. S. Ilton and H. Idriss, *Surf. Sci.*, 2016, **643**, 142.
21. A. Kotani and H. Ogasawara, *J. Electron Spectrosc. Relat. Phenom.*, 1992, **60**, 257.
22. K. Siegbahn, Nobel Lecture, 1981, in *Nobel Lectures, Physics 1981–1990*, World Scientific Publishing Co., Singapore, 1993.
23. A. Kotani and Y. Toyozawa, *J. Phys. Soc. Jpn.*, 1974, **37**, 912.
24. R. P. Vasquez, *Surf. Sci. Spectra*, 1993, **2**, 155.
25. R. P. Vasquez, *Surf. Sci. Spectra*, 1993, **2**, 160.
26. R. P. Vasquez, *Surf. Sci. Spectra*, 1993, **2**, 165.
27. M. Barreau, D. Salusso, J. Zhang, M. Haevecker, D. Teschner, A. Efimenko, F. Bournel, J.-J. Gallet, E. Borfecchia, K. Sobczak, C. Petit and S. Zafeiratos, *J. Mater. Chem. A*, 2024, **12**, 7605.
28. Y. A. Teterin, A. Y. Teterin, A. M. Lebedev and K. E. Ivanov, *J. Electron Spectrosc. Relat. Phenom.*, 2004, **137–140**, 607.
29. I. Nagakura, T. Ishii and T. Sagawa, *J. Phys. Soc. Jpn.*, 1972, **33**, 754.
30. J. L. Campbell and T. Papp, *At. Data Nucl. Data Tables*, 2001, **77**, 1.
31. M. Iwan and C. Kunz, *Phys. Lett. A*, 1977, **60**, 345.
32. M. J. Edgell, R. W. Paynter and J. E. Castle, *J. Electron Spectrosc. Relat. Phenom.*, 1985, **37**, 241.
33. A. Jablonski and C. J. Powell, *J. Electron Spectrosc. Relat. Phenom.*, 1999, **100**, 137.
34. S. Tougaard, *J. Electron Spectrosc. Relat. Phenom.*, 2010, **178–179**, 128.
35. M. Vicanek, *Surf. Sci.*, 1999, **440**, 1.
36. S. Gusenleitner, E. Handick, D. Hauschild and F. Reinert, *J. Appl. Phys.*, 2014, **115**, 123513.
37. S. Tougaard and I. Chorkendorff, *Phys. Rev. B*, 1987, **35**, 6570.
38. D. R. Penn, *Phys. Rev. B*, 1987, **35**, 482.
39. S. Tanuma, C. J. Powell and D. R. Penn, *Surf. Interface Anal.*, 1991, **17**, 927.
40. S. Tougaard, *J. Vac. Sci. Technol., A*, 2020, **39**, 011201.
41. J. E. Castle, H. Chapman-Kpodo, A. Proctor and A. M. Salvi, *J. Electron Spectrosc. Relat. Phenom.*, 2000, **106**, 65.
42. H. Tillborg, A. Nilsson and N. M»rtensson, *J. Electron Spectrosc. Relat. Phenom.*, 1993, **62**, 73.
43. C. R. Brundle and B. V. Crist, *J. Vac. Sci. Technol., A*, 2020, **38**, 041001.
44. A. G. Shard, *J. Vac. Sci. Technol., A*, 2020, **38**, 041201.
45. R. F. Reilman, A. Msezane and S. T. Manson, *J. Electron Spectrosc. Relat. Phenom.*, 1976, **8**, 389.
46. A. Jablonski and C. J. Powell, *J. Phys. Chem. Ref. Data*, 2020, **49**, 033102.

47. J. R. Shallenberger, N. J. Smith and J. Banerjee, *Surf. Interface Anal.*, 2021, **53**, 569.
48. J. H. Scofield, *J. Electron Spectrosc. Relat. Phenom.*, 1976, **8**, 129.
49. C. D. Wagner, L. E. Davis, M. V. Zeller, J. A. Taylor, R. H. Raymond and L. H. Gale, *Surf. Interface Anal.*, 1981, **3**, 211.
50. L. I. Johansson, I. Lindau, M. Hecht, S. M. Goldberg and C. S. Fadley, *Phys. Rev. B*, 1979, **20**, 4126.
51. M. H. Engelhard, D. R. Baer, A. Herrera-Gomez and P. M. A. Sherwood, *J. Vac. Sci. Technol., A*, 2020, **38**, 061203.
52. N. Fairley, V. Fernandez, M. Richard-Plouet, C. Guillot-Deudon, J. Walton, E. Smith, D. Flahaut, M. Greiner, M. Biesinger, S. Tougaard, D. Morgan and J. Baltrusaitis, *Appl. Surf. Sci. Adv.*, 2021, **5**, 100112.
53. J. E. Castle and A. M. Salvi, *J. Vac, Sci. Tech. A*, 2001, **19**, 1170.
54. D. A. Shirley, *Phys. Rev. B*, 1972, **5**, 4709.
55. B. Moeini, M. R. Linford, N. Fairley, A. Barlow, P. Cumpson, D. Morgan, V. Fernandez and J. Baltrusaitis, *Surf. Interface Anal.*, 2022, **54**, 67.
56. J. Végh, *J. Electron Spectrosc. Relat. Phenom.*, 2006, **151**, 159.
57. P. J. Cumpson and M. P. Seah, *Surf. Interface Anal.*, 1992, **18**, 361.
58. A. G. Shard, *Surf. Interface Anal.*, 2014, **46**, 175.
59. D. Matsakis, A. Coster, B. Laster and R. Sime, *Phys. Today*, 2019, **72**, 10.
60. M. Çopuroğlu, H. Sezen, R. L. Opila and S. Suzer, *ACS Appl. Mater. Interfaces*, 2013, **5**, 5875.
61. J. Cazaux, *J. Electron Spectrosc. Relat. Phenom.*, 2000, **113**, 15.
62. J. Cazaux, *J. Electron Spectrosc. Relat. Phenom.*, 1999, **105**, 155.
63. D. R. Baer, K. Artyushkova, H. Cohen, C. D. Easton, M. Engelhard, T. R. Gengenbach, G. Greczynski, P. Mack, D. J. Morgan and A. Roberts, *J. Vac. Sci. Technol., A*, 2020, **38**, 031204.
64. L. H. Grey, H.-Y. Nie and M. C. Biesinger, *Appl. Surf. Sci.*, 2024, **653**, 159319.
65. D. J. Morgan, *Surf. Interface Anal.*, 2025, **57**, 28.
66. G. Greczynski and L. Hultman, *Vacuum*, 2022, **205**, 111463.
67. G. Greczynski and L. Hultman, *Sci. Rep.*, 2021, **11**, 11195.
68. G. Greczynski and L. Hultman, *ChemPhysChem*, 2017, **18**, 1507.
69. A. Lippitz, O. Boese, E. Kemnitz and W. E. S. Unger, *Surf. Sci. Spectra*, 2002, **8**, 214.
70. S. Kohiki, T. Ohmura and K. Kusao, *J. Electron Spectrosc. Relat. Phenom.*, 1983, **28**, 229.
71. G. Greczynski and L. Hultman, *Appl. Surf. Sci.*, 2023, **635**, 157598.
72. P. Le Fèvre, H. Magnan, D. Chandesris, J. Jupille, S. Bourgeois, A. Barbier, W. Drube, T. Uozumi and A. Kotani, *Nucl. Instrum. Methods Phys. Res., Sect. A*, 2005, **547**, 176.
73. C. Kalha, N. K. Fernando, P. Bhatt, F. O. L. Johansson, A. Lindblad, H. Rensmo, L. Z. Medina, R. Lindblad, S. Siol, L. P. H. Jeurgens, C. Cancellieri, K. Rossnagel, K. Medjanik, G. Schönhense, M. Simon, A. X. Gray, S. Nemšák, P. Lömker, C. Schlueter and A. Regoutz, *J. Phys.:Condens. Matter*, 2021, **33**, 233001.

74. S. Ueda and Y. Sakuraba, *Sci. Technol. Adv. Mater.*, 2021, **22**, 317.
75. C. Kalha, L. E. Ratcliff, J. J. G. Moreno, S. Mohr, M. Mantsinen, N. K. Fernando, P. K. Thakur, T. L. Lee, H. H. Tseng, T. S. Nunney, J. M. Kahk, J. Lischner and A. Regoutz, *Phys. Rev. B*, 2022, **105**, 045129.
76. R. Fantin, A. Van Roekeghem, J.-P. Rueff and A. Benayad, *Surf. Interface Anal.*, 2024, **56**, 353.
77. P. S. Bagus, C. J. Nelin, X. Zhao, S. V. Levchenko, E. Davis, X. Weng, F. Späth, C. Papp, H. Kuhlenbeck and H.-J. Freund, *Phys. Rev. B*, 2019, **100**, 115419.
78. J. M. Seymour, E. Gousseva, R. A. Bennett, A. I. Large, G. Held, D. Hein, G. Wartner, W. Quevedo, R. Seidel, C. Kolbeck, C. J. Clarke, R. M. Fogarty, R. A. Bourne, R. G. Palgrave, P. A. Hunt and K. R. J. Lovelock, *Faraday Discuss*, 2022, **236**, 389.
79. P. S. Bagus, *Phys. Rev.*, 1965, **139**, A619.
80. D. R. Jennison, P. Weightman, P. Hannah and M. Davies, *J. Phys. C: Sol. State Phys.*, 1984, **17**, 3701.
81. N. Pueyo Bellafont, P. S. Bagus and F. Illas, *J. Chem. Phys.*, 2015, **142**, 214102.
82. K. Hirao, T. Nakajima, B. Chan and H.-J. Lee, *J. Chem. Phys.*, 2023, **158**, 114112.
83. C. Daul, *Int. J. Quantum Chem.*, 1994, **52**, 867.
84. P. S. Bagus, C. J. Nelin, C. R. Brundle, B. V. Crist, N. Lahiri and K. M. Rosso, *Phys. Chem. Chem. Phys.*, 2022, **24**, 4562.
85. N. Pueyo Bellafont, G. Álvarez Saiz, F. Viñes and F. Illas, *Theor. Chem. Acc.*, 2016, **135**, 35.
86. J. M. Kahk and J. Lischner, *Phys. Rev. Mater.*, 2019, **3**, 100801.
87. J. M. Pi, M. Stella, N. K. Fernando, A. Y. Lam, A. Regoutz and L. E. Ratcliff, *J. Phys. Chem. Lett.*, 2020, **11**, 2256.
88. I. Tolbatov and D. M. Chipman, *Theor. Chem. Acc.*, 2014, **133**, 1560.
89. J. M. Kahk and J. Lischner, *Faraday Discuss.*, 2022, **236**, 364.
90. N. A. Besley, *J. Chem. Theory Comput.*, 2021, **17**, 3644.
91. J. M. Kahk, G. S. Michelitsch, R. J. Maurer, K. Reuter and J. Lischner, *J. Phys. Chem. Lett.*, 2021, **12**, 9353.
92. D. Golze, L. Keller and P. Rinke, *J. Phys. Chem. Lett.*, 2020, **11**, 1840.
93. D. Golze, J. Wilhelm, M. J. van Setten and P. Rinke, *J. Chem. Theory Comput.*, 2018, **14**, 4856.
94. J. Li, Y. Jin, P. Rinke, W. Yang and D. Golze, *J. Chem. Theory Comput.*, 2022, **18**, 7570.
95. M. Annegarn, J. M. Kahk and J. Lischner, *J. Chem. Theory Comput.*, 2022, **18**, 7620.
96. B. G. Janesko, *J. Chem. Theory Comput.*, 2023, **19**, 5112.
97. T. Zarrouk, R. Ibragimova, A. P. Bartók and M. A. Caro, *J. Am. Chem. Soc.*, 2024, **146**, 14645.
98. H. Höchst, S. Hüfner and A. Goldmann, *Phys. Lett. A*, 1976, **57**, 265.
99. R. Lee, R. Quesada-Cabrera, J. Willis, A. Iqbal, I. P. Parkin, D. O. Scanlon and R. G. Palgrave, *ACS Appl. Mater. Interfaces*, 2023, **15**, 39956.

100. A. E. Hughes, C. D. Easton, T. R. Gengenbach, M. C. Biesinger and M. Laleh, *J. Vac. Sci. Technol., A*, 2024, **42**, 053205.
101. G. Beamson, D. T. Clark, J. Kendrick and D. Briggs, *J. Electron Spectrosc. Relat. Phenom.*, 1991, **57**, 79.
102. P. S. Bagus, C. J. Nelin, C. R. Brundle, B. V. Crist, E. S. Ilton, N. Lahiri and K. M. Rosso, *Inorg. Chem.*, 2022, **61**, 18077.
103. D. J. Morgan, *Surf. Interface Anal.*, 2023, **55**, 567.
104. P. Bargiela, V. Fernandez, C. Cardinaud, J. Walton, M. Greiner, D. Morgan, N. Fairley and J. Baltrusaitis, *Appl. Surf. Sci.*, 2021, **566**, 150728.
105. T. G. Avval, N. Gallagher, D. Morgan, P. Bargiela, N. Fairley, V. Fernandez and M. R. Linford, *J. Vac. Sci. Technol., A*, 2022, **40**, 063206.
106. S. Chatterjee, B. Singh, A. Diwan, Z. R. Lee, M. H. Engelhard, J. Terry, H. D. Tolley, N. B. Gallagher and M. R. Linford, *Appl. Surf. Sci.*, 2018, **433**, 994.
107. B. Moeini, N. Gallagher and M. R. Linford, *Surf. Interface Anal.*, 2023, **55**, 853.
108. B. Moeini, T. G. Avval, N. Gallagher and M. R. Linford, *Surf. Interface Anal.*, 2023, **55**, 798.
109. P. Bhatt, M. Isaacs, Y. Liu and R. G. Palgrave, *Appl. Surf. Sci.*, 2024, **672**, 160808.
110. P. S. Bagus, F. Illas, G. Pacchioni and F. Parmigiani, *J. Electron Spectrosc. Relat. Phenom.*, 1999, **100**, 215.
111. H. Idriss, *Surf. Sci.*, 2021, **712**, 121894.
112. P. S. Bagus, C. J. Nelin, C. R. Brundle, B. Vincent Crist, N. Lahiri and K. M. Rosso, *J. Chem. Phys.*, 2020, **153**, 194702.
113. R. Zimmermann, P. Steiner, R. Claessen, F. Reinert, S. Hüfner, P. Blaha and P. Dufek, *J. Phys.: Condens. Matter*, 1999, **11**, 1657.
114. M. Oku, H. Matsuta, K. Wagatsuma, Y. Waseda and S. Kohiki, *J. Electron Spectrosc. Relat. Phenom.*, 1999, **105**, 211.
115. H. W. Nesbitt and D. Banerjee, *Am. Mineral.*, 1998, **83**, 305.
116. P. S. Bagus, E. S. Ilton and C. J. Nelin, *Surf. Sci. Rep.*, 2013, **68**, 273.
117. N. Pauly, F. Yubero, J. P. Espinós and S. Tougaard, *Surf. Interface Anal.*, 2019, **51**, 353.
118. P. Vasconcelos Borges Pinho, A. Chartier, J.-B. Moussy, D. Menut and F. Miserque, *Materialia*, 2020, **12**, 100753.
119. F. Parmigiani and L. Sangaletti, *J. Electron Spectrosc. Relat. Phenom.*, 1994, **66**, 223.
120. A. N. Mansour, *Surf. Sci. Spectra*, 1994, **3**, 231.
121. M. C. Militello and S. J. Simko, *Surf. Sci. Spectra*, 1994, **3**, 395.
122. G. Greczynski, D. Primetzhofer and L. Hultman, *Appl. Surf. Sci.*, 2018, **436**, 102.
123. D. J. Morgan, *J. Carbon Res.*, 2021, **7**(3), 512021.
124. R. Al-Gaashani, A. Najjar, Y. Zakaria, S. Mansour and M. A. Atieh, *Ceram. Int.*, 2019, **45**, 14439.
125. G. Beamson and D. Briggs, *High Resolution XPS of Organic Polymers: The Scienta ESCA300 Database*, John Wiley & Sons, Chichester, 1992.

126. B. R. Strohmeier, *Surf. Interface Anal.*, 1990, **15**, 51.
127. R. Rowe, K. R. J. Lovelock and P. A. Hunt, *J. Chem. Phys.*, 2021, **155**, 014501.
128. S. Men, K. R. J. Lovelock and P. Licence, *Phys. Chem. Chem. Phys.*, 2011, **13**, 15244.
129. R. Lee, T. S. Nunney, M. Isaacs, R. G. Palgrave and A. Dey, *ACS Appl. Mater. Interfaces*, 2024, **16**, 35675.
130. V. L. Vorob'ev, L. V. Dobysheva, A. Y. Drozdov, P. V. Bykov and V. Y. Bayankin, *J. Electron Spectrosc. Relat. Phenom.*, 2021, **252**, 147124.
131. C. Li, Q. Liu, J. A. Boscoboinik and G. Zhou, *Phys. Chem. Chem. Phys.*, 2020, **22**, 3379.
132. A. Pashusky, *Vacuum*, 1993, **44**, 997.

CHAPTER 3

Photoemission Spectroscopy Instrumentation

TIM S. NUNNEY

Thermo Fisher Scientific, Unit 1, The Felbridge Centre, East Grinstead, RH19 1PE, UK
E-mail: tim.nunney@thermofisher.com

3.1 Introduction

Photoemission spectroscopy (PES) instruments have been commercially available for over 50 years, and although a lot of the fundamental principles of instrument design have changed very little since then, the technology has advanced to greatly improve the speed and sensitivity of the technique. That first wave of PES instruments could load one sample at a time, and had limited sensitivity which made experiments slow, but were customised to focus on the specific end application. Over the years the technology has advanced to meet the increasing range of applications that PES is being applied to investigate. These advances have encompassed all of the fundamental components of a PES system, including the vacuum system, which brings the internal pressure down to an ultra-high vacuum range, X-ray sources for exciting photoemission from the sample, and the electron optics and analyser, which detect the signal. This is also not to forget the other key components dedicated to particular experimental modes that have been introduced over the years as the applications of the technique have expanded.

Inorganic Materials Series No. 19
Photoelectron Spectroscopy in Inorganic Materials Science
Edited by Robert G. Palgrave and Philip R. Davies

Published by the Royal Society of Chemistry, www.rsc.org

Following on from the instrument design and hardware, this chapter will then focus on the intricacies of PES operation, discussing how the system's components work together to measure elemental composition, empirical formula, chemical state, and electronic state. We will also discuss the challenges that can arise during PES operation, such as sample contamination and charging effects, and how these can be mitigated. The different types of experiments that can be conducted using PES will also be explored, including depth profiling, which allows for the analysis of successive layers of a sample, and PES imaging, which reveals the distribution of chemistries across a surface. We will also discuss other specialist implementations such as angle-resolved PES, and recent interest in near-ambient pressure PES and higher energy source PES (HAXPES).

3.2 Vacuum System Considerations

3.2.1 Why Is Ultra-high Vacuum Required?

Photoemission spectroscopy (PES) relies on ultra-high vacuum (UHV) conditions for optimal operation, where UHV is defined as a pressure less than 1×10^{-8} mbar. One of the primary reasons that UHV is essential for PES is to maintain the cleanliness of the sample surface. In an environment with normal atmospheric pressure, gas molecules would rapidly adsorb onto the sample surface, potentially contaminating it and skewing the results of the analysis. In UHV conditions, however, the number of residual gas molecules is drastically reduced, minimising the risk of such contamination. This ensures that the PES analysis accurately reflects the properties of the sample itself, rather than those of an adsorbed layer of gas. More critically however, UHV conditions are necessary to provide an unobstructed path for the photoelectrons ejected from the sample surface to the detector. In the PES process, a beam of photons is directed at the sample, causing it to emit photoelectrons. The energies of these photoelectrons are then measured to provide information about the sample's elemental composition and chemical state. If the chamber were not under UHV conditions, these photoelectrons could collide with gas molecules enroute to the detector, which would cause them to lose kinetic energy, and thus compromise the accuracy of the measurements.

In order to maintain UHV, the instrument will need to be designed to enable samples to be transferred from atmosphere without compromising the vacuum level in the analytical part of the system. This requires careful consideration of pumping methods, chamber materials, sealing methods and operating procedures which will be discussed in the following sections.

3.2.2 Vacuum Chamber Design

Modern PES systems usually comprise a minimum of two vacuum chambers and a common schematic of this arrangement is shown in Figure 3.1: the main, or analysis, chamber which contains the spectrometer, and an

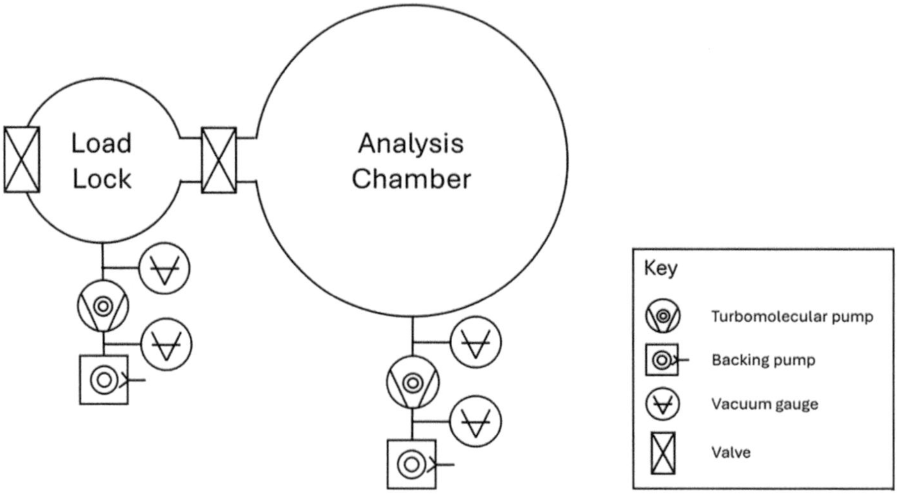

Figure 3.1 Schematic of the vacuum system for a typical PES instrument.

ancillary chamber which is usually referred to as the fast-entry airlock or load-lock chamber. The two-chamber design means that samples can be brought into the instrument through the load-lock chamber, where they can be pumped down into a high-vacuum pressure regime before transferring into the main chamber. This arrangement is to ensure that the UHV condition in the main chamber is protected as much as possible, by limiting contamination from components of the air which would affect the vacuum, such as oxygen or water vapour. These gases adsorb on the walls of the vacuum chamber, slowing progress to UHV, and so limiting their presence in the analysis chamber is crucial. In fact, after service events when the chamber has been exposed to the atmosphere it has to be heated to remove these adsorbed gases from the system, typically for more than 10 hours, in a process referred to as 'bake-out'.

Chambers are constructed from multiple components, making seals between parts very important to maintain vacuum. The most common way of sealing vacuum flanges is by using raised areas on the flanges which cut into soft metal seals, typically gaskets made from copper or silver, when compressed against them. These all-metal sealing systems work well, though can introduce variation in the alignment of the part being attached if thicker gaskets are used which may need to be corrected. In recent times it has become possible to use polymer seals or differentially pumped O-ring seals to get equivalent vacuum performance but with better alignment of the flanges. The chambers themselves need to be made from materials that will not outgas into the vacuum or otherwise affect the system pressure. For the analysis chamber, the material it is manufactured from also needs to also be able to shield the instrument from magnetic fields, such as the Earth's natural field, which will otherwise deflect the photoelectrons and adversely affect the experiment. Alloys with a high nickel–iron content such as mu-metal that are effective at screening magnetic fields are used for this purpose, either by manufacturing the chamber from that material, or by using plates of the material to form a shield either internally or externally to the vacuum system. If the shielding plates are used, they have to make a tight seal to prevent field penetration into the chamber.

3.2.3 Pumping

Each chamber will require a pumping system to achieve the necessary vacuum condition. For the load-lock chamber this will need to be a series of pumps that can routinely pump down from atmosphere to high vacuum quickly and be vented back to atmosphere when samples need to be exchanged. A two-stage system is usually used which comprises a main pump and a backing pump. The standard pump used for the main pump is a turbomolecular pump, which has the pumping speed and reliability to fit into this operating regime. This is supported by a backing pump which operates in the low vacuum region and will either be an oil-based rotary vane pump or an oil-free pump such as a dry scroll pump. PES systems are not very sensitive to vibration, so turbomolecular pumps are now routinely used for pumping the main chamber too (with vibration isolation if a small spot size electron source for Auger electron spectroscopy is to be used), whereas in the past ion pumps or diffusion pumps may have been used. The concerted move to turbomolecular pumps has in part been due to the large amounts or argon that need to be pumped when using gas cluster ion sources, which have grown in prominence in the last decade. This is not pumped well by ion pumps in particular, and this requirement combined with the increased reliability and performance of turbomolecular pumps has driven this shift in the pumping system for the main chamber. To achieve UHV it is necessary to add an additional pumping method, which will remove oxygen-containing gas molecules from the chamber background gas. The most common way of doing this is with a getter system such as a titanium sublimation pump which works by periodically heating a filament to evaporate a film of titanium onto a section of the chamber wall. This is very reactive and any gas molecules that will react with the fresh surface will be retained in the metal film. The instrument software usually controls this process to prevent the pump from operating during experiments, where a burst in pressure might affect the system.

If sources on the system use high gas loads, such as ion sources or UV sources, then there may be additional pumping required to differentially pump these to maintain the vacuum level in the analysis chamber. These will be either turbomolecular pumps or backing pumps depending on the gas load and required vacuum level for the process. Near-ambient pressure PES systems have very specific pumping requirements, and these will be discussed in the section concerning those instruments.

3.3 Sample Handling

3.3.1 Sample Holders and Sample Mounting

Where the early PES instruments had a single sample mounted in a fixed position to just analyse one place on the surface, modern PES instruments are now generally built to be able to load a range of different sample types and sizes. When choosing the correct sample holder and mounting method there are five main things to consider:

(i) The sample dimensions.
(ii) The sample form factor.
(iii) How the sample will be attached to the holder.
(iv) The vacuum compatibility of the sample.
(v) The desired experimental procedure.

The first three items are interlinked; if the sample is a powder, then the sample dimensions are less of an issue, but the mounting method will be different than if the sample was a single, 1 cm^2 piece of silicon wafer, and the type of sample holder may change. After considering the first three points, the final steps before deciding how to proceed are understanding any potential issues with introducing the sample into vacuum, and exactly what experiments are to be undertaken.

The design of the PES instrument will have the largest impact on the size of the sample that can be introduced. This determines the maximum sample thickness, either by the dimensions of the transfer valve between the load-lock and the analysis chamber, or by the arrangement of equipment around the lens, for example magnetic immersion lenses that tend to be located beneath the sample. The other sample dimensions, the length and width, will largely be affected by the transfer valve dimensions, but since that will be designed to admit the sample holder, the sample will have to be large to have an issue.

If the sample is a single piece, for example a metal coupon, a piece of polymer film or a silicon wafer, then it can be mounted onto the sample holder either using clips, clamps, or using vacuum-compatible tape. Conductive tape is preferred if tape used, though care needs to be taken to make sure that it is suitable for UHV and does not out-gas adhesive that might contaminate the sample. Most brands for use with electron microscopes are suitable. Powder samples can also be mounted on tape or pressed into wells on special powder sample holders. Care needs to be taken with powders to ensure that they are firmly affixed to the sample holder to prevent the chance of loose grains being distributed around the instrument during pump down. Sample mounting examples are shown in Figure 3.2.

Aside from the physical aspect of fixing the sample to the sample stage, the other key consideration with regards to mounting is the electrical properties of the sample. For samples that are conductors or insulators, this is not

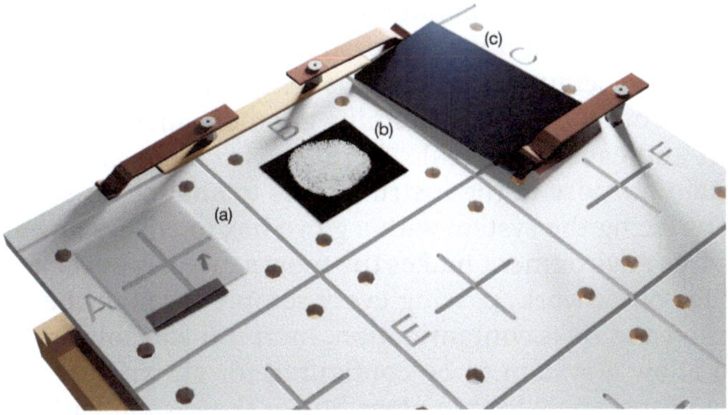

Figure 3.2 Examples of sample mounting for a PES analysis using (a) conductive carbon tape for a polymer film, (b) conductive carbon tape for a powder sample, and (c) clips for securing a wafer sample.

too much of an issue, but if the sample has regions that could end up with different electrical environments, this can cause differential charging of the surface during analysis, which affects the data quality. In this case, being able to electrically isolate the sample either by the mounting method, or by being disconnected from ground during data acquisition, is important to ensure that the whole sample charge can be maintained at the same level.

Once the sample is mounted, it can be loaded into the instrument and pumped down. To protect the analysis chamber vacuum, samples need to be pumped down into the high vacuum regime (typically better than 5×10^{-6} mbar) before transferring. Materials that are not vacuum compatible are unlikely to reach this level (although samples that can degrade under X-ray exposure could appear to be fine before transferring and only become problematic once the experiment commences). Porous materials, for example catalyst supports such as zeolites, can reach an almost acceptable level quite quickly but then take a long time to remove all the adsorbed gas from inside the pores. Where a flat, solid sample could be ready to transfer in less than 30 minutes, porous materials can take several hours to pump to an acceptable level to transfer. This needs to be considered when determining the length of an experimental protocol for analysing these types of materials.

Finally, once the samples have been mounted, pumped down, and transferred into the analysis chamber, the experiment can be designed. The steps in determining the requirements for the common experimental protocols will be discussed later in this chapter.

3.3.2 Inert or Vacuum Transfer

One special case for sample mounting and loading is the increasingly important requirement to analyse air-sensitive materials. This is driven to a large extent by the need to analyse several materials that are key to clean energy research, such as new Li-ion battery and solar cell devices. In order to analyse these samples, it must be possible to move them from the sample container and into vacuum without exposure to air (specifically oxygen and water vapour and other oxygen-containing gas molecules such as CO_2, which can also react and cause changes). There are two main methods for doing this. If the main purpose of the instrument is to analyse air-sensitive materials, then interfacing the system with a glove box to ensure an easy transfer in an oxygen-free environment makes the most sense. However, if the system will be used for other work, passing every sample through the glove box is time-consuming and risks contamination. Inert transfer solutions have been developed to allow a system to be configured for all types of samples and allow loading of materials from a glove box to the system without air exposure. These can be either vacuum transfer devices, where the sample is transported under vacuum either with or without continuous pumping, or inert transfer systems where the sample is kept in the vessel under the glove box atmosphere, and which is pumped away during the loading process. Both methods work well and ensure that air-sensitive materials can be analysed using PES.

3.4 Analytical Instrumentation

Once the samples have been pumped down in the entry chamber and transferred into the analysis chamber, the analytical work can begin. The analysis chamber will have components for positioning samples, monitoring the vacuum and other tasks that support the analytical workflows, but in this section, we will concentrate on the key parts for PES and related surface analysis experimentation, starting with photon sources.

3.4.1 Photon Sources

The first consideration for an instrument for photoelectron spectroscopy is to decide on the photon source or sources that will be used. The most common source for the laboratory instruments commercially available at the time of writing is a monochromated X-ray source with an aluminium anode, but alternative sources may also be used should additional information be required from the experiment. These could include UV photon sources for analysis of the valence band (the low binding energy electrons that are predominantly involved in bonding) or X-ray sources with higher photon energies to excite the emission of photoelectrons from deeper core levels. Therefore, when choosing a photon source to use in an experiment, the analyst needs to first consider if the energy range of the source is correct to identify the spectral features needed, if the energy resolution of the source is sufficient to be able to resolve the spectral features, and whether the photon flux from the source will facilitate obtaining the required signal level in a reasonable amount of time. The Kα X-ray line from aluminium has been adopted as the default choice because it nicely meets these requirements, giving access to transitions from the elements which are likely to be encountered in the laboratory, being easy to monochromate to give a narrow line width, and delivering sufficient flux to result in good signal-to-noise in collected data in acceptable amounts of time.

3.4.1.1 Monochromated X-ray Sources

The standard X-ray source for a modern X-ray photoelectron spectroscopy (XPS) instrument is conceptually quite a simple device, comprising an electron source (the cathode) to supply a stream of high energy electrons which bombard a target material (the anode), and a material to act as the monochromator to energy filter the emitted X-rays. The high energy electrons incident on the target cause core electrons to be ejected from the atoms in the anode, leaving a 'hole' in the electronic structure that needs to be refilled. One of the ways that this can happen is through an electron moving from a higher energy level in the atom to fill the hole, and the excess energy can be released as fluorescence: the emission of a characteristic X-ray. Electrons will be ejected from a range of orbitals in the atom, which in turn will result in the emission of X-ray photons with a range of energies. Along with these characteristic X-rays that are emitted, the changes to the vector of the electrons as they interact with the nuclei of the anode material causes the emission of Bremsstrahlung, or braking, radiation. This is a broad band of X-rays which effectively acts as a background to the stronger emission lines.

The cathode is usually just a thermal electron source, either an electrically heated filament material such as tungsten, or a higher brightness lower temperature material such as lanthanum hexaboride (LaB_6). LaB_6 is used when the source is designed to be able to vary the analysis area on the sample by changing the size of the X-ray spot, as its emission characteristics make it more suitable to produce enough flux at the lower emission levels necessary to create focused electron beams down to diameters of a few microns on the anode. This small spot focusing is performed by using electrostatic lenses and/or apertures between the cathode and the anode, and in systems where the X-ray spot can be scanned across the sample, the motion is created by scanning the electron beam across the anode.

A significant potential difference is required between the cathode and the anode to ensure that the incident electrons have sufficient energy to ionise atoms in the target material and generate enough photon flux to be used. For aluminium, which is the standard laboratory source, this is usually in the range of 10–15 kV which results in a suitable level of X-ray flux, with either the anode or the cathode being at high potential, and the other electrode being at or near ground potential. A large amount of heat is generated at the anode during the generation of X-rays, and so the assembly is water-cooled. Aluminium does not have the best thermal characteristics for easily removing the heat from the area bombarded by the electron beam, and so the anode is manufactured by depositing a thin layer of aluminium onto a substrate that is more thermally conductive such as copper or silver. Even with cooling, the anode material will degrade slowly over time and will need replacement once the aluminium becomes so thin that the X-ray flux starts to decrease. For this reason, manufacturers usually design the anode to be movable so a fresh area can be used without having to open the vacuum system, which results in several years of operation without the need for replacement.

The next step following the generation of X-rays is to monochromate the beam using a single crystal (or a set of single crystals) to diffract the specific X-ray wavelength required and direct it towards the sample. By using a monochromator, the Bremsstrahlung radiation background can be removed, and the energy dispersion in the width of the X-ray line will be reduced; for Al Kα this is from ~1.0 eV to ~0.25 eV, which will result in a significant improvement in spectral resolution. The other benefit is that the other unwanted parts of the X-ray emission spectrum are filtered out, which will eliminate signals due to X-ray lines other than Al Kα. Figure 3.3 shows these benefits for a spectrum of Ag with equivalent analyser conditions.

The monochromator does this by diffracting the X-ray beam using a single crystal (or a set of single crystals) which has a suitable lattice spacing for the X-ray wavelength.

In commercially available Al Kα X-ray sources the $[10\bar{1}0]$ face of quartz single crystals is usually used as the monochromator material. The Bragg angle is effectively defined by the angle between the anode and the sample, but the positioning and shape of the crystal is also a crucial consideration.

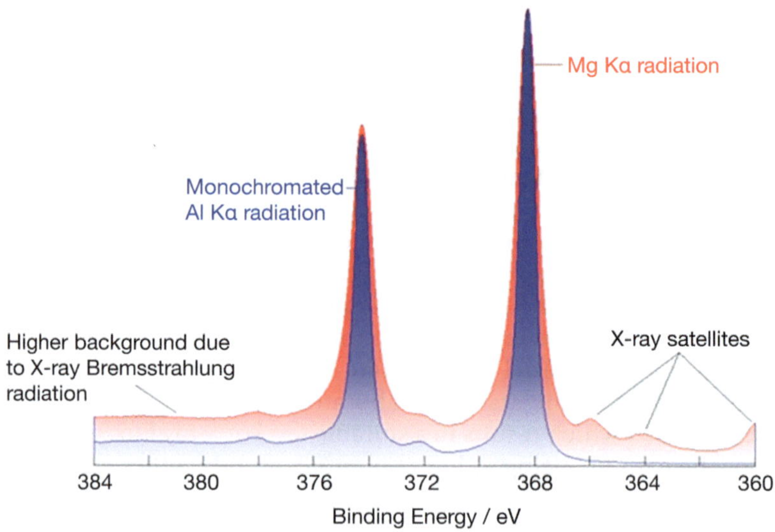

Figure 3.3 XP spectra for the Ag 3d region collected using a non-monochromated Mg Kα X-ray source (red) and a monochromated Al Kα X-ray source (blue).

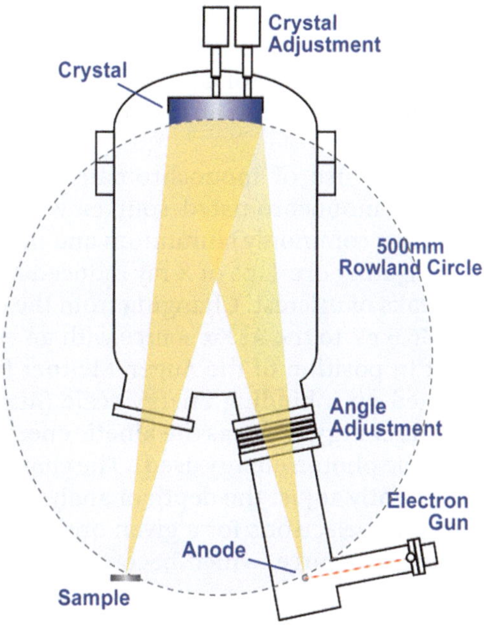

Figure 3.4 An example of a monochromated X-ray source showing how the anode, crystal and sample positions need to lie on the Rowland circle for effective X-ray focusing following diffraction from a shaped crystal.

The anode, sample and crystal positions are arranged in a 'Rowland circle', as shown in Figure 3.4, which ensures that the distance between the anode and crystal and the sample and crystal are the same. The crystal is usually also shaped to refocus the X-ray beam at the sample. The curvature of the crystal is usually a compromise from the ideal geometry where the radius of curvature matches the Rowland circle radius, to having a curvature where the radius is twice the radius of the Rowland circle. This is achieved in manufacturing by creating a curved substrate and bonding the crystal wafer onto its surface. By doing this, the curved diffractor will create an image spot of X-rays at the sample which is almost equivalent to the dimensions of the electron beam at the anode, making it possible to adjust the area illuminated by the X-rays on the sample by adjusting the focus of the electron beam in the source.

3.4.1.2 Alternative X-ray Sources

For the majority of analysis, the Al Kα monochromated source will usually be sufficient to meet the experimental requirements. However, there are occasions where having an alternative X-ray source is beneficial. This can be to increase or decrease the X-ray photon energy, which can be used to:

(i) Remove overlaps of features resulting from Auger–Meitner emission processes with photoelectron peaks.
(ii) Change the depth of analysis for a given photoelectron peak.
(iii) Access deeper core levels within an atom.

Prior to the widespread use of monochromated sources, instruments were equipped with non-monochromated sources which were fitted with two anode materials, most commonly aluminium and magnesium. This was largely to help with mitigating overlaps of X-ray induced Auger–Meitner electron peaks with XPS peaks of interest. Changing from the Mg Kα source with a photon energy of 1256.6 eV to the Al Kα source with an energy of 1486.6 eV would result in a shift in position of the Auger–Meitner features by 230 eV, when the data is plotted on a binding energy scale (Auger–Meitner peaks being constant in kinetic energy, whereas the kinetic energy of photoelectron peaks is dependent on the photon energy used). The change of source energy could also be used to slightly adjust the depth of analysis, as the attenuation length of the emitted photoelectrons for a given orbital would be increased by using the higher energy source, which becomes much more significant when harder X-ray sources are used.

A schematic of such a twin anode source is shown in Figure 3.5. Switching between the two X-ray lines is achieved by using the filament nearest the desired anode to produce the electron beam. As with the monochromated source, there needs to be a high potential between the cathode and the anode to produce sufficient X-ray flux, and a significant amount of heat is generated so the source needs to be water-cooled. Due to the lack of focusing, the source needs to be moved very close to the sample when being used. For this reason, sources are mounted on adjustable bellows that allow motion to and from this position. This proximity of the source to the sample can result in the sample being heated during the use of the source. A window of thin aluminium foil is fitted to the front of the source to act as a simple barrier to limit the electrons from the filament reaching the sample, and as a high-pass filter to try and remove as much of the Bremsstrahlung X-ray emission as possible.

As analyser design improved to allow measurement of higher kinetic energy electrons, the range of anode materials for non-monochromated sources increased to generate higher energy X-ray photons. Anode materials such as zirconium, titanium, silver, chromium, and copper to name but a few have all been available commercially as one side of a twin anode X-ray source. These sources produce a lot of flux, and so high-count rates can be measured, but there are two main drawbacks. One is that the measured data will have poor energy resolution compared to that from a monochromated source. The second complication is that there will be the presence of satellite peaks in the measured data due to all created X-ray photon energies causing photoelectron emission. Along with the need to get the source close to the sample, these weaknesses have resulted in non-monochromated sources

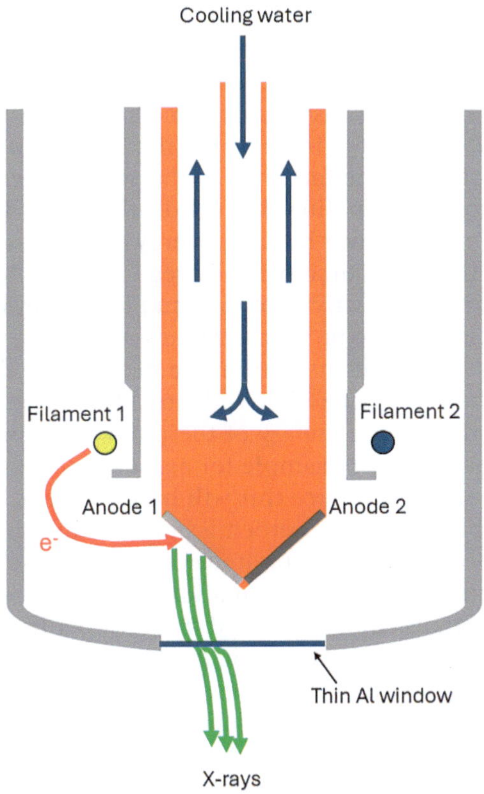

Figure 3.5 Schematic of a twin anode non-monochromated X-ray source. Only one side of the source is used at any one time. The figure shows anode 1 being used to generate X-rays.

being used less as the performance of the Al Kα monochromators improved for standard analysis, but the requirement to be able to remove overlaps or adjust the analysis depth remained. Other monochromated X-ray sources were developed to meet these needs but without the resolution or satellite feature limitations.

Ag Lα X-ray photons have almost double the energy of Al Kα photons, and so can, with a little adjustment, be monochromated using the same crystal assembly, using second-order rather than first-order diffraction. This idea was first introduced in the late 1970s, early 1980s,[1] and has increased in use over the last ten years as motorisation of the monochromator crystal assembly and anode position has made it possible to easily switch between the two sources. The fact that second-order diffraction is used does result in lower intensity, necessitating longer acquisition times. Other commercially available monochromated X-ray sources, which along with the Ag source are conventionally referred to as sources for hard X-ray photoelectron spectroscopy or HAXPES, use dedicated crystal diffractors rather than higher order diffraction from the standard system. The two most commonly encountered are Cr Kα and Ga Kα which offer photon energies of 5.5 keV and 9.2 keV respectively. The sources work in the same way as the standard monochromator, although the Ga source uses a liquid metal anode to enable higher levels of electron bombardment to increase the flux by running at higher power. This is to compensate for the fact that as the source energy increases, the photoionisation cross-section, (in other words, the ability to cause the emission of electrons from the lower binding energy orbitals), decreases. Spectra collected using a HAXPES source typically require much longer acquisition times, and therefore the sensitivity of the sample to X-rays needs to be considered more than it needs to be for the Al Kα source.

HAXPES sources are mainly used to probe deeper into the sample and to access core levels that are inaccessible with an Al Kα source. With an increase in photon energy, the kinetic energy of the photoelectrons for a particular transition is increased, so for example for a photoelectron emitted from an Al 2p orbital, with an Al Kα source this will have a kinetic energy of approximately 1414 eV, whereas if the photoemission is induced by a Cr Kα source the kinetic energy of the emitted photoelectron is 5343 eV. This difference in kinetic energy means that the attenuation length of the photoelectron is increased, and so signals can be detected from deeper in the material using the higher energy source as shown in Figure 3.6. The other benefit of using a higher energy source is to access core levels that cannot be accessed with Al Kα, such as Si 1s. This provides the opportunity to use different transitions for analysis which can help with simplification of the spectrum, for example considering a spectrum from a less complex s-orbital compared to one from a p-, d- or f-orbital with the addition of spin–orbit splitting. Auger–Meitner lines may also be accessed, which can be used to help discern chemical states through the calculation of the Auger parameter. Generally, it is easier to work with deeper core levels with hard X-ray sources as the variation in photoionisation cross section at the higher binding energy part of the spectrum is less than in the low binding energy region below 1.5 keV that can be accessed with an Al Kα source.

3.4.1.3 UV Sources for Ultraviolet Photoelectron Spectroscopy (UPS)

Ultraviolet photoelectron spectroscopy (UPS) is one of the most well-known surface analysis techniques complementary to XPS. The two main applications for the technique are investigation of the valence electronic structure of surfaces and the measurement of the workfunction or ionisation potential of a sample. UPS operates on the same principles as XPS, with the main difference being that ionising radiation is utilised at energies that are a fraction of that used for XPS. Since UV photons have a much lower energy than X-rays, most core-level photoemissions are not accessible using UPS, and spectral acquisition is limited to the valence band region. Additionally, UPS

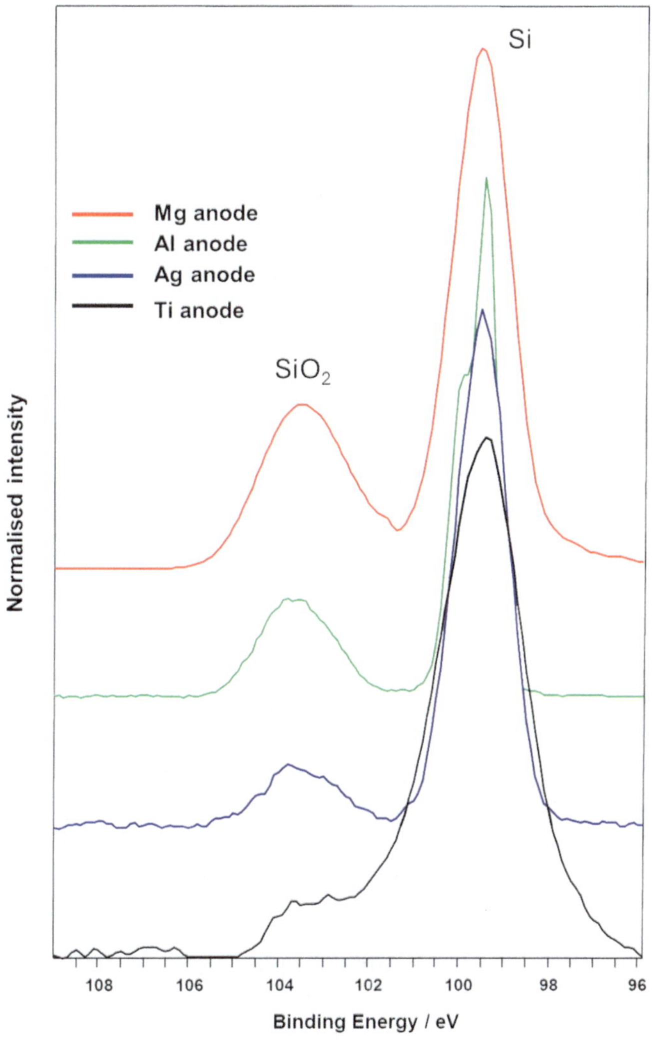

Figure 3.6 Si 2p spectra collected from a sample of thin SiO_2 on Si using four different X-ray source energies. The variation in the measured oxide percentage is due to the increasing analysis depth with increasing source energy.

has a much greater surface sensitivity than XPS, because the lower energy UV photons generate significantly lower energy photoelectrons, which have a much shorter inelastic mean free path than observed in XPS. UPS has an approximate sampling depth of 2–3 nm compared to ~10 nm for PES with an Al Kα source. This makes UPS much more sensitive to surface contamination than XPS, and careful cleaning of the surface for analysis is often required; this could be a much larger area than for XPS since the irradiated area from a UV source is often a few mm^2 in size compared to a few hundred microns

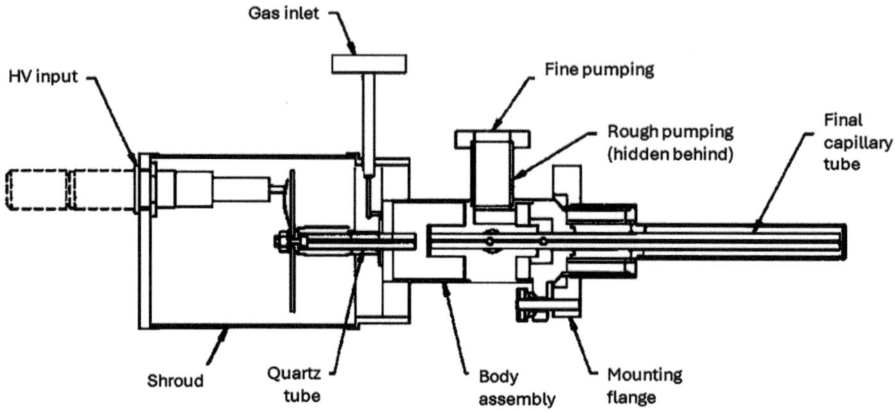

Figure 3.7 Schematic of a gas discharge UV lamp. The discharge is created in the quartz tube and guided into the analysis chamber through the final capillary.

for an X-ray source, and this needs to be considered during sample cleaning to obtain good results.

In a laboratory setting, ultraviolet photons are typically produced using a gas discharge lamp, as shown in Figure 3.7. The gas used is usually helium, although other noble gases can be used to give other emission lines. The principal photons used in spectroscopy which are emitted by a helium gas discharge have energies of 21.2 eV (He I) and 40.8 eV (He II). Each photon energy has a different ionisation cross-section when interacting with the atoms in the sample, which means that spectra collected with He(I) emission can have a different appearance to those collected with He(II) emission, which tend to more closely resemble valence band spectra collected using X-ray photons.

The operation of the source is fairly simple. Gas is admitted into a region which contains an anode at high voltage (the typical starter voltage is 5 kV) and a cathode at ground potential. The two electrodes are positioned at either end of a quartz capillary tube to contain the discharge and guide the created UV light from the source towards a second quartz capillary which directs the beam to the sample. Once a discharge has been created, the operating voltage is typically lowered to between 0.5 and 1 kV to maintain a current of approximately 50 mA measured in the power supply. As the pressure required for the discharge is in the range of 10^{-2} mbar, the source has two stages of differential pumping to ensure that the vacuum condition in the analysis chamber is maintained.

Although the source is not complex, successful obtaining a discharge with helium is very dependent upon the purity of the incoming gas and the cleanliness of the electrodes in the lamp. Flushing the gas lines to remove any contamination following replacement of the gas supply is essential, and if there are still issues with starting a source or maintaining a consistent flux it can be advantageous to use the lamp with argon which is easier to ionise and can successfully clean the internal surfaces to improve operation with helium.

3.4.2 Photoelectron Detection

Following photoemission from the sample, the emitted photoelectrons need to be collected and characterised in order to create a spectrum. This is done using a multi-component part of the instrument which is usually referred to as 'the analyser', although that is actually just one part of the assembly. It can be broken down into three sections. The first section is the electron optics which collect the photoelectrons emitted from the sample and focus them onto the entrance of the analyser. The analyser is a band-pass energy filter, which determines the kinetic energy of the photoelectrons that can make it to the final component which is the electron detector. Depending on the type of instrument, these three elements can be simple or complex, and this section will describe the considerations in their design and operation.

3.4.2.1 Electron Optics

To begin, the instrument needs to be able to successfully collect and focus the photoelectrons emitted from the surface of the sample. In early instruments the analyser was located very close to the sample, but as the PES technique matured, and additional access to the sample surface was necessary, there was a need to use a transfer lens to move the analyser entrance away from the sample. In its simplest form the electron transfer optics comprise two lenses, an objective lens to gather the electrons, and a second element to refocus the electron beam onto the entrance of the analyser. This would be an electrostatic system, where the lens elements are simple electrodes or meshes, which can have a potential applied to give the desired focus. Even in a simple system the shape of the lens is crucial – small deformations (which could be caused by vacuum accidents, for example) can greatly reduce the efficiency of electron transmission. The objective lens also defines the parameter known as the acceptance angle, which is the range of emission angles from a point of the surface from which a signal will be detected. A large acceptance angle will increase the sensitivity of the instrument but will limit the resolution available for angle-dependent studies. One alternative to the electrostatic objective lens is to use a magnetic immersion lens as the first element. This approach is most commonly used for instruments which are designed for high spatial resolution parallel imaging detection (where the lower aberration of the magnetic lens allows for improved magnification and resolution), but also has benefits for standard spectroscopy by increasing the collection angle. Utilisation of a magnetic element can increase sensitivity by around a factor of three compared to an electrostatic element, due to the improved collection efficiency, which is largely due to electron trajectories outside the collection range in electrostatic mode being within the larger acceptance angle of the magnetic element (up to $\pm 90^\circ$ compared to $\pm 30^\circ$).

The electron optics can also be used to define the analysis area on the sample. In the previous section it was described how the analysis area can be adjusted by changing the size of the X-ray spot size (in effect by changing the electron beam size on the anode material), which is called source-defined small area analysis. The alternative to this is to use apertures in the electron lenses to define the analysis area, which is called lens-defined small area analysis. Two apertures are required, one to define the field of view of the analyser, and one to act as a restriction on the acceptance angle, which is used to maintain the resolution of the edges of the measured region by making the transmitted electron beam more parallel and reducing spherical aberrations. This approach is not able to give small areas that are as small as source-defined methods for spectroscopy directly. However, when this method is used as part of a parallel imaging system then the ultimate spatial resolution for spectroscopy by averaging pixels in an image can be brought to similar levels. The main drawback with lens-defined analysis is that the apertures do result in the sensitivity being reduced by some of the emitted signal being rejected and so acquisition times for spectroscopy can be longer.

The final role for the electron transfer lens is to determine the kinetic energy of the electrons entering the analyser. The kinetic energy of the electrons across the majority of the energy range is too large to be easily resolved by the analyser, and to eliminate this problem it is necessary to retard them prior to them reaching the entrance. This is usually performed just before the entrance to the analyser. There are two approaches to how the retardation is applied, which depend on the mode of operation of the analyser and will be described in the next section. An example lens and analyser layout is shown in Figure 3.8.

3.4.2.2 Analysers

The standard analyser used for PES is the concentric hemispherical analyser (CHA), also known as the hemispherical sector analyser (HSA), which consists of two electrodes separated by a gap for electrons to pass between. A schematic of the arrangement is shown in Figure 3.8. A potential difference is applied across the two hemispherical electrodes, with the outer hemisphere being more negative (to repel electrons) and the inner hemisphere being more positive (to attract electrons). This arrangement will allow electrons to follow a path from the entrance to exit if their energy satisfies the following equation:

$$E = e\Delta V\left(\frac{R_1 R_2}{R_2^{\,2} - R_1^{\,2}}\right)$$

where E is the kinetic energy of the electrons, e is the charge on the electron, ΔV is the potential difference between the hemispheres, R_1 is the inner hemisphere radius, and R_2 is the outer hemisphere radius.

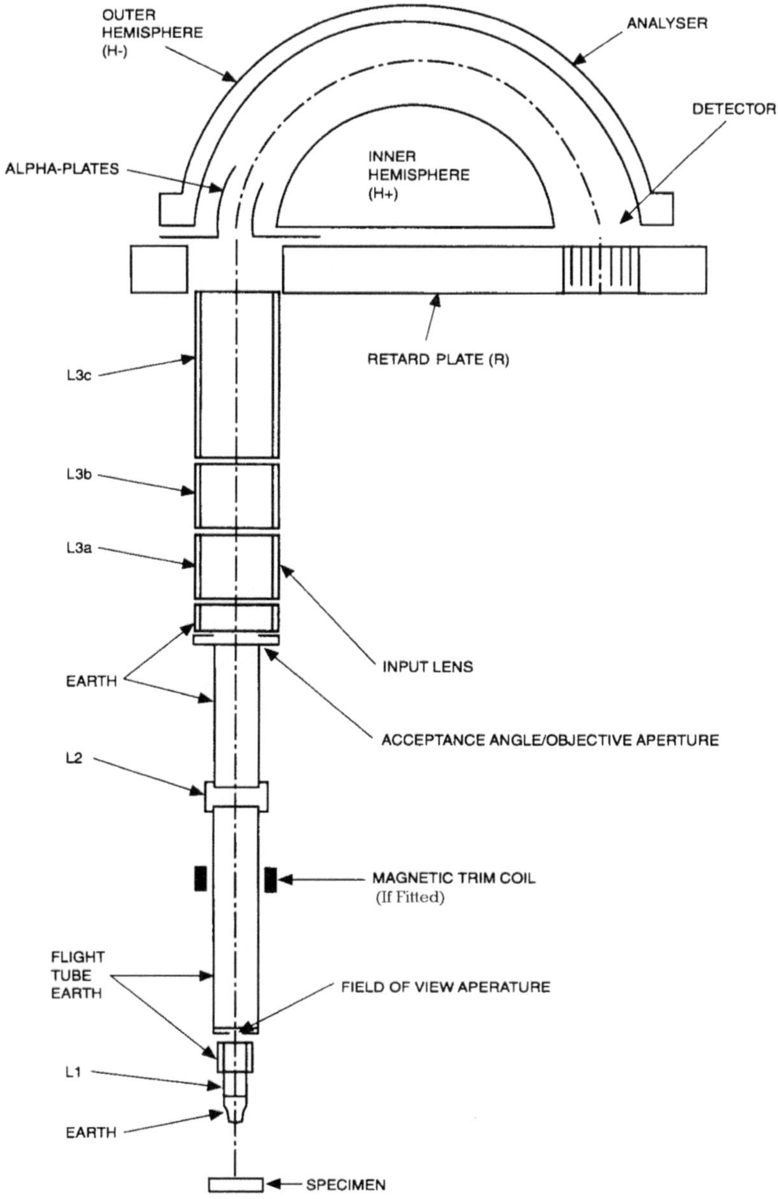

Figure 3.8 Schematic of a PE spectrometer lens and analyser assembly.

As the hemisphere radii are fixed and the charge on the electron is constant, then the transmission of the analyser for a given kinetic energy is defined by the potential difference between the two electrodes. If ΔV is fixed for an acquisition, then whether an electron passes through the analyser is dependent on its kinetic energy, and so the analyser acts as a filter, allowing electrons of a specific energy to reach the detector at the exit. As mentioned in the previous section, the kinetic energies across the whole energy range are too large to allow good spectral resolution, and so the electron optics retard the electrons prior to entering the analyser. There are two modes of operation of the analyser depending on the retardation strategy used. By far and away the most common mode for PES data acquisition is constant analyser energy (CAE), which is also called fixed analyser transmission (FAT). In CAE mode, the electrons are retarded to a value defined by the operator in the experimental set-up known as the pass energy, which is the kinetic energy required to pass through the analyser. The voltages on the hemispheres are scanned to vary the kinetic energy that will pass though, but crucially the potential difference between the hemispheres is kept constant. This maintains consistent resolution across the energy range and meaning that the spectral resolution is chosen by setting the pass energy value. A low pass energy gives high spectral resolution, but lower sensitivity. A high pass energy will give better sensitivity through higher analyser transmission but will result in lower spectral resolution. Examples of varying the pass energy for a Si 2p spectrum are shown in Figure 3.9.

The alternative operating mode of the analyser, constant retard ratio (CRR), allows the value of ΔV to change across the kinetic energy range. The pass energy is therefore proportional to the kinetic energy, and so the transmission and resolution changes across the spectrum. This mode is rarely

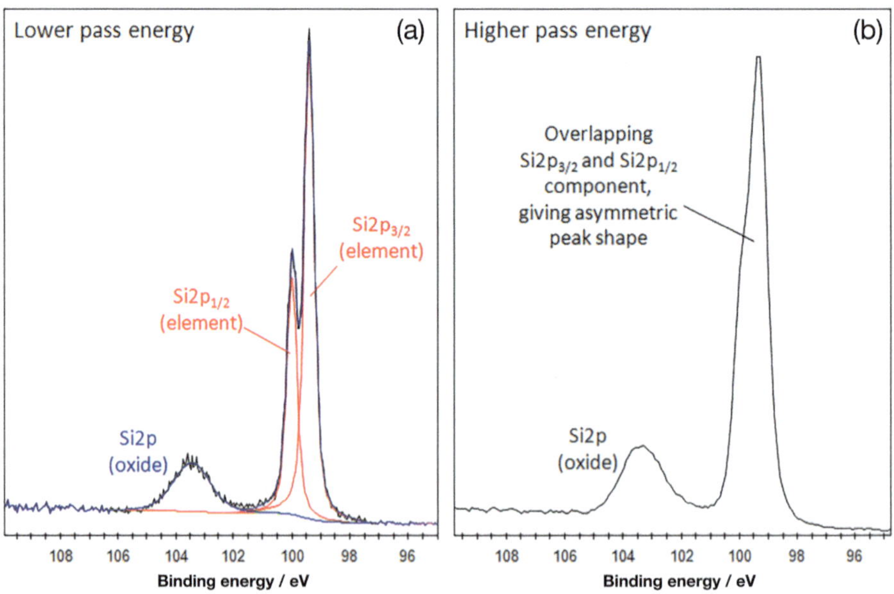

Figure 3.9 Si 2p spectra collected at (a) low pass energy and (b) high pass energy showing the change in spectral resolution which is especially evident in the Si 2p doublet for the elemental chemical state.

used for photoelectron spectroscopy, but is commonly used for Auger electron spectroscopy, where CRR mode helps to reduce the impact of the large secondary electron tail on the appearance of the spectrum. CRR mode makes quantification much more difficult, and therefore another reason why CAE mode is the most common mode for PES.

3.4.2.3 Detectors

On a modern instrument the detector is usually a multi-element device to accurately and efficiently facilitate the measurement of the number of photoelectrons that have passed through the analyser. All detector designs can be broken down into two main components. The first element amplifies the incoming electron signal to make it easier to measure, while the second element acts as the electron counter. Channel electron multipliers ('channeltrons') were the original electron detectors used and they are still found in applications where a high dynamic range is required. They work as an effective electron signal amplifier, where an incoming electron strikes the surface inside the channeltron, initiating a cascade of electron emission and resulting in a final signal that can be measured at the channeltron output. Arrays of up to nine channeltrons have been used in commercially available systems to amplify and count the incoming signal.

For even greater numbers of detection channels, there has been a move in recent times to larger area detectors. These use channel plates, which are in effect an area with many small holes with a similar coating to the inside of a channeltron, and act as the electron amplifier. Usually at least two plates are used with the channels placed with the opposite orientation which maximises the electron cascade. At the exit to the channel plate stack there needs to be a device that can detect the signal. This can be a phosphor screen with a camera that measures the light flashes and intensities as electrons strike the screen, which was the original way of using this type of system, but over the last twenty years the preferred method has developed to use either a resistive anode or a delay line to collect the signal. These are both position-sensitive detectors and can measure the point in two dimensions where the electron signal has hit the surface. In both cases this is by measuring the time that the pulse takes to reach the opposite corners of the assembly.

Being able to do this allows for the collection of 'snapshot spectra' where the analyser energy is fixed, and by using a relatively high pass energy a range of energies can be collected by the detector resulting in a spectrum being acquired in less than a second. This is useful in experiments where a large number of spectra are to be collected such as depth profiling or spectrum mapping. If the detector is operated in a two-dimensional mode, then it can be used for parallel imaging, where the positional information from where the photoelectron was emitted from the sample is maintained through the analyser and projected onto the detector, or parallel angle-resolved PES where the emission angle information is maintained. These experiment modes will be explained in more detail later in this chapter.

Each element of the detector needs to be calibrated to ensure that differences in the response of each channel do not affect the measured data. For channeltron-based systems this is by tuning the operating voltage to ensure an even response across the array. Multi-channel instruments using channel plates can have the signature of the detector measured to remove it from the measured signal, analogous to the background in an FTIR spectrometer. This is one aspect of the calibration of the system, but there are other things that need to be checked before using the system to collect data.

3.4.3 Essential Analytical Accessories

Photon sources and the spectrometer are not the only essential components on an instrument. The standard equipment list for a PES system will also include a method for analysing insulating samples, which is normally referred to as charge compensation, and a device for cleaning samples or removing material to extend the analysis from the surface and into the bulk of the material, which is the ion source.

3.4.3.1 Ion Sources

Samples, even when stored in very clean conditions can acquire a thin layer of surface contamination which can attenuate the measured PES signal from the surface. This is usually carbonaceous in nature, often referred to as 'adventitious carbon contamination' (see Chapter 2) and may need to be removed once the sample has been introduced into the analysis chamber, although measuring the contamination on the sample 'as-received' is usually best practice. To facilitate sample cleaning, instruments are usually configured with an ion source, which is used to direct a beam of energetic, ionised gas atoms towards the surface in order to remove material from the surface. The ionised gas is typically argon, which will not chemically react with the sample, nor the internal components of the source once it has been ionised, but has the characteristics such as mass which will result in an appropriate sputter yield of material from the surface. As well as being used for sample cleaning, ion sources can also be used to remove material from the sample to extend the analysis into the material and investigate composition changes with increasing depth or at interfaces. This experimental procedure is called depth profiling and involves alternate cycles of PES analysis followed by the use of an ion beam to remove thin layers from the sample surface.

Most ion sources use a hot filament to create energetic electrons which interact with gas leaked into the source. The ions that are created can then be extracted, accelerated and focused onto the sample through electrostatic elements. The beam of high-energy ions is rastered across the surface and collides with the surface atoms, leading to their ejection and revealing the underlying layers. This is very effective for removing surface contamination such as adventitious carbon, or oxides that may have formed during storage of the sample. The major advantage of monatomic ion sources is their high sputtering yield, allowing for a relatively fast removal of material even at lower beam energies. However, the typical operating energy range of the ion beam, from a few hundred to several thousand electron volts, can cause significant damage to the sample surface, potentially altering the original composition and structure of the material through preferential sputtering of lighter elements, or by causing chemical changes as a consequence of the bond breaking induced by the ion–surface interaction.

To try to minimise the damage caused by monatomic ion sources, cluster ion sources have been adopted as an alternative method for sample cleaning and depth profiling. An example of such a source is shown in Figure 3.10. Cluster ion sources generate a beam of weakly-bound clusters of gaseous ions, which are directed at the sample surface at a low velocity. Unlike monatomic ion sources, the energy from the impact of the cluster ions is concentrated in the surface and very near surface region of the bombarded material, limiting the potential for chemical damage of the remaining sample. This makes cluster ion sources particularly suitable for PES depth profiling of fragile and soft materials, such as polymers. However, the sputtering yield of cluster ion sources is lower than that of monatomic ion sources for inorganic materials, which can result in longer analysis times, but with the benefit of less chance of preferential sputtering or chemical modification.

The cluster ion source is very similar to the monatomic ion source in a lot of respects, but the main differences are for the creation of the gas clusters, and the need to be able to filter the clusters once ionised to have control of the mass-charge of the particles directed towards the sample. Clusters are created by cooling the incoming gas which is achieved by expanding gas at a

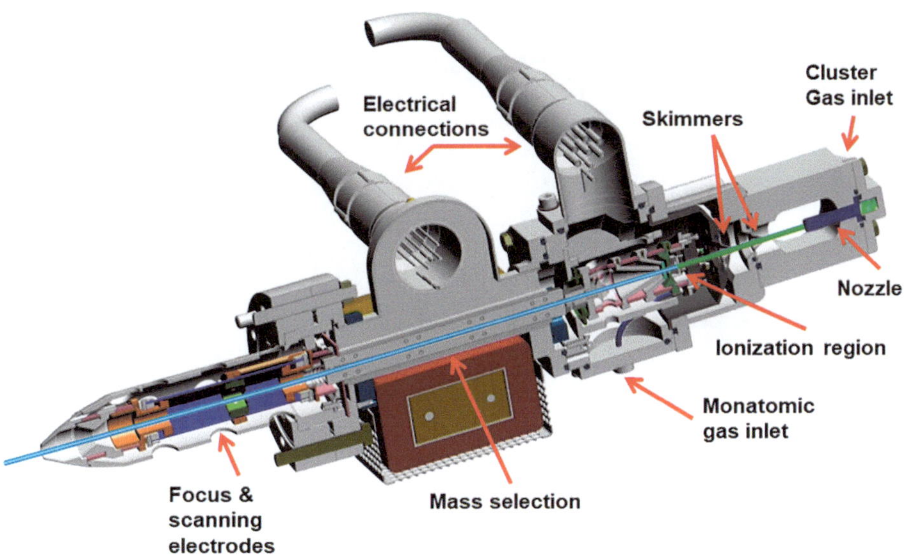

Figure 3.10 Schematic of a dual mode monatomic and gas cluster ion source.

relatively high pressure through a specially shaped nozzle into a low pressure region. This creates a supersonic expansion which rapidly cools the argon gas atoms, and causes them to condense into clusters. The gas expansion is accelerating to supersonic speed, and has to be extracted through skimmer elements to prevent shockwaves breaking the clusters apart. The expansion region also needs to be differentially pumped to prevent the excess gas raising the pressure of the analysis chamber. The extracted clusters are then bombarded by electrons in the same way as in the monatomic ion source, to ideally ionise one atom in the cluster. The ionised clusters can then be focused into a region of the source to select singly-ionised clusters of the correct size. This is usually by using a curved region of the source with an applied magnetic field which will only allow clusters of the requested size to pass through. The beam of mass-selected, singly charged clusters is then focused onto the sample.

Cluster ion sizes range from a few 10s to 1000s of atoms, and the beam energies range from 2–20 keV. The consideration for the choice of ion beam is the energy per atom in the cluster, where a low energy per atom will typically result in less chemical damage, but also generally a lower sputter rate. Generally low energy per atom cluster ion beams, such as 4 keV 2000 atoms, *i.e.* 2 eV per atom, are good for polymers, where the low energy per atom prevents chemical damage to the polymer, but the high mass difference between the cluster ion and the polymer still results in a high sputter rate. Inorganic materials require much higher energy per atom values of around 50–100 eV per atom to result in good removal rates. A comparison of the difference in monatomic and cluster ion beam sputtering of polyimide is shown in Figure 3.11.

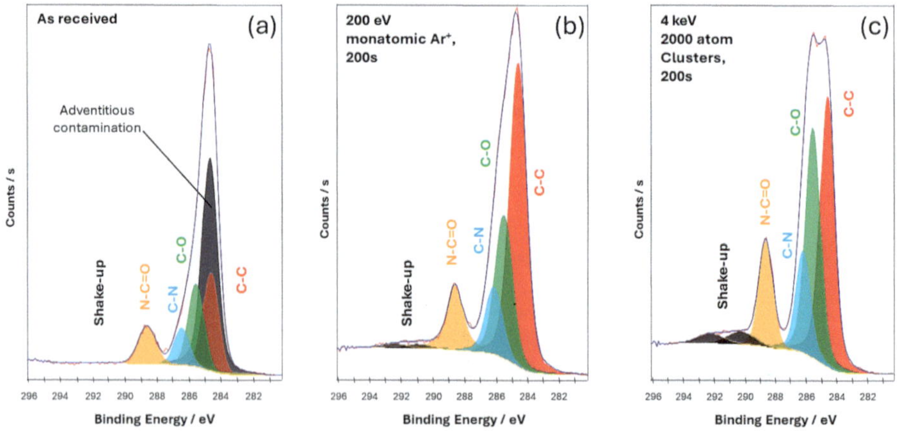

Figure 3.11 C 1s spectra for (a) as-received polyimide, (b) polyimide cleaned using a low energy monatomic ion source, and (c) polyimide cleaned using a gas cluster ion source. The change in the ratio of the intensity of the C–C and C–O peaks and the loss of the shake-up satellites due to damage to the chemistry is clear in spectrum (b).

3.4.3.2 Charge Compensation Systems

One of the fundamental challenges in PES is the consistent analysis of insulating samples. This is due to the build-up of positive charge on the surface of the sample during PES measurements, where unlike in a conducting sample the emitted photoelectrons cannot be replaced by conduction from the system ground. This phenomenon can distort the energy of the emitted photoelectrons, thereby affecting the quality and accuracy of the PE spectrum, as shown in Figure 3.12. To overcome this issue, a technique known as charge compensation is employed.

The goal of charge compensation is to maintain a stable surface potential during the PES measurement, enabling the acquisition of accurate and

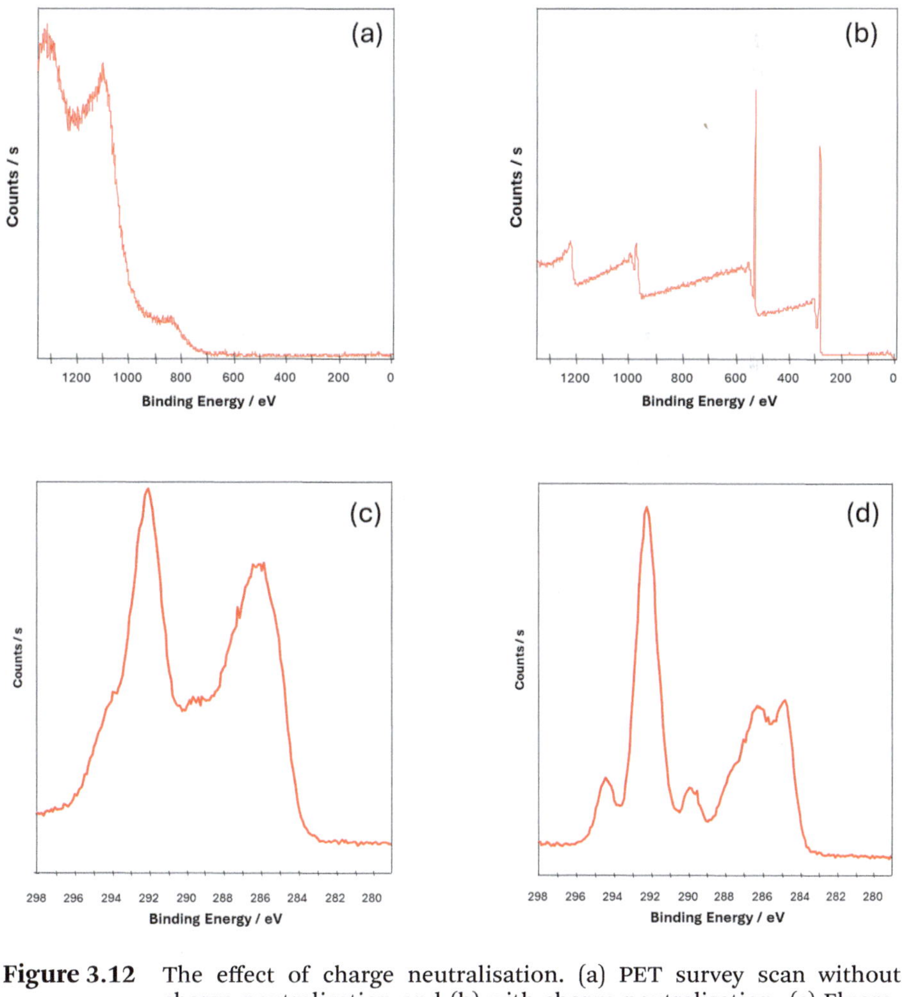

Figure 3.12 The effect of charge neutralisation. (a) PET survey scan without charge neutralisation and (b) with charge neutralisation. (c) Fluoropolymer-treated fabric analysed with poor charge neutralisation and (d) with effective charge neutralisation.

reliable data. This is achieved by providing a supply of low-energy electrons to the analysis area that can neutralise the positive charge and maintain a stable potential at the surface.

The common method of charge compensation involves the use of an electron flood gun. The flood gun is aligned to the same position as the X-ray source on the sample and emits electrons with energies typically less than five electron volts. These electrons neutralise the positive charge on the sample surface, thereby preventing the distortion of the photoelectron energy and the PE spectrum. This may not be enough to achieve complete neutralisation of the analysis position, however. There may also be static negative charges present on the sample which could deflect the low energy electron beam, and so co-axial flood sources have been developed that use combination beams of low energy electrons and low energy argon ions. These co-axial beams can act to balance the surface charge by supplying either positive or negatively charged particles as required by the surface potential, effectively self-regulating.

In both methods, it is critical to carefully balance the amount of charge compensation to avoid overcompensation, which will require post-processing of the data to correct the peak positions as overcompensation will move the spectrum to lower binding energy. This can be challenging on samples where the exact peak position is not known, and potentially lead to misassignment of features. On modern instruments, the charge compensation systems are now designed to be turn-key in operation, removing the need to optimise for every sample, and make insulator analysis a routine proposition. The only real consideration needs to be ensuring that any materials that may have a mix of insulating and conducting regions are electrically isolated so that they can have a uniform surface charge created through the charge compensation system, rather than having differentially charging areas due to the possible variation in surface and near surface charges. It is also worth noting that with the increase in the use of hard X-ray sources, the charge compensation requirements due to differential charging can be more significant if the charging area is sub-surface due to the increase in analysis depth.

3.4.4 Other Techniques (AES, ISS, IPES, REELS, Raman, *etc.*)

Although PES instruments are primarily for PES, it is very common for systems to also have additional analytical techniques available for use with the samples under investigation. In this section, the methods that are commonly seen on PES systems will be briefly discussed to explain the additional information they can add to a PES analysis.

3.4.4.1 Auger Electron Spectroscopy (AES)

Auger Electron Spectroscopy (AES) is another analytical technique that is used primarily for surface analysis. Named after Pierre Auger, a French physicist who was one of the pioneers of the technique, AES provides detailed information about the elemental composition and some information on the chemical state of surfaces. In AES, the sample is irradiated with a focused beam of electrons rather than the photons used in PES. This electron beam can be focused to a much smaller spot size than photons, which makes it possible to analyse much smaller features. The probe spatial resolution can be less than 100 nm for a source mounted on a PES system (although dedicated AES systems can achieve much smaller spatial resolutions) compared to 5–10 μm for PES.

When the focused beam of electrons strikes the sample surface, if the beam energy is high enough, electrons are emitted from the core levels of the atoms, similar to photoelectron emission with PES. The remaining core hole can then be filled by an electron from a higher energy level, and at this point, two competing processes can occur. First, as the higher lying electron fills the core hole, there can be emission of a photon, which is the process behind energy dispersive X-ray spectroscopy (EDS). Alternatively, instead of radiative relaxation, as the higher lying electron fills the core hole, an additional electron, also from a higher lying level, can be emitted. This is the Auger–Meitner electron. Unlike PES, where the photoelectron energy depends directly on the energy of the incident photon, the energies of Auger–Meitner electrons depend only on the sample electronic configuration and not on the energy of the incident electron beam. The spectrum is collected using the PES analyser in largely the same way as a PE spectrum, although it is common for AES spectra to be collected in the constant retard ratio (CRR) operating mode rather than the constant analyser energy (CAE) mode used for PES. The overall shape of the spectrum, typical of AES, is of a large, decreasing, background of electrons at low kinetic energy followed by a rising background that continues to increase up to the energy of the primary beam. The Auger–Meitner peaks are superimposed on this background.

AES has the advantage of a high spatial resolution compared to PES, which makes it particularly useful for complementing PES surface analysis. However, it is more limited concerning chemical bonding state information. For certain elements, such as copper and zinc, Auger spectroscopy is quite rich in chemical information, but with most elements, such as carbon, for example, the information is very limited. Furthermore, the use of a high energy electron beam can be quite damaging to sensitive materials and for insulating samples a method of charge compensation will need to be used, commonly a flood of low energy argon ions to prevent surface charging.

3.4.4.2 Low Energy Ion Scattering (or Ion Scattering Spectroscopy) (LEIS)

Low Energy Ion Scattering (LEIS), also known as Ion Scattering Spectroscopy (ISS), is another analytical technique that provides insight into the elemental composition of the surface of a sample. At its core, LEIS operates based on the principle of momentum transfer. When a beam of ions is directed at a surface, the ions are scattered by atoms at the surface. The kinetic energy of the scattered ions is measured, and peaks are observed that correspond to elastic scattering of ions from atoms at the surface. Each element at the sample surface produces a peak at a different measured kinetic energy due to the momentum transfer between the incident ion and atom.

The process begins by directing a beam of ions, usually noble gases like helium or neon, at the sample surface. The ions scatter upon collision with the surface atoms, and the kinetic energy of these scattered ions is then measured using the PES analyser, but operating with inverted polarity compared to PES as positively charged particles are now being detected rather than electrons. The ion's energy loss depends on the relative masses of the surface atom and incident ion, with heavier atoms causing more energy loss. This also means that only surface atoms with a mass greater than the incident ion mass can be detected, which makes helium a popular choice for the incident ion. A schematic of the LEIS experiment is shown in Figure 3.13.

This energy loss is the basis for the spectral peaks observed in LEIS. Each element present on the surface will produce a unique peak in the spectrum, with the energy separation between peaks relating to the relative atomic masses of the elements. Thus, the ion scattering spectrum serves as a fingerprint of the surface's elemental composition. In some cases, different isotopes of the same element can even be resolved, although this usually requires a heavier primary ion than helium.

One of the main strengths of LEIS is its extreme surface sensitivity. The strong scattering signals are essentially limited to the topmost layer of atoms, making LEIS an ideal technique for analysing the outermost atomic layer of a sample. However, this high surface sensitivity also means that the sample must be very clean, as even a small quantity of surface contaminant can

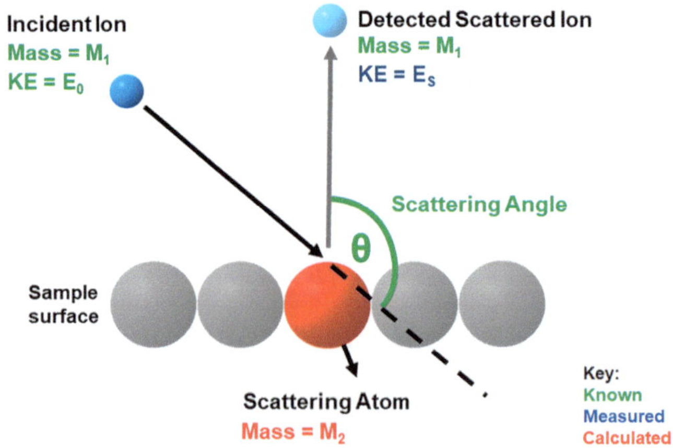

Figure 3.13 Schematic showing the basic principle of low energy ion scattering. The same process is also the basis of REELS, although the incident particle is an electron rather than an ion.

strongly influence the ion scattering spectrum. Fortunately, the use of an ion beam, albeit a beam of He ions, means that the sample will be gently cleaned during the experiment, which also needs to be remembered when acquiring data to ensure that spectra are averaged that are not dynamically changing.

In LEIS, the absolute kinetic energy (E_s) of each peak also depends on the energy of the scattered ion beam (E_0). The ratio of the scattered and incident ion energies, E_s/E_0, is the relevant quantity when dealing with LEIS. The angle through which the ions are scattered is also important. In cases where the ion source and detector are fixed, the angle is simply one parameter that must be included in the calculations. It is also worth noting that the primary ion energy (E_0) is not always known with sufficient accuracy at the start of an experiment. The solution is to calibrate the energy using a standard sample, preferably of a pure, metallic element such as gold. Once the scattered peak energy from the standard sample is measured, the only remaining unknown is E_0, which can be calculated using the relevant equations.

LEIS can be a valuable tool to complement PES, particularly for the understanding of problems such as segregation, where the difference between the PES analysis depth and the extreme surface selectivity of LEIS can help confirm which elements are actually at the surface. The slow sputtering of the material by the ion beam needs to be accounted for, and the scattering nature of the method makes it more suitable for flat samples compared to rougher materials, but overall the additional information provided can be very useful for materials where the outermost layer is of significant importance.

3.4.4.3 Reflected Electron Energy Loss Spectroscopy (REELS)

REELS operates based on the interaction of a primary electron beam with a sample in a very similar way to the ion beam used in LEIS described in the previous section. In the case of REELS a beam of electrons, rather than ions, with an energy typically in the range of 250–1000 eV is directed at the sample surface, causing the electrons to scatter and lose energy due to interactions with the atoms at the surface. By measuring the energy of these scattered electrons, REELS can provide information about the sample's surface composition and electronic structure.

The process begins by firing a beam of electrons at the sample surface. Most of these electrons reflect directly from the surface without losing any energy and are detected at a kinetic energy that is the same as the beam energy by the PES analyser. These elastically scattered electrons form the most intense part of the REEL spectrum, leading to a strong peak which defines the 0 eV position on the energy loss scale. However, some electrons from the beam do interact with the surface which results in losing some of their kinetic energy. This energy loss can occur due to various processes, such as the excitation of valence electrons into unoccupied electron levels or the excitation of plasmons. These inelastically scattered electrons then have a lower kinetic energy than the initial beam and appear at a higher energy loss value in the REEL spectrum. The energy loss of the scattered electrons is directly related to the energy required for various excitation processes within the sample. Therefore, by analysing the energy loss spectrum, REELS can provide information about the electronic structure of the sample. For instance, it can measure the band gap of a metal oxide or the relative energies of the lowest unoccupied energy levels in an organic material. It can also be used to identify loss features in a PE spectrum, since they will appear in the REEL spectrum too, helping with the identification of PE spectral features.

One other aspect of REELS is its ability to detect and quantify hydrogen, something which is not possible with PES. This is effectively the same process as LEIS, with the hydrogen atom being light enough that there is a small momentum transfer when struck by the incoming electron which can be seen as a small energy loss of around 1.5–2.5 eV. By use of known standard samples it can be possible to create a calibration plot to be able to quantify the amount of hydrogen in an unknown sample. An example is shown in Figure 3.14.

Adjusting the beam energy can allow REELS to provide depth information about the sample too, essentially by changing the interaction depth with

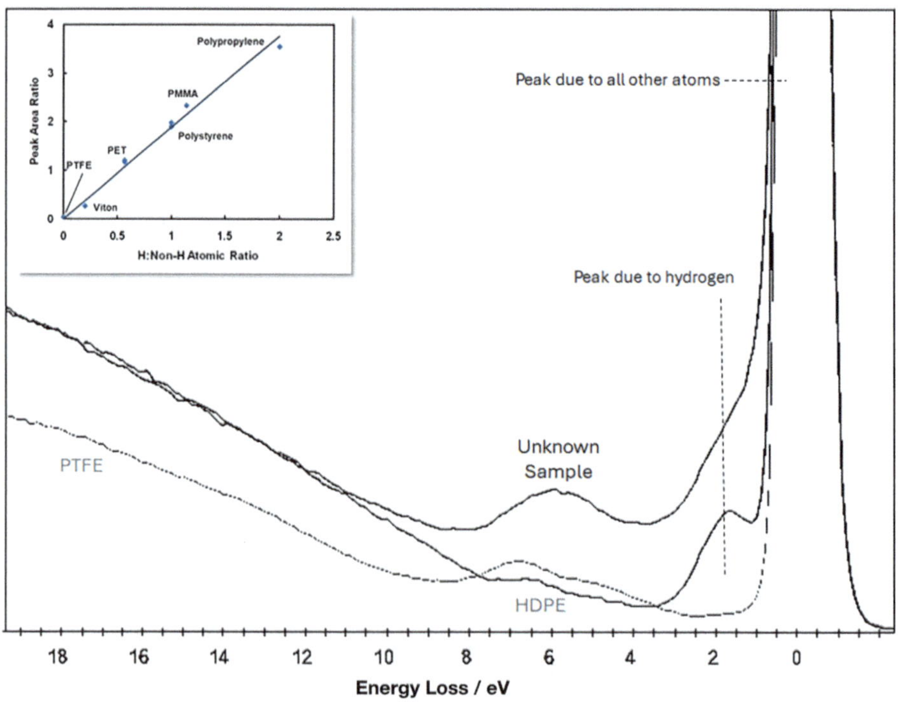

Figure 3.14 The measurement of the presence of hydrogen using REELS for PTFE, HDPE and an unknown sample. The intensity of the peak arising due to the presence of hydrogen can be plotted onto a control chart (inset) to establish the amount of hydrogen present.

electron energy in a similar way to the change in analysis depth by changing the photon energy in PES. The information obtained from REELS typically comes from a similar range of depths as PES, which is about 0–10 nm for the typically used beam energies. It's important to note that REELS requires careful calibration and interpretation. The energy loss spectrum can be influenced by various factors, including the angle and energy of the incident electron beam, the sample temperature, and the sample's surface roughness. Therefore, the accurate interpretation of REEL spectra often requires comparison with theoretical calculations or reference spectra from known materials to establish the nature of the observed features. It is a very simple experiment to perform, however, and so the potential use for probing the unoccupied levels of a materials energy diagram can make it a very useful tool for the analysis of materials for electronic and energy applications.

3.4.4.4 Inverse Photoemission Spectroscopy (IPES)

An alternative method to REELS for investigating the unoccupied states of a material is inverse photoemission spectroscopy (IPES). As the name suggests, the technique is, at a basic level, the opposite of photoemission spectroscopic methods such as PES and UPS, involving an incident beam of electrons directed towards the sample, and the detection of the emitted photons. The instrumentation required is therefore a source of electrons and a method of detecting the emitted photons. To achieve good resolution of the spectrum, the source of electrons needs to be operated with a very low energy spread on the kinetic energy of the electron beam, which is scanned through a range from 0 eV up to a maximum of around 20–30 eV. When the electron beam interacts with the sample, the electrons can be absorbed in a random fashion into any of the unoccupied states which are below the kinetic energy of the electron beam. This results in a continuous emission spectrum, again referred to as Bremsstrahlung, which is representative of the density of states of the unoccupied states of the sample. This is quite complex, and there are two main ways to measure the emission spectrum. The more expensive method is to use a UV monochromator to energy select and detect the emitted photons, but a less costly, yet still effective approach can also be employed using a band pass filter and a detector. In this arrangement, the detector is configured to only detect photons with a specific energy, often between 9 and 10 eV, by using a band pass filter in front of the detector to only admit photons with that energy, and with a resolution on the centre energy of the filter of around 0.6 eV. The kinetic energy of the electron source is scanned across the energy range, and the detector will only measure a narrow band of states at a binding energy above the Fermi level defined by the energy of the band pass filter below the electron kinetic energy. Scanning the beam energy changes the detection energy and allows measurement of the photon intensity for a particular energy, which is proportional to the density of states of the unoccupied states in the sample.

The challenge with IPES is that the cross-section for inverse photoemission is several orders of magnitude lower than for photoemission in the UV region. Where UP spectra typically have high intensities, IPE spectra can take substantial amounts of time to collect due to the low count rates caused by the low cross-section for photon emission. This necessitates the stability of the sample under the electron beam to need to be considered, although the low energies do mitigate the long exposures times to an extent. This complexity and low sensitivity make REELS a more popular choice for probing the unoccupied states, but there are situations, largely where the separation between the Fermi level and the unoccupied states is very low, that REELS is unable to be used for the measurement, and IPES is the technique that needs to be used.

3.4.4.5 Raman Spectroscopy

Raman Spectroscopy is an analytical technique that has been widely utilised in various scientific fields, from chemistry and physics to materials science and biochemistry. Recently it has become possible to integrate a Raman spectrometer onto a PES instrument, which adds the possibility of correlative molecular spectroscopy with PES analysis. Raman spectroscopy is based on the principle of inelastic scattering of light, also known as the Raman effect. When a sample is illuminated with monochromatic light, such as from a laser, the majority of the light is scattered elastically, meaning its frequency remains the same. This is known as Rayleigh scattering. However, a small fraction of the light (approximately 1 in 10 million photons) is scattered inelastically, meaning its frequency changes. This inelastic scattering is the Raman effect. In the Raman effect, the incident photon interacts with the vibrational states of the molecules in the sample. This interaction results in a shift in the frequency of the scattered light, which is either increased (Stokes Raman scattering) or decreased (anti-Stokes Raman scattering) compared to the frequency of the incident light. The magnitude of this shift is directly related to the vibrational energy levels of the molecules in the sample.

A Raman spectrometer consists of a laser source to illuminate the sample, a lens to collect the scattered light, a spectrometer to separate the light into its component frequencies, and a detector to measure the intensity of the light at each frequency. The resulting spectrum of intensity *versus* frequency shift provides a unique 'fingerprint' of the sample, revealing information about its molecular structure and composition.

One of the main strengths of Raman spectroscopy is its sensitivity to vibrational modes, which makes it an excellent tool for identifying molecules and studying their structures. This can be very helpful when analysing polymers with PES, where a definitive identification of the precise material can be more difficult.

Despite its many advantages, Raman spectroscopy also has its challenges. The Raman effect is inherently weak, which can make it difficult to detect in the presence of strong fluorescence or background noise. Its depth of analysis is generally much larger than PES too, depending on the optical properties of the material being analysed. The analysis depth could be more than an order of magnitude deeper than the PES measurement at the same point, requiring care to understand the difference between effectively bulk and surface measurements. There is one exception to this, where the analysis regime is similar between the two methods and that is for 2D materials such as graphene and transition metal chalcogenides. For these types of materials, the Raman effect is strong, and the nature of the materials means that the data for both techniques are associated with similar volumes. This emerging area of study has meant that the combination of Raman with PES has developed into a much more common combination over the last decade.

3.5 Near Ambient Pressure XPS (NAP-XPS)

The origin of a large amount of the initial interest in XPS was in understanding catalytic processes, where an understanding of the nature of the active site could have benefits in optimising and improving the catalyst. As explained at the start of this chapter, XPS is largely considered a UHV technique, but early pioneers of the method such as Siegbahn,[2] the scientist who was awarded the Nobel prize for the invention of Electron Spectroscopy for Chemical Analysis (ESCA), and Joyner *et al.*[3] investigated how it could also be used to bridge the 'pressure gap' from UHV to analyse liquid samples. This initial interest was not maintained after these first investigations however but has re-emerged since the early part of the 2000s as instrument design has made it possible to create more efficient systems for operating at higher pressures.

The key to operating an XPS experiment at higher pressures is the effectiveness of the differential pumping arrangement for the analyser. A schematic of a typical arrangement is shown in Figure 3.15. As was discussed in Section 3.2.1, one of the main reasons why UHV is required is so that the emitted photoelectrons do not collide with gas molecules and lose energy which will degrade the measured spectrum. Therefore, a successful NAP-XPS experiment needs to extract the photoelectrons from the high-pressure region around the sample and transfer them into vacuum as quickly as possible. To do this, the entrance of the NAP-XPS analyser is close to the sample, with an aperture that is designed to maximise the acceptance angle but minimise the transmission of gas into the first section of the electron transfer optics. The aperture dimensions and its distance from the sample are closely related to the gas pressure that the sample is to be exposed to, since the attenuation length of the photoelectrons will reduce and the transmission of gas into the transfer optics will increase with increasing pressure, making shorter distances and smaller apertures imperative. Once the photoelectrons have entered the transfer lens region, significant differential pumping is used to keep removing gas from each stage of the lens, with a typical analyser requiring four stages of differential pumping before the entrance to

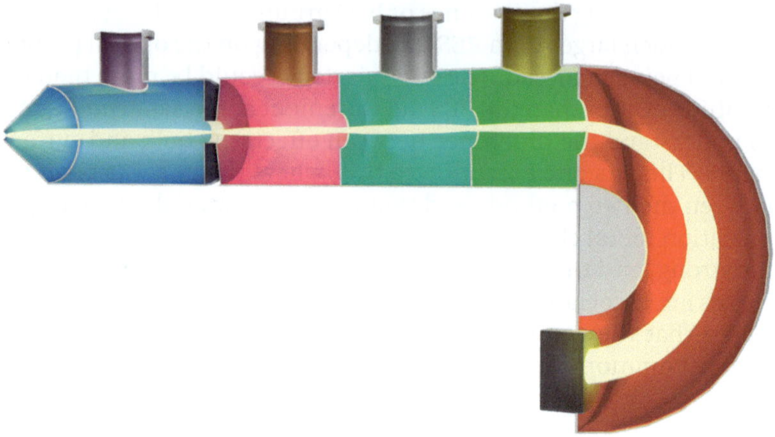

Figure 3.15 Schematic of a NAP-XPS analyser showing the arrangement of apertures and differential pumping prior to the entrance to the hemispherical analyser.

the hemispherical analyser. To further assist with controlling the pressures throughout the focusing stages in the transfer optics, additional apertures may be added to the lens to restrict the ingress of gas into the lens stages closer to the hemispherical analyser. This can add further complexity to the lens design, but also acts as an effective way to maximise the signal being measured.

As well as changes to the electron lens and analyser, the X-ray source design will also need modification. The high voltages associated with the generation of X-rays for XPS are not compatible with higher pressure environments and so NAP-XPS systems are designed to ensure that the X-ray source can be kept in UHV. To do this an X-ray transparent window needs to be used between the X-ray source and the sample region. Silicon nitride has the necessary properties to be able to allow acceptable transmission of X-rays and also remain intact with the substantial pressure difference between the two areas of the system.

NAP-XPS experiments can take substantially longer than UHV XPS experiments due to the considerations listed above. As the pressure in the sample region increases, the acquisition time required to achieve acceptable signal-to-noise will increase substantially. This can be mitigated by increasing the X-ray flux directed towards the sample, which is why a lot of NAP-XPS work is performed at synchrotron light sources which have much brighter sources than lab-based systems. As long as the samples are not too sensitive to X-ray exposure, lab-based NAP-XPS is now becoming a much more routine experimental method. Some care does need to be taken with insulating samples, as the method for charge neutralisation is markedly different to a UHV system and is more in common with an environmental scanning electron microscope (E-SEM). To measure insulating samples, the ionisation of the gas in the region around the sample by the incoming X-rays is used to generate both positive ions and electrons which can then travel to the analysis region and act to neutralise any charge build-up. As in an E-SEM, regulating the pressure of gas is the key to achieving effective charge compensation, which may influence the possible pressure that can be used for experiments.

3.6 Calibration

Calibrating the detector ensures that the measured intensity of the peaks is correct, but the other key aspect of a PE spectrum is the peak position, which is a critical piece of information to identify the chemistry of the material under investigation. The energy scale of the spectrometer is what is needed to be calibrated to ensure the best accuracy for the measurement of peak positions. It is performed by measuring peak positions from known materials across the energy range of the spectrum and adjusting the operating voltages of the spectrometer to correct for deviations in position (see also Section 2.8). With modern instruments the error in the peak position can be typically less than 50 meV following calibration, which is less than the typical precision of a PES measurement. Commonly used materials for calibrating the spectrometer are copper, silver and gold foils, which are easy to obtain in a pure form, and easy to clean in the vacuum system. The principal peaks for the three metals are spaced between 0 and 1000 eV when excited by Al Kα, ensuring that the majority of the energy scale can be checked. By measuring the peak positions of Cu $2p_{3/2}$, Ag $3d_{5/2}$ and Au $4f_{7/2}$ the energy scale linearity can be corrected. Other peaks can be used, for example the adventitious carbon that is found on every sample when introduced into a vacuum system[4] (as described in Chapter 2). The peak position of the Cu LMM X-ray induced Auger–Meitner peak can also be measured so that the X-ray photon energy difference to the nominal 1486.6 eV can be calculated, which can be used to optimise the monochromator to improve its efficiency and resolution.

3.7 Experimental Design

Beyond an understanding of the experimental hardware and the considerations that need to be made with regards to sample mounting, acquiring PES data is a reasonably straightforward process on modern instruments which are largely completely computer controlled, and in a lot of cases can even be operated remotely once the samples have been introduced into the system. In this final section of this chapter, the basics of the common experimental methods will be introduced along with some notes on the common things to check for during data acquisition.

3.7.1 Point Analysis

The majority of PES analysis carried out in laboratories across the globe is point analysis, where the surface composition is measured at one or several points on the sample. The experiment can be set up by considering just a few key areas. Firstly, how large are the areas to be analysed. PES instruments analysis areas can vary from a few microns to several hundred microns for point analyses, and if a feature of specific dimensions is to be measured rather than points on a uniform surface then the analysis area will need to be adjusted to fit. Depending on the instrument type this may involve changing the X-ray spot size (source-defined analysis) or the field of view of the analyser (lens-defined analysis). After the analysis area has been determined, the next decision is whether the charge compensation system will be required, *i.e.* whether the sample is an insulator. On most modern instruments this is a simple task to enable, and the only implication is if the data will be shifted by use of the neutralisation system which will need to be accounted for in post-acquisition processing.

The remaining task before data acquisition can start is to define the spectrum acquisition parameters. Most analyses will start by acquiring a survey scan: a spectrum that covers a large spectral range (for Al Kα X-ray source instruments this will be a binding energy range of 0–1350 eV) which will enable the elemental composition to be measured. System parameters that give high sensitivity at the expense of spectral resolution, *i.e.* using a high pass energy, are usually used to ensure that weak signals are not missed. In addition to the pass energy, spectral acquisition parameters also include the energy step size, the dwell time (the time that the spectrometer collects at a particular energy before moving to the next), and the number of scans or sweeps that the spectrometer will average together. Survey scans usually use larger step sizes to speed up the acquisition, and because the higher pass energy would make using smaller step sizes redundant due to over-sampling, acquiring data that cannot be resolved due to the pass energy used.

The survey scan is used to understand the chemical composition of the surface, and based on the results high-resolution or narrow scans are collected for the elements of interest. Care needs to be taken in compensating for overlapping features in regions of interest, where collecting the peak from an orbital that is not the strongest may be advantageous if it is not complicated by an overlap that the major transition is affected by. Region scans are collected at lower pass energies to make the resolution of spectral features easier, and with smaller step sizes. As long as the sample is not sensitive to X-rays, the experiment can be designed to acquire for sufficient time to obtain good statistics on the data and make the data processing to quantify the sample compositions as straightforward as possible.

3.7.2 Line Scans and Imaging

Extending the data acquisition from a single point to a line or an area can be done for a couple of reasons. The most obvious reason is to measure either across a feature or collect a PES image of the surface by mapping or parallel imaging. However, as touched upon in the previous section, some samples are sensitive to damage by X-rays, and by collecting at multiple points but for shorter acquisition times, the resulting data set can be averaged to produce spectra that are representative of the material. As an example, rather than collecting ten scans at one point, ten points where one scan has been collected at each position can be averaged together to achieve the same result but with 90% less X-ray exposure per point.

For line scans or images to measure features on the surface, the experimental parameters need to be chosen to efficiently collect the data in an acceptable amount of time. Modern instruments are able to collect region spectra in extremely fast times, through either rapid scanning or so-called snapshot acquisition where the analyser energy is fixed rather than scanned. The step size for line scans or the pixel size for spectrum maps is determined by the analysis area, either the X-ray spot size or the analyser field of view. These experiments can cover large areas, defined by the stage movement of the system, and the only limit is the time it will take to collect the data based on the number of spectral regions that will be acquired and the number of points that make up the line or area. Spectrum mapping can also be speeded up by scanning the X-ray source or the stage at speed and acquiring data very quickly.

Parallel imaging, as mentioned earlier in this chapter, works differently. In spectrum mapping, a spectrum is collected at each point in the image sequentially, moving point to point to acquire all the required data. Parallel imaging collects data for every pixel in the image simultaneously, but only at one binding energy position. Once the first image is complete, the next image at the next binding energy is collected, and by stepping through the range of energies, a series of images are collected which are the equivalent of the spectrum map, with each pixel being representative of a spectrum at that point. Better spatial resolution can be achieved with parallel imaging, although the gap to spectrum mapping is getting smaller, and the high number of data points but lower signal levels lend themselves to multi-variate statistical processing as a way of getting good insights.

3.7.3 Depth Profiling

PES systems are not generally regarded as imaging tools as the spatial resolution is less useful than other imaging techniques. The most common experiment performed after point analysis is depth profiling. As briefly described in the ion source section, this is an experiment mode where material is removed from the surface after each data acquisition step, which enables a composition depth profile of the sample to be created. As with line scans and PES imaging, the main consideration from the spectrum acquisition perspective is the amount of time it will take to collect the required spectra, and the pass energy, step size and number of scans need to be considered to ensure that the experiment time does not become too large.

There are also some things that are specific to depth profiling. The ion source mode to be used is very important to control the level of disruption to the remaining surface after etching, and also the speed of material removal and therefore the depth resolution that will be achieved in the profile. This will be deciding whether to use monatomic or gas cluster ions, if available, and what beam energy to use. The analysis area that the data will be collected from also needs to be considered. It is considered good practice to use an analysis area that is five times smaller than the area that the ion beam is being rastered over during etching. This is to ensure that the depth resolution of the profile is not degraded by the edges of the etched area being included in the analysis. If the sample to be depth profiled is an insulator it may be necessary to build some extra time into the experiment for the surface potential to stabilise after etching, as a large positive potential could be created at the surface during the sputtering process. Finally, the order of the spectra to be acquired needs to be optimised. Even in UHV some materials can oxidise through interaction with the background gas in the chamber, and for this reason it is good practice to acquire the oxygen region first so that any adsorption of O-containing gas is kept to a minimum. Reactive metals should be collected next to limit any oxidation that may happen during acquisition.

The slight complication with depth profiling is that different materials etch at different rates, and so there is rarely a linear correlation between etch time and depth into the sample if the layer stack in the sample under investigation is complex. Etch rates can be calibrated using standard materials of known thickness and if this is not possible it is quite common to report depths based on calibration using a standard of known thickness such as silicon oxide on silicon or tantalum pentoxide on tantalum.

References

1. K. Yates and R. H. West, *Surf. Interface Anal.*, 1983, **5**(4), 133.
2. H. Siegbahn and K. Siegbahn, *J. Electron Spectrosc. Relat. Phenom.*, 1973, **2**(3), 319.
3. R. Joyner, M. Roberts and K. Yates, *Surf. Sci.*, 1979, **87**, 501.
4. D. R. Baer, K. Artyushkova, H. Cohen, C. D. Easton, M. Engelhard, T. R. Gengenbach, G. Greczynski, P. Mack, D. J. Morgan and A. Roberts, *J. Vac. Sci. Technol. A*, 2020, **38**, 031204.

CHAPTER 4

Analysis of Battery Materials by X-ray Photoelectron Spectroscopy

ETHAN CRUMLIN[a,b] AND ELISABETTA ARCA[c*]

[a]Chemical Sciences Division, Lawrence Berkeley National Laboratory, Berkeley, CA, 94720, USA; [b]Advanced Light Source, Lawrence Berkeley National Laboratory, Berkeley, CA, 94720, USA; [c]Stephenson Institute for Renewable Energy and Department of Physics, University of Liverpool, Liverpool L69 7ZF, UK
*E-mail: Elisabetta.Arca@liverpool.ac.uk

4.1 Introduction

X-ray photoelectron spectroscopy (XPS) has become an essential technique for the characterisation of battery materials. Much of its success arises from the ability of XPS to identify lithium compounds in a quantitative manner, providing simultaneously information about the chemical species present in the samples and their atomic ratio. This has made XPS an extremely powerful tool to characterise battery materials and their evolution during cycling, providing crucial information about the evolution of the overall performance of the electrode materials and the devices.[1–5] For example, correct data analysis can be used to establish a correlation between the presence of specific chemical species and the observed overpotentials.[6] Very few techniques can enable this deep a level of atomistic understanding of a battery device.

Inorganic Materials Series No. 19
Photoelectron Spectroscopy in Inorganic Materials Science
Edited by Robert G. Palgrave and Philip R. Davies

Published by the Royal Society of Chemistry, www.rsc.org

While extremely powerful, XPS is also a rather complex technique, which over the years has had to adapt to overcome some of its own limitations and constantly deepen the level of information acquired.

For instance, XPS has been traditionally a surface-sensitivity technique, which limits the probing depth to a few nanometres when using soft X-ray sources, and initially required samples to be good electrical conductors. The technological advancement of XPS instruments has enabled some of these limitations to be overcome, or at least be minimised. The development of hard X-ray photoelectron spectroscopy (HAXPES) has allowed an increase in the probing depth in a non-destructive manner.[7–10] Likewise, charge neutralisers have been optimised to enable measurements of insulating materials.[11] Yet scientists measuring battery materials by XPS must be aware of some of the peculiar behaviours of battery samples, such as differential charging or built-in potential due to several physical phenomena, including battery voltage, overpotentials, presence of dipole moments and electrochemical double layers, which are specific to this field. Awareness of these effects is necessary not only to avoid incorrect interpretation of the data, but also to give access to a deeper level of understanding of how batteries work.

In this chapter we will examine various aspects of XPS applied to the analysis of battery materials and devices. Post-mortem analysis will be discussed in Sections 4.2, 4.3 and 4.4 signposting potential artefacts, limitations or care in sample preparation and analysis. Section 4.5 will focus on the current state of the art of specific battery components such as interphases, anode, cathode and solid electrolyte materials. Finally, in Section 4.6, we will examine state of the art *operando* methods, highlighting strengths, weakness, and opportunities for future development.

Overall, this chapter aims to provide the reader with a clear understanding of the versatility and power of XPS for the analysis of battery materials, as well as highlighting common mistakes and potential pitfalls and strategies to minimise them.

4.2 Sample Preparation for Post-mortem Analysis

To date, much battery characterisation happens using so-called post-mortem analysis, which consists of analysing the electrodes recovered from cycled cells after careful preparation. This applies to Li, Na, and K ion batteries using liquid electrolytes and all-solid-state batteries (ASSB). While this is one of the most used methods, the detail of the sample preparation can affect the overall results of this kind of measurement, leading sometimes to erroneous conclusions. Sample preparation has always had a major influence on the quality and reliability of experimental measurements. In the case of XPS applied to battery materials, this is even more crucial due to the combination of surface sensitivity of the XPS measurements and the delicate nature of battery components, particularly after cycling. Whereas post-mortem analysis is widely used in the battery community, variations in the approach used to prepare and handle the samples introduce large uncertainties, particularly when data are compared across laboratories, which might adopt intrinsically different handling practises. Ultimately researchers should remember that the surface they are measuring is the result of all the processes carried out including cell disassembly, washing protocols, aging, and transport across facilities and into the instrument.[12] Definition of a unique approach to battery disassembly and sample preparation is unlikely to be adoptable as the protocols can vary depending on which scientific questions must be answered. Thus, a minimum requirement is that papers report the preparation method in detail, as well as highlighting the modification and possible artefacts that this creates. In the following paragraphs, different aspects of sample handling will be examined, highlighting their role in enabling reliable measurements of electrode surfaces.

The first step of post-mortem analysis is the battery disassembly process. The procedure can be slightly different depending on the battery assembly type: T-cells (also called Swagelok cells), coin cells or pouch cells.[13] T-cells are intrinsically easier to disassemble, minimising the risk of short-circuit and damage to the electrode. Coin cells and pouch cells present some additional challenges, and care needs to be taken to avoid short-circuit or physical damage to the electrodes. However, coin cell disassembling tools are nowadays available commercially, making the whole process faster and more reliable, whereas for pouch cells, insulating ceramic scissors are needed.[13] Once the electrode is recovered without damage, the next step might involve the use of solvents to clean the electrode surface from excess electrolytes. Washing procedures are by far the most controversial aspect of post-mortem analysis. There are clear net differences in the electrode chemical composition prior to and after washing, even for a very short amount of time, as shown in Figure 4.1. The washing procedure not only removes the residual salt, most commonly $LiPF_6$, from the surface, but potentially it removes some of the most soluble components of the Solid Electrolyte Interphase (SEI) or Cathode Electrolyte Interphase (CEI), and it might potentially alter the whole composition and structure of the interphase layers.[14,15] Both solubility and rate of dissolution play a role in this. If the study aims to determine the composition of the SEI or CEI, it is recommended that both unwashed and washed electrodes are measured. This inevitably comes at the expense of a longer measurement time. At a minimum, the procedure should report details about the chemical composition and volume of the washing solvent used, washing time and temperature, and drying processes (both conditions and time).

Loading the disassembled battery sample into the spectrometer must necessarily happen using air-free methods, as exposure of the electrode to air even for a few seconds, particularly after cycling, causes severe changes of its composition.[15,17,18] This is true for both ASSB and batteries comprising liquid electrolytes. The extent of the effects of air exposure is different in the two cases, with the latter mostly experiencing oxidation/degradation of the organic components, whereas in ASSB it is common to see carbonation of the surface due to the presence of highly electropositive elements (lithium or lanthanum for example).[19] Nowadays, air-free transfer systems

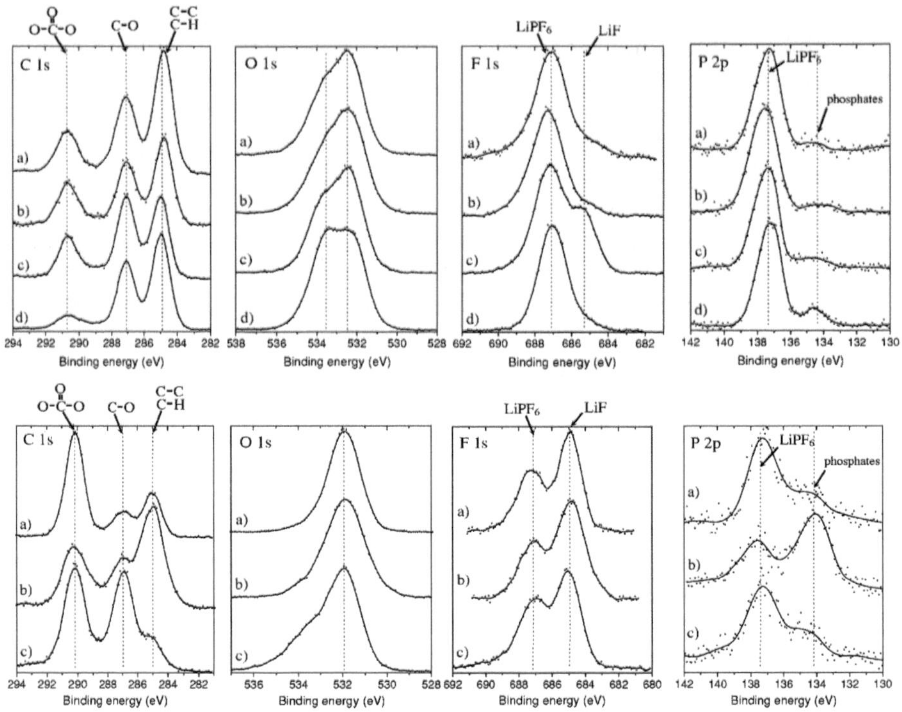

Figure 4.1 Effect of washing on the composition of the solid electrolyte layer. The top row shows three samples cycled under different conditions (a)–(c) in comparison to the same after washing, reported in the bottom row. Reproduced from ref. 16 with permission from IOP Publishing, Ltd, Copyright 2005.

are commonly available in modern XPS instruments and are fundamental to preserve the chemical integrity of the samples. This is equally valid for ambient pressure systems (AP-XPS, often also called near-ambient pressure systems, NAP-XPS) or ultra-high vacuum systems (UHV-XPS). Whenever spectroscopic facilities are not available in-house, shipping of samples is another factor to consider, as this is another potential source of contamination, oxidation, and uncertainty for the measurements. Glass vials or sample boxes need to be carefully chosen to avoid migration of residues (*e.g.* silicone) onto the sample surfaces, with consequent contamination. Air sealing is another crucial aspect to avoid sample degradation. If the facility has available one of the commercial cell openers, repacking the electrode inside a clean coin-cell case is often the best solution.

4.3 Sample Measurements

Electrical conductivity of battery materials varies all the way from good, such as in graphite anodes, to poor such as in some solid electrolytes or close to insulating such as the SEI layers, see Table 4.1. In the case of poorly

Table 4.1 Some conductivities of important battery materials.

Material	Electrical conductivity σ, (S cm^{-1})	Reference
LiF	Insulating	
Li_2CO_3	10^{-2}–10^{-1}	20
PEO	10^{-4}–10^{-7} [a]	21
Graphite	10^3	22
Black carbon	1	23
$LiFePO_4$	10^{-9}–10^{-3} [b]	24–26
$LiCoO_2$	10^{-1}–10^{-4}	27

[a]Modified with various additives.
[b]Doping, structural properties and defects can have substantial influence.

conducting or insulating samples, charging effects can arise during measurement (see Chapter 2, Section 2.8). These are caused by the positive charge left on the sample surface after the photoemission process, which, in the case of poor conductors, is not discharged through grounding to the spectrometer. If no charge neutraliser is available, poorly conducting samples will accumulate positive charges at the surface, causing the peaks to shift to higher binding energies, increase peak width and, in extreme cases, deformation of the peak line-shape can occur, making the data unusable. In the case of heterogenous mixtures, such as in SEI and solid-state electrolytes, the different components can have very different electrical conductivity, giving rise to differential charging. Correct grounding is a crucial aspect of sample preparation. Electronic conductivity of battery materials might be lower or more inhomogeneous in comparison to samples from other research fields. When the thickness of the insulating layer is thin and the sample has a good electrical grounding, it might be possible to conduct the measurements without the need of a charge neutraliser. In this case, a small shift in the peaks might still be visible with different state of charge or for more insulating components (differential charging) and they need to be accounted for (Figure 4.2). When the charging becomes excessive, leading to severe shift from the expected position and an increase in the full-width half maximum of the peak, neutralisers are necessarily used. While a thorough discussion of the technical aspects of different charge neutralisers is beyond the scope of this chapter, here we will briefly discuss some aspects which are relevant to the context of batteries. A thorough discussion of charging effects and neutralisation techniques can be found in ref. 11.

Broadly, charge neutralisers can make use of pure electron compensation, by flooding the surface with low-energy electrons, or by Ar-ion-assisted methods (see Chapter 3, Section 3.4.3.2). Depending on the manufacturer, UHV systems might have either of the two methods, whereas AP-XPS instruments use the high pressure of gas close to the sample to assist with charge neutralisation. The Ar-assisted method is often considered advantageous for samples displaying high surface roughness, by helping a better distribution of

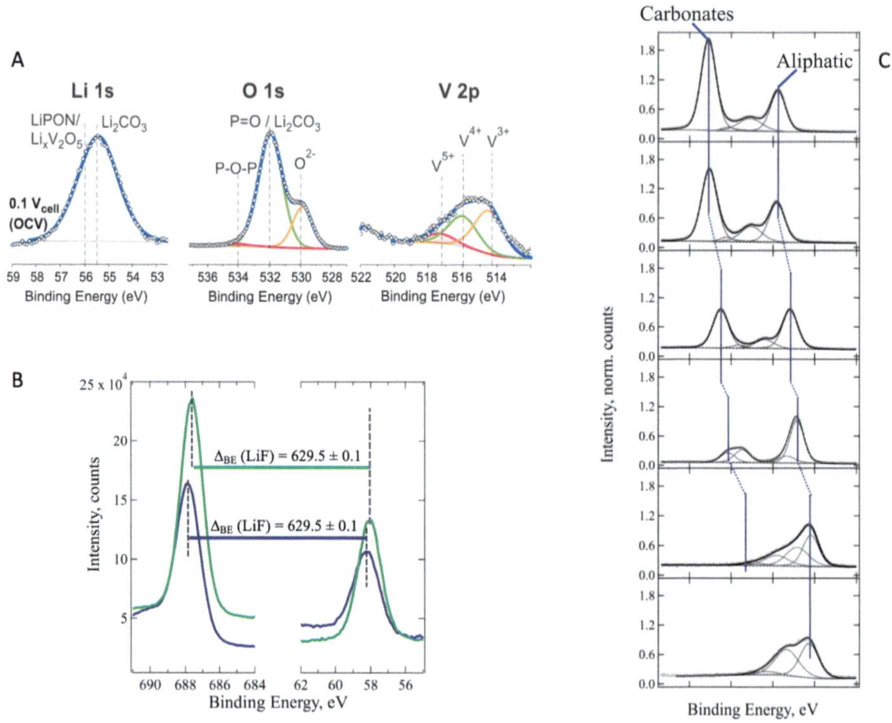

Figure 4.2 Peak shifting in XP spectra. (A) Different components are present where the shift is due to a change in the chemical environment of the atom (*e.g.* O 1s core level) or a change in the oxidation state (*e.g.* V 2p core level). Reproduced from ref. 31 with permission from Springer Nature, Copyright 2012. (B) Li 1s and F 1s core levels recorded for two different solid electrolyte interfaces grown on Si anodes. While there is a shift in the absolute position of the Li and F respective core-level due to a small amount of charging, the energy separation between the two peaks is a constant characteristic of the compounds. (C) Energy shift as a function of carbonate present in the material. As can be seen, the higher the carbonate content the larger the shift to higher binding energy. Reproduced from ref. 32 with permission from American Chemical Society, Copyright 2022.

the charges and avoiding shadowing effects. However, the energies of the Ar ions are not negligible, and the possibility of sample damage arising from charge neutralisation must be checked. Highly reactive samples are likely to degrade during this process. In the context of battery materials, and if one can work with a smooth enough surface, the use of an electron-only neutraliser might be preferable (not available in AP-XPS). Additionally, reducing the X-ray power or defocusing the beam might help in mitigating charging effects. One of the consequences of charge neutralisation is the loss of the absolute reference for the binding energy scale. The use of the adventitious C 1s peak is often applied and is still under debate (see also Chapter 2, Section 2.8.3),[28,29] although this method can be highly unreliable and prone to be misleading depending on the nature of the sample and the degree of contamination present in it. This is particularly the case for heterogenous samples where the differential charging behaviour of different phases is an important aspect of the measurements, such as in the case of batteries. Additionally, most battery electrodes comprise different types of carbon (*e.g.* carbon black, binders, graphite *etc.*), making the concept of charge correction using a well-established adventitious carbon line challenging. Using the method of binding energy separation rather than absolute peak positions on the binding energy scale might be the preferred approach.[30] This method is based on the concept that charging will equally affect all components of a given species – *e.g.* in LiF both the Li 1s peak and F 1s peak will be shifted upwards by some meV or eV. However, since they are both shifted by the same amount, their relative position Δ(F 1s–Li 1s) will be a constant. This approach is recommended not only for samples that require charge neutralisation, but also for samples that can be measured without it. Indeed, battery samples that do not require charge neutralisation might still present shifts in some of their elemental peaks due differential charging, thickness beyond tunnelling current for insulating components, built-in potentials, or differences because of the electrode state of charge.

While charging phenomena need to be carefully considered when using the absolute peak position on the binding energy scale, the approach based on the binding energy separation between components is not affected by these phenomena, enabling a reliable identification of chemical compounds.[30,32,33]

An important aspect of XPS is its surface sensitivity. Different X-ray sources can lead to different information depths for an individual core level. However, when comparing different elements, core lines may have substantially different kinetic energies and hence the signal will be coming from slightly different depths. To keep the probing depth the same for all elements, one should look at core levels with similar comparable kinetic energy, as demonstrated, for example, in ref. 34 (Figure 4.3).

This can be obtained by a careful selection of excitation source and the core levels. Different X-ray excitation sources are nowadays available even in laboratory-based tools, the most common being Al Kα (~1.5 eV), Ag (~3 keV), Cr (~5 keV) to Ga Kα (~9 keV), however rarely all on the same instrument. Although dual anodes are becoming more common, the need for three or more sources is becoming increasingly compelling even in laboratory-based systems. The use of multiple X-ray sources is itself non-destructive, but the user still needs to be mindful of artefacts and technical difficulties. Firstly, the sensitivity of the buried layers decreases exponentially with probing depth, while keeping in mind that top surface layers will still account for most of the signal.[35–37] Secondly, the elemental cross section decreases significantly with increasing excitation energy for a given core level, meaning that the time and costs of the measurements will be increased substantially. For some elements, alternative approaches are possible, such as using a different core level – for example Ti 2p is commonly used with Al Kα sources, whereas Ti 1s is preferable for Cr or Ga sources, which have the higher photon

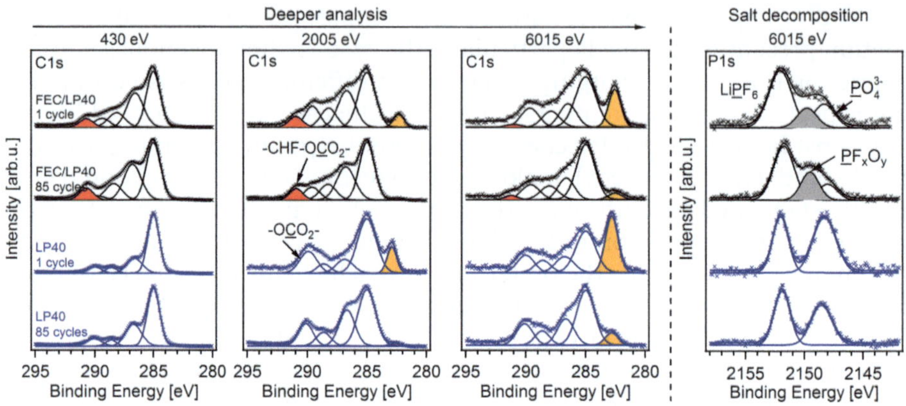

Figure 4.3 An example of how different excitation energies can lead to an increase in the information depth for a given core level, whereas for different core levels, one needs to consider the kinetic energy of the escaping electron and use an excitation source which would lead to the same probing depth. Reproduced from ref. 34 with permission from American Chemical Society, Copyright 2015.

energy necessary to access it.[35] However, this is not always possible and elements commonly present in battery materials, such as Li, O and C, will inevitably require longer integration times. Furthermore, while the resolution of measurements performed using hard X-rays is constantly improving,[38] it still tends to be lower than those using soft X-rays. Some of these shortcomings can be addressed using synchrotron-based sources instead of laboratory-based XPS systems. For example, a wider selection of wavelength is usually available at synchrotron end-stations, with the micro-focused beam which enables faster measurements and the different optics mitigating the issue of high resolution at higher energies. However, care needs to be taken with power density when using the micro-focused beam, as it can exacerbate charging effects and potentially cause degradation of sensitive compounds.

A final note must be made about the correlation between roughness and probing depth. For atomically flat surfaces, one can calculate the probing depth of the measurement accurately. This is very rarely the case for battery materials. Even model electrodes can develop substantial roughness through modification of the surface electrode itself or the growth of a CEI or SEI layer. In many cases, the battery samples consist of composite electrodes, whose roughness is orders of magnitude larger than the probing depth of XPS measurements. This means that calculating an accurate probing depth is very challenging in these conditions as the plane of focus will be at an average height, with grains protruding away and valleys running through the focus planes. It is highly advisable to get an estimate of the surface roughness of the measured samples before the measurements take place to at least be able to understand how reliable the probing depth estimation will be.

4.4 Quantitative Analysis

One of the greatest strengths of XPS is the ability to couple the determination of the chemical species present in a sample with the ability to quantify their atomic ratio. Quantitative analysis by XPS relies on a number of parameters. Most lab-based systems use pre-programmed scaling factors, often referred to as Relative Sensitivity Factors (RSFs)[39–42] which can be calculated based on the variables upon which the photoelectron intensities depend, including impingent X-ray flux, transmission function of the analyser, inelastic mean free path, photoionisation cross-section for the relevant subshell, and dipole asymmetric parameter,[42] or they can be determined empirically from the analysis of standard references (see Chapter 2, Section 2.6). Several additional experimental parameters such as the electron take-off angle or homogeneity of the samples, as well as data analysis procedure, such as the determination of the correct background,[43] will heavily affect the accuracy of the quantification. This is particularly important when comparing results across facilities. A thorough discussion about these parameters can be found in ref. 44–47 whereas here we highlight some key points.

- Sensitivity factors are determined for homogenous samples. This is rarely the case in battery materials. When analysing samples that are inhomogeneous particularly across their thickness, one needs to evaluate if the sensitivity factors need to be corrected for this.
- When comparing elements whose binding energies are very different, the accuracy of the transmission function is very important for quantitative analysis. Furthermore, when using a monochromatic X-ray source, elements with different inelastic mean free paths will be emitted from different depths within the sample. In synchrotron-based experiments, one can tune the energy of the excitation source to ensure all data are collected from the same depth, however this is far more challenging in laboratory-based settings, where usually only one or two sources are available. We can foresee that instruments equipped with more than two sources will find more widespread use even for laboratory settings.
- To improve accuracy, one would ideally calibrate the sensitivity factors for a given instrument using a standard with known atomic composition. This is rarely done, as reference samples might not be readily available, and the process is time-consuming.

Once data are collected, the next step is the data analysis. This is indeed one of the major sources of error in the quantitative analysis published in the literature. In addition to appropriate use of the sensitivity factors, the choice of a correct background, line shape of the peaks and including satellite features where appropriate, are all important factors in determining the correctness of the data analysis. Several papers are available to guide the reader through a more appropriate choice of fitting parameters.[46,48–51]

4.5 Analysis of Battery Materials: Interphases, Cathodes, Anodes and Solid Electrolytes

Battery materials comprise both interfaces and interphases.[52] The former are defined as a 2-dimensional plane of contact between two materials, whereas the latter are defined as a tri-dimensional structure, potentially comprising several materials and their interfaces (Figure 4.4).[52] Both interfaces and interphases play a crucial role in the stability and performance of battery devices and as such, they have and will be under intense scrutiny from an experimental standpoint. Traditionally, a major application of XPS analysis has been the identification of chemical compounds, spanning from the composition of the SEI and respective analogue at the cathodes (CEI), to the chemical changes of the cathode, anode and solid electrolyte.

SEI and CEI layers form because most batteries operate at potentials that are larger than the electrolyte stability window.[54] This is particularly true for carbonate-based electrolytes and the $LiPF_6$ salt, the *current* commercial standard. As a result, the electrolyte will undergo a decomposition process with the production of soluble and insoluble decomposition products. The latter will deposit on the surface of the cathode and the anode forming respectively the CEI and SEI.[55–59] A properly formed interphase will constitute a barrier that will prevent further electrolyte decomposition, while still allowing Li ions to migrate between the two electrodes. Compositional analysis of the SEI has been widely applied to anode materials including graphite, silicon, and Li-metal anodes. Detailed analysis of the composition of the SEI and CEI is beyond the scope of this chapter and the reader is directed to relevant literature on the subject.[9,10,32,33,60–67] Here we will only highlight some key aspects. CEI and SEI are most often dynamic structures, whose composition and physical properties changes over the charging cycle.[32,53,60,68] Qualitative

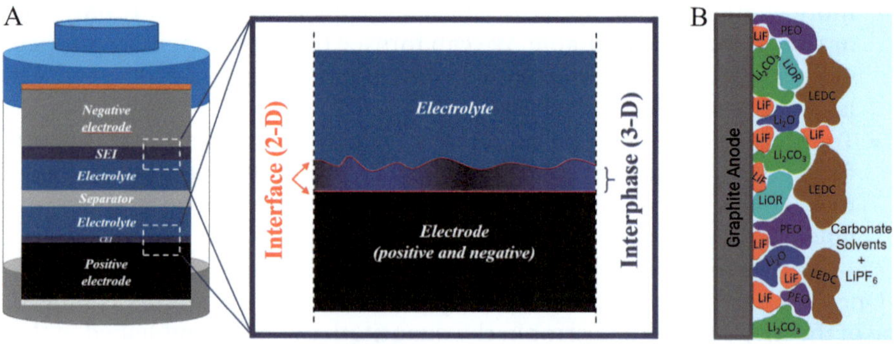

Figure 4.4 (A) difference between the interphase and the interface in a battery device. Reproduced from ref. 53 with permission from John Wiley & Sons, Copyright 2022. (B) mosaic representation of the SEI grown on a graphite anode showing the complexity of its composition. Reproduced from ref. 74 with permission from Elsevier, Copyright 2019.

and quantitative analysis is limited to the probing depth, usually much lower than the thickness of the layer itself.[32] These phases comprise a variety of inorganic and organic compounds, most noticeably lithium fluoride, carbonates, and phosphates. While inorganic compounds can be precisely identified, identification of specific organic molecules is more complex, and often not possible at all. In principle, the ratio between carboxyl, carbonates and aliphatic carbons should aid the identification of the organic molecule, however in reality, the complexity of the different carbon species, coupled with the mixture of organic and inorganic species, makes this process virtually impossible. For this reason, complementary FT-IR is necessary to distinguish between different organic species,[69–72] leaving to XPS still the ability to identify the presence of organic functional groups, rather than specific molecules. While this type of analysis will provide information about the interphase, the information about the evolution of the SEI/electrode or CEI/electrode interfaces is far more elusive. The growth of a thick layer impedes probing of these buried interfaces in most cases, with scientists currently developing approaches to access them – see Figure 4.5 and Sections 4.6 and 4.7.

An important benefit of understanding SEI and CEI composition and physical properties is the ability to control their formation and composition through cycling conditions or additives or even create artificial SEI/CEI to improve battery stability and performance. The anecdotic case is the SEI on graphite and silicon anodes respectively. While compositionally these look very similar,[73,74] the SEI on graphite is far more efficient in preventing further electrolyte decomposition, whereas for Si-anodes the SEI instability is the root cause of their limited battery lifetime.[32,33,75] This has been proven to be

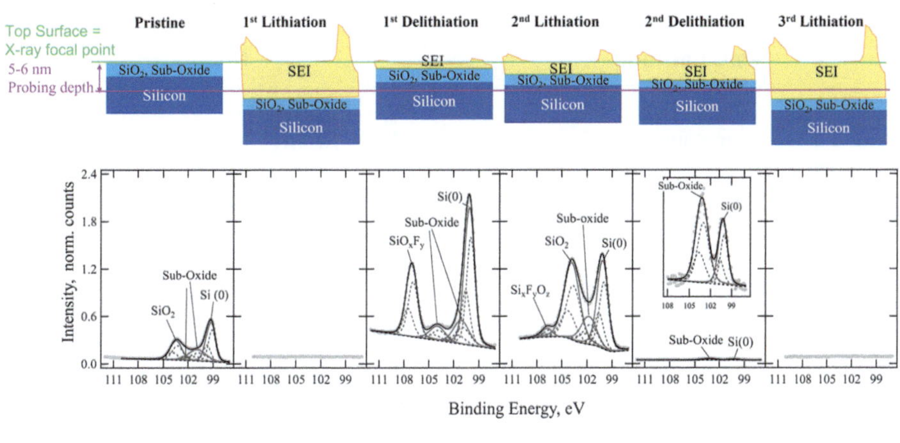

Figure 4.5 Effect of the SEI grown on silicon anodes as a function of state of charge. Since the plane of focus remains on the surface, when the SEI grows, it pushes the interface between the SEI/anode outside of the measurement probing depth. The value indicated in the figure (5–6 nm) is valid for an Al Kα source. Reproduced from ref. 32 with permission from American Chemical Society, Copyright 2022.

the case even for electrode compositions whose morphology has been engineered to manage the volumetric expansion, in addition to fully expanded electrodes.[75] Despite more than a decade of research, the question as to why the SEI kinetic control role is still so different remains open, and more information about both the interphase and the interfaces is needed.

Another important application of XPS for the analysis of battery materials is the determination of oxidation states, particularly for cathodes where this is intrinsically connected to their redox activities. Determination of oxidation states based on XPS data is far from a trivial process, particular for materials comprising transition metal elements, whose line-shape can be complicated by final states effects and the presence of satellite peaks. Identification of the oxidation position on the basis of absolute peak position is highly discouraged. Many cathode materials and solid electrolytes are oxides or sulfides, and in many cases can present semiconducting, Mott-insulator[77] or highly correlated oxide[78] characteristics. In these cases, doping plays a major role and therefore shifts the peak positions due to a change of the position of the Fermi level within the band gap of the materials. In other words, the attribution of a change in oxidation state simply due to a peak position can be incorrect. Also, more than one oxidation state can be present at once, depending on the state of charge of the material. For compounds where the redox activity is linked to the transition metal oxide, it is crucial to perform a full peak fitting analysis which includes final states effects (Figure 4.6).[79,80] While a detailed analysis of common oxides of the first row transition metals exists in the literature,[76] the process becomes more complicated for new materials, particularly those involving more than one anion, which can influence the crystal field splitting and hence the makeup of the final state. The

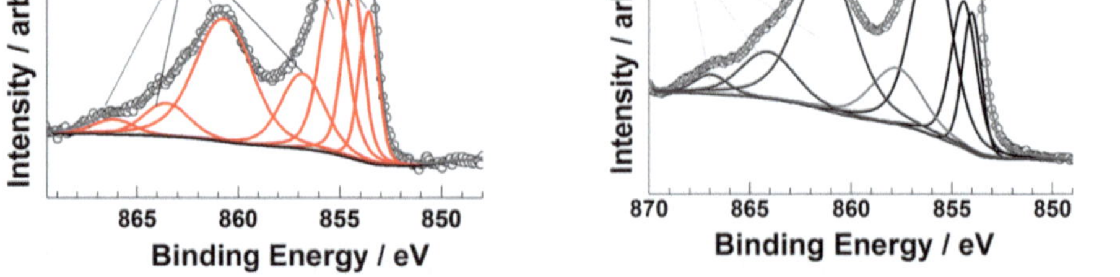

Figure 4.6 XP spectra of the Ni ion in a NMC cathode compared to the same ion in NiO, showing the different satellite structures and multiplet splitting arising from the crystal field splitting. Reproduced from ref. 80 with permission from John Wiley and Sons, Copyright 2017.

problem does not have an easy solution and largely will require molecular orbital (MO) theory[81,82] and density functional theory calculations[83–85] to aid with the interpretation of spectral features or product formation. Additional information about the significance of binding energy shifts can be found in ref. 86.

4.6 Sputtering Depth Profiling for the Analysis of Battery Materials

Sputtering methods are often used during the XPS characterisation of battery materials primarily for two purposes: (1) cleaning of the surface to remove the residual salts and (2) depth profiling to determine the composition of a given layer as a function of probing depth. Depth profiling tends to be a less expensive and more widely used method than the combination of soft/hard X-ray, yet it is important to understand the caveats and limitations that come with this approach.

Depth profiling is a destructive technique consisting of removing layers of materials by bombarding the surface with high energy ions or molecules (see Chapter 3, Section 3.4.3.1 Ion Sources). In a depth profiling process, the user mills down the surface for a brief period of time (seconds to minutes depending on the experiment), followed by a collection of high-resolution spectra.[87,88] By physically removing layers in a stepwise process, the user can look at the compositional profile of chemical species in the layer in question or remove a thick layer to gain access to a buried interface or a buried layer.[89–93] This is a useful process in many instances, such as to access the electrode underneath the CEI or SEI layers. However, depth profiling can already introduce severe artefacts when applied to hard materials, which are exacerbated in the field of batteries due to the mobility of lithium ions, and the presence of delicate organic species.[89,94,95] Polymeric or organic materials are profiled very poorly by atomic Ar sources,[96] and therefore, depth profiling of SEI and CEI should not be performed using this type of source, and cluster sources should be used instead.

Modern instruments allow control of the sputtering parameters and in many cases there is more than one type of sputter source available. Monoatomic Ar is commonly present even in old instruments. However, due to preferential sputtering of lighter elements[88,97] and the highly localised nature of the Ar^+ charge, this method can often lead to boat-shape profiling of the lithium content in the materials, which can cause incorrect interpretations. Furthermore, the energetics of the profiling process might activate reaction processes between species present on the surface. Hence conclusions about chemical species and their concentration need a careful consideration of possible artefacts.

Cluster sources such as cluster Ar or carbon clusters are less energetic than monoatomic Ar, since the charge is spread over a wider volume. However, this method is more time-consuming as the sputtering rates are more than one order of magnitude lower than for the monoatomic sources, requiring longer instrument time and higher costs. Furthermore, potential artefacts might still be present, whose extent still needs to be assessed.[88]

Finally, profiling works properly for smooth surfaces (roughness on the order of a few nm), but for battery materials where roughness can be more than several tens of nm, depth profiling can be further affected by shadowing effects and inhomogeneous sputtering rates. This is another aspect of the process that needs careful consideration.

In summary, depth profiling is a widely used technique with the advantage of accessing buried layers. However, the artefacts created during the profiling process can be very extensive and difficult to predict or account for, creating the risk of severely misleading conclusions. A better option might be to design experiments that avoid the use of sputtering altogether if possible.

4.7 *In Situ* and *Operando* Approaches

In situ and *operando* approaches are becoming progressively more widespread as powerful tools to overcome many of the limitations plaguing post-mortem analysis methods. *In situ* methods consist of acquiring XP spectra during the evolution of a battery material when, for example, it is in contact with one another, or exposed to gas or temperature cycles, without necessarily the ability to carry out electrochemical protocols. Precious information can still be derived by this method, particularly about the chemical reactivity of battery components with one another or their temperature stability.[98–102] *Operando* XPS instead implies that the materials/devices are studied under operational conditions. This allows information provided by XPS, such as chemical composition, to be coupled with a device's electrochemical properties, providing a means to correlate device characteristics such as overpotential, capacity fade, and efficiency, with an atomistic understanding of their origin.[63,103–109] While this is an incredibly powerful approach, able to overcome many of the limitations of post-mortem analysis, it comes with the drawbacks of highly complex experimental arrangements at present. Some of the drawbacks include having to develop specific electrochemical cells and high costs for the experiments, due to long acquisition times. Different types of battery present different challenges. All solid-state batteries (ASSB) tend to be easier to handle, due to the different complexity of managing liquid *vs.* solids in an ultra high vacuum-XPS system. For ASSB, electrochemical stage holders are already available, which are integrated with traditional high vacuum systems. Testing can be performed by connecting the electrochemical stage holder to an external potentiostat, enabling a large range of electrochemical protocols to be conducted while acquiring XP spectra.[110] One aspect to consider is how samples will be mounted, as this can differ depending on the range of information that needs to be extracted. It has to be noticed that only the top layer will be measured, hence a choice has to be made on how to prepare the cells – for example a composite cathode might allow the experimentalist to probe both cathode and solid electrolyte (SE) materials, whereas in a traditional cathode/SE/anode configuration, the cathode (or anode) would need to be extremely thin to gain any information about the SE. Because of the limited probing depth when soft X-ray sources are used, often dual source or synchrotron radiation is used to probe deeper. This however might increase the chance of charging phenomena, particularly within the buried layers if the materials are very resistive. Homogeneity of the current distribution is also another point of careful planning. Due to the need to photo-eject electrons, the electrode can't be traditional Cu or Al. In this case, graphene membranes can be used to provide a continuous and homogenous current distribution,[111,112] or alternatively, electro-less approaches are adopted.[6,113] Effects such as current inhomogeneities, very limited contact pressure, or limited C-rates (the rate at which a battery charges/discharges), can affect the results of the electrochemical process. It is highly recommended to compare the electrochemical performance of the *operando* cells to those of a standard electrochemical system (*e.g.* coin cells) to validate the findings and avoid potential artefacts. Although these tests might have been conducted, they are rarely reported in the literature. Other effects, such as the limitation of the C-rates, might be hard to avoid particularly in the virtual electrode approach.[6] While this method has the advantage of not requiring any hardware, and thus being widely applicable, the discharge process is usually extremely slow. The latter is an important issue, as the C-rates have a profound effect on the electrochemical performance of a battery,[114] and it needs to be considered during data interpretation.

Batteries which make use of liquid electrolytes are far more complex in terms of *operando* measurements. Here two approaches are possible: measurement through a thin layer of liquid (a droplet approach or a thin meniscus approach)[7,115–118] or the design of a full sample holder.[3,104,108,116,119]

The thin meniscus approach is largely the most commonly used for liquid aqueous based electrolytes, however careful consideration about the safety of the experiments is required when handling commercial electrolytes made of $LiPF_6$ in carbonate solvents. This method consists in pulling out the electrode from the liquid reservoir, until the meniscus is thin enough that electrons can travel through the liquid phase. Most solvents have vapour pressures that are too high to be used in high-vacuum systems, and this method often requires the use of AP-XPS. For those liquid cells where measurements are safe to conduct with this method, the signal from the liquid phase is usually of high quality, whereas the solid–liquid interface and the electrode are more difficult to probe. This is because the electrons need to travel through the liquid before being able to enter the analysers, which causes the signal to be strongly attenuated.[120] Despite these limitations, precious information can be gained using this method. In particular its ability to probe the Helmholtz layer in a water solution has been demonstrated.[117] XPS on the carbonates' electrolytes has been reported,[121] showing very limited X-ray induced damaged during measurements and proving the potential to expand the use of AP-XPS systems to commercial electrolytes with adequate safety conditions in place.

Electrochemical XPS sample holders for *operando* measurements using liquid electrolytes are mostly based on graphene membrane technology,[112,122] although recently some alternatives have been proposed.[123] While a properly constructed membrane can withstand the pressure differential even inside a UHV instrument,[123,124] any defect in the membrane will lead to its delamination. Therefore, this home-made set up still remains quite delicate, and the practical challenges are substantial. Wettability, viscosity, and the type of meniscus that the liquid forms are often key limitations, and because they are different properties of the liquids, designs that work with water as a solvent might not necessarily work equally well with carbonate electrolytes. The study of solid–liquid interfaces and of the electrode in contact with the solution is far from trivial. For a membrane-based sample holder, a reverse configuration (with respect to that used for the thin meniscus approach) is used. In the thin meniscus approach, the photo-ejected electrons travel through the liquid phase. In the membrane approach, a very thin electrode (5–10 nm) is deposited on the graphene membrane and electrons must travel through this thin layer of material.[123] This approach would inevitably require hard X-rays and it presents the further challenge of deposition of battery materials on graphene. While the latter is possible in principle, there is a limited range of materials that can be grown with deposition methods compatible with graphene membrane technology. Such methods include for example chemical vapour deposition (CVD) or atomic layer deposition (ALD).

Studies using *operando* approaches have clearly demonstrated the benefit of this technique and the need to develop it further. Indeed, determination of the chemical species formed as a function of the applied potential, and the ability to correlate this to parameters such as overpotential, can provide crucial pieces of information for the further development of battery materials; information that is largely inaccessible especially in such a direct manner with other techniques. In terms of future development, the ability to get femtosecond time-resolved X-ray photoemission spectroscopy (TR-XPS), already available at some synchrotron end-stations,[125] will enable studies on the kinetics of the process happening in battery materials and devices, opening up the possibility to correlate information about the chemical changes and device properties in a time domain relevant for the electrochemical process; a game-changing ability for understanding battery degradation processes.

4.8 Conclusion

In summary, this chapter provides an overview of different applications of X-ray Photoelectron Spectroscopy (XPS) to the field of battery materials development. The widespread use of post-mortem analysis to characterise a material's composition both qualitatively and quantitatively has been reviewed, highlighting potential issues and the care that needs to be taken during both data collection and subsequent analysis. The evolution of the technique, and particularly the development of *in situ* and *operando* methods, represents the necessary follow up to overcome some of the limitations and drawbacks of the traditional post-mortem approach, but also to enable a deeper understanding of the correlation between material properties and device parameters. Technical challenges are still limiting the widespread application of *operando* methods, with great effort from the scientific community to overcome them. Prospective areas of development include the wide adoption of time-resolved methods, which will enable the study of the kinetics of the process in a time-frame relevant to this process to gain an even deeper understanding of some of the key issues of battery performance and, in particular, degradation processes.

References

1. I. Källquist, R. Le Ruyet, H. Liu, R. Mogensen, M.-T. Lee, K. Edström and A. J. Naylor, *J. Mater. Chem. A*, 2022, **10**, 19466.
2. J. G. Connell, X-ray photoelectron spectroscopy for batteries, in *Batteries: Materials principles and Characteriszation Methids*, ed. C. Liao, IOP Pulbishing Ltd, Bristol, 2021, DOI: 10.1088/978-0-7503-2682-7ch8.
3. A. T. S. Freiberg, S. Qian, J. Wandt, H. A. Gasteiger and E. J. Crumlin, *ACS Appl. Mater. Interfaces*, 2023, **15**, 4743.
4. J. Auvergniot, A. Cassel, D. Foix, V. Viallet, V. Seznec and R. Dedryvère, *Solid State Ionics*, 2017, **300**, 78.
5. S. Wi, V. Shutthanandan, B. M. Sivakumar, S. Thevuthasan, V. Prabhakaran, S. Roy, A. Karakoti and V. Murugesan, *J. Vac. Sci. Technol., A*, 2022, **40**, 010808.
6. K. N. Wood, K. X. Steirer, S. E. Hafner, C. Ban, S. Santhanagopalan, S.-H. Lee and G. Teeter, *Nat. Commun.*, 2018, **9**, 2490.
7. S. Axnanda, E. J. Crumlin, B. Mao, S. Rani, R. Chang, P. G. Karlsson, M. O. M. Edwards, M. Lundqvist, R. Moberg, P. Ross, Z. Hussain and Z. Liu, *Sci. Rep.*, 2015, **5**, 9788.
8. K. Ciosek Högström, S. Malmgren, M. Hahlin, M. Gorgoi, L. Nyholm, H. Rensmo and K. Edström, *Electrochim. Acta*, 2014, **138**, 430.
9. S. Malmgren, K. Ciosek, M. Hahlin, T. Gustafsson, M. Gorgoi, H. Rensmo and K. Edström, *Electrochim. Acta*, 2013, **97**, 23.
10. B. T. Young, D. R. Heskett, C. C. Nguyen, M. Nie, J. C. Woicik and B. L. Lucht, *ACS Appl. Mater. Interfaces*, 2015, **7**, 20004.
11. D. R. Baer, K. Artyushkova, H. Cohen, C. D. Easton, M. Engelhard, T. R. Gengenbach, G. Greczynski, P. Mack, D. J. Morgan and A. Roberts, *J. Vac. Sci. Technol., A*, 2020, **38**, 031204.
12. F. A. Stevie, R. Garcia, J. Shallenberger, J. G. Newman and C. L. Donley, *J. Vac. Sci. Technol., A*, 2020, **38**, 063202.
13. T. Waldmann, A. Iturrondobeitia, M. Kasper, N. Ghanbari, F. Aguesse, E. Bekaert, L. Daniel, S. Genies, I. J. Gordon, M. W. Löble, E. De Vito and M. Wohlfahrt-Mehrens, *J. Electrochem. Soc.*, 2016, **163**, A2149.

14. L. Somerville, J. Bareño, P. Jennings, A. McGordon, C. Lyness and I. Bloom, *Electrochim. Acta*, 2016, **206**, 70.
15. S. Malmgren, K. Ciosek, R. Lindblad, S. Plogmaker, J. Kühn, H. Rensmo, K. Edström and M. Hahlin, *Electrochim. Acta*, 2013, **105**, 83.
16. R. Dedryvère, S. Laruelle, S. Grugeon, L. Gireaud, J. M. Tarascon and D. Gonbeau, *J. Electrochem. Soc.*, 2005, **152**, A689.
17. R. Zhang, S. Yang, H. Li, T. Zhai and H. Li, *InfoMat*, 2022, **4**, e12305.
18. L. Cheng, E. J. Crumlin, W. Chen, R. Qiao, H. Hou, S. Franz Lux, V. Zorba, R. Russo, R. Kostecki, Z. Liu, K. Persson, W. Yang, J. Cabana, T. Richardson, G. Chen and M. Doeff, *Phys. Chem. Chem. Phys.*, 2014, **16**, 18294.
19. L. Cheng, M. Liu, A. Mehta, H. Xin, F. Lin, K. Persson, G. Chen, E. J. Crumlin and M. Doeff, *ACS Appl. Energy Mater.*, 2018, **1**, 7244.
20. S.-F. Wang, Y.-L. Liao, Y.-F. Hsu, T. Y. Weng and P. Jasinski, *J. Power Sources*, 2023, **557**, 232560.
21. F. Gucci, M. Grasso, S. Russo, G. J. T. Leighton, C. Shaw and J. Brighton, *Polymers*, 2023, **15**, 2581.
22. W. M. E. Haynes, *CRC Handbook of Chemistry and Physics (95th edn)*, CRC Press, Boca Raton, 2014.
23. B. Marinho, M. Ghislandi, E. Tkalya, C. E. Koning and G. de With, *Powder Technol.*, 2012, **221**, 351.
24. C. Wang and J. Hong, *Electrochem. Solid-State Lett.*, 2007, **10**, A65.
25. A. Kulka, A. Milewska, W. Zając, K. Świerczek, E. Hanc and J. Molenda, *Solid State Ionics*, 2012, **225**, 575.
26. J. Molenda, A. Kulka, A. Milewska, W. Zając and K. Świerczek, *Materials*, 2013, **6**, 1656.
27. E. Antolini, *Solid State Ionics*, 2004, **170**, 159.
28. M. C. Biesinger, *Appl. Surf. Sci.*, 2022, **597**, 153681.
29. G. Greczynski and L. Hultman, *Prog. Mater. Sci.*, 2020, **107**, 100591.
30. K. N. Wood and G. Teeter, *ACS Appl. Energy Mater.*, 2018, **1**, 4493.
31. Y.-C. Lu, E. J. Crumlin, G. M. Veith, J. R. Harding, E. Mutoro, L. Baggetto, N. J. Dudney, Z. Liu and Y. Shao-Horn, *Sci. Rep.*, 2012, **2**, 715.
32. E. Arca, G. M. Veith, R. Satish, T. Lin, G. Teeter and R. Kostecki, *J. Phys. Chem. C*, 2022, **126**, 14058.
33. Y. Yin, E. Arca, L. Wang, G. Yang, M. Schnabel, L. Cao, C. Xiao, H. Zhou, P. Liu, J. Nanda, G. Teeter, B. Eichhorn, K. Xu, A. Burrell and C. Ban, *ACS Appl. Mater. Interfaces*, 2020, **12**, 26593.
34. C. Xu, F. Lindgren, B. Philippe, M. Gorgoi, F. Björefors, K. Edström and T. Gustafsson, *Chem. Mater.*, 2015, **27**, 2591.
35. A. Regoutz, M. Mascheck, T. Wiell, S. K. Eriksson, C. Liljenberg, K. Tetzner, B. A. D. Williamson, D. O. Scanlon and P. Palmgren, *Rev. Sci. Instrum.*, 2018, **89**, 073105.
36. S. Siol, J. Mann, J. Newman, T. Miyayama, K. Watanabe, P. Schmutz, C. Cancellieri and L. P. H. Jeurgens, *Surf. Interface Anal.*, 2020, **52**, 802.

37. Z. W. Lebens-Higgins, H. Chung, M. J. Zuba, J. Rana, Y. Li, N. V. Faenza, N. Pereira, B. D. McCloskey, F. Rodolakis, W. Yang, M. S. Whittingham, G. G. Amatucci, Y. S. Meng, T.-L. Lee and L. F. J. Piper, *J. Phys. Chem. Lett.*, 2020, **11**, 2106.
38. J. F. Watts and J. E. Castle, *Surf. Interface Anal.*, 2024, **56**, 408.
39. D. J. H. Cant, B. F. Spencer, W. R. Flavell and A. G. Shard, *Surf. Interface Anal.*, 2022, **54**, 442.
40. M. P. Seah, *Surf. Interface Anal.*, 1980, **2**, 222.
41. C. D. Wagner, Handbook of X-ray photoelectron spectroscopy : a reference book of standard data for use in X-ray photoelectron spectroscopy, *Phys. Electron. Div.*, 1979.
42. B. F. Spencer, S. Maniyarasu, B. P. Reed, D. J. H. Cant, R. Ahumada-Lazo, A. G. Thomas, C. A. Muryn, M. Maschek, S. K. Eriksson, T. Wiell, T. L. Lee, S. Tougaard, A. G. Shard and W. R. Flavell, *Appl. Surf. Sci.*, 2021, **541**, 148635.
43. S. Tougaard, *J. Vac. Sci. Technol., A*, 2020, **39**, 011201.
44. J. C. Green and P. Decleva, *Coord. Chem. Rev.*, 2005, **249**, 209.
45. G. Greczynski and L. Hultman, *J. Appl. Phys.*, 2022, **132**, 011101.
46. A. G. Shard, *J. Vac. Sci. Technol., A*, 2020, **38**, 041201.
47. G. H. Major, T. G. Avval, B. Moeini, G. Pinto, D. Shah, V. Jain, V. Carver, W. Skinner, T. R. Gengenbach, C. D. Easton, A. Herrera-Gomez, T. S. Nunney, D. R. Baer and M. R. Linford, *J. Vac. Sci. Technol., A*, 2020, **38**, 061204.
48. D. J. Morgan, *Surf. Interface Anal.*, 2023, **55**, 567.
49. G. H. Major, N. Fairley, P. M. A. Sherwood, M. R. Linford, J. Terry, V. Fernandez and K. Artyushkova, *J. Vac. Sci. Technol., A*, 2020, **38**, 061203.
50. M. H. Engelhard, D. R. Baer, A. Herrera-Gomez and P. M. A. Sherwood, *J. Vac. Sci. Technol., A*, 2020, **38**, 063203.
51. A. Herrera-Gomez, M. Bravo-Sanchez, O. Ceballos-Sanchez and M. O. Vazquez-Lepe, *Surf. Interface Anal.*, 2014, **46**, 897.
52. C. Villevieille, *Adv. Mater. Interfaces*, 2022, **9**, 2101865.
53. S. P. Kühn, K. Edström, M. Winter and I. Cekic-Laskovic, *Adv. Mater. Interfaces*, 2022, **9**, 2102078.
54. P. Peljo and H. H. Girault, *Energy Environ. Sci.*, 2018, **11**, 2306.
55. F. Jeschull, J. Maibach, K. Edström and D. Brandell, *J. Electrochem. Soc.*, 2017, **164**, A1765.
56. H. Nakai, T. Kubota, A. Kita and A. Kawashima, *J. Electrochem. Soc.*, 2011, **158**, A798.
57. K. Edström, M. Herstedt and D. P. Abraham, *J. Power Sources*, 2006, **153**, 380.
58. R. Dedryvère, S. Leroy, H. Martinez, F. Blanchard, D. Lemordant and D. Gonbeau, *J. Phys. Chem. B*, 2006, **110**, 12986.
59. A. Kotronia, H. D. Asfaw, C.-W. Tai, M. Hahlin, D. Brandell and K. Edström, *ACS Appl. Mater. Interfaces*, 2021, **13**, 3867.
60. B. S. Parimalam, A. D. MacIntosh, R. Kadam and B. L. Lucht, *J. Phys. Chem. C*, 2017, **121**, 22733.

61. A. V. Cresce, S. M. Russell, D. R. Baker, K. J. Gaskell and K. Xu, *Nano Lett.*, 2014, **14**, 1405.
62. A. L. Michan, B. S. Parimalam, M. Leskes, R. N. Kerber, T. Yoon, C. P. Grey and B. L. Lucht, *Chem. Mater.*, 2016, **28**, 8149.
63. A. Benayad, J. E. Morales-Ugarte, C. C. Santini and R. Bouchet, *J. Phys. Chem. A*, 2021, **125**, 1069.
64. M. Schnabel, S. P. Harvey, E. Arca, C. Stetson, G. Teeter, C. Ban and P. Stradins, *ACS Appl. Mater. Interfaces*, 2020, **12**, 27017.
65. A. M. Andersson, A. Henningson, H. Siegbahn, U. Jansson and K. Edström, *J. Power Sources*, 2003, **119–121**, 522.
66. K. Edström, T. Gustafsson and J. O. Thomas, *Electrochim. Acta*, 2004, **50**, 397.
67. G. M. Veith, M. Doucet, R. L. Sacci, B. Vacaliuc, J. K. Baldwin and J. F. Browning, *Sci. Rep.*, 2017, **7**, 6326.
68. S. Fang, D. Jackson, M. L. Dreibelbis, T. F. Kuech and R. J. Hamers, *J. Power Sources*, 2018, **373**, 184.
69. I. Hasa, A. M. Haregewoin, L. Zhang, W.-Y. Tsai, J. Guo, G. M. Veith, P. N. Ross and R. Kostecki, *ACS Appl. Mater. Interfaces*, 2020, **12**, 40879.
70. F. Shi, P. N. Ross, H. Zhao, G. Liu, G. A. Somorjai and K. Komvopoulos, *J. Am. Chem. Soc.*, 2015, **137**, 3181.
71. D. Streich and P. Novák, *Electrochim. Acta*, 2016, **190**, 753.
72. J. Bitenc, A. Vizintin, J. Grdadolnik and R. Dominko, *Energy Storage Mater.*, 2019, **21**, 347.
73. M. Nie, D. P. Abraham, Y. Chen, A. Bose and B. L. Lucht, *J. Phys. Chem. C*, 2013, **117**, 13403.
74. S. K. Heiskanen, J. Kim and B. L. Lucht, *Joule*, 2019, **3**, 2322.
75. J. D. McBrayer, M.-T. F. Rodrigues, M. C. Schulze, D. P. Abraham, C. A. Apblett, I. Bloom, G. M. Carroll, A. M. Colclasure, C. Fang, K. L. Harrison, G. Liu, S. D. Minteer, N. R. Neale, G. M. Veith, C. S. Johnson, J. T. Vaughey, A. K. Burrell and B. Cunningham, *Nat. Energy*, 2021, **6**, 866.
76. M. C. Biesinger, B. P. Payne, A. P. Grosvenor, L. W. M. Lau, A. R. Gerson and R. S. C. Smart, *Appl. Surf. Sci.*, 2011, **257**, 2717.
77. C. A. Marianetti, G. Kotliar and G. Ceder, *Nat. Mater.*, 2004, **3**, 627.
78. A. S. Menon, B. J. Johnston, S. G. Booth, L. Zhang, K. Kress, B. E. Murdock, G. Paez Fajardo, N. N. Anthonisamy, N. Tapia-Ruiz, S. Agrestini, M. Garcia-Fernandez, K. Zhou, P. K. Thakur, T. L. Lee, A. J. Nedoma, S. A. Cussen and L. F. J. Piper, *PRX Energy*, 2023, **2**, 013005.
79. Z. Fu, J. Hu, W. Hu, S. Yang and Y. Luo, *Appl. Surf. Sci.*, 2018, **441**, 1048.
80. R. Azmi, V. Trouillet, M. Strafela, S. Ulrich, H. Ehrenberg and M. Bruns, *Surf. Interface Anal.*, 2018, **50**, 43.
81. M. Okubo and A. Yamada, *ACS Appl. Mater. Interfaces*, 2017, **9**, 36463.
82. R. Dedryvère, L. Gireaud, S. Grugeon, S. Laruelle, J. M. Tarascon and D. Gonbeau, *J. Phys. Chem. B*, 2005, **109**, 15868.
83. A. Urban, D.-H. Seo and G. Ceder, *npj Comput. Mater.*, 2016, **2**, 16002.
84. M. Ebadi, A. Nasser, M. Carboni, R. Younesi, C. F. N. Marchiori, D. Brandell and C. M. Araujo, *J. Phys. Chem. C*, 2019, **123**, 347.

85. L. Dahéron, H. Martinez, R. Dedryvère, I. Baraille, M. Ménétrier, C. Denage, C. Delmas and D. Gonbeau, *J. Phys. Chem. C*, 2009, **113**, 5843.
86. P. S. Bagus, C. J. Nelin and C. R. Brundle, *J. Vac. Sci. Technol., A*, 2023, **41**, 068501.
87. D. Li, Y. Chen, C. Zhou, C. Shi, Z. Xu, Z. Miao, Z. Xi and J. Han, *Mater. Chem. Front.*, 2024, **8**, 715.
88. A. G. Shard and M. A. Baker, *J. Vac. Sci. Technol., A*, 2024, **42**, 050801.
89. P. Niehoff, S. Passerini and M. Winter, *Langmuir*, 2013, **29**, 5806.
90. J. E. Coyle, M. T. Brumbach, G. M. Veith and C. A. Apblett, *J. Phys. Chem. C*, 2020, **124**, 8153.
91. V. Eshkenazi, E. Peled, L. Burstein and D. Golodnitsky, *Solid State Ionics*, 2004, **170**, 83.
92. F. Lin, D. Nordlund, I. M. Markus, T.-C. Weng, H. L. Xin and M. M. Doeff, *Energy Environ. Sci.*, 2014, **7**, 3077.
93. M. Hekmatfar, A. Kazzazi, G. G. Eshetu, I. Hasa and S. Passerini, *ACS Appl. Mater. Interfaces*, 2019, **11**, 43166.
94. M. Seo, Y. Lee, H. Shin, E. Kim, H.-S. Kim, K.-B. Chung, G. Kim and B. S. Mun, *ACS Appl. Mater. Interfaces*, 2024, **16**, 26922.
95. E. Lewin, J. Counsell and J. Patscheider, *Appl. Surf. Sci.*, 2018, **442**, 487.
96. Y. J. Hofstetter and Y. Vaynzof, *ACS Appl. Polym. Mater.*, 2019, **1**, 1372.
97. J. B. Malherbe, S. Hofmann and J. M. Sanz, *Appl. Surf. Sci.*, 1986, **27**, 355.
98. S. Wenzel, T. Leichtweiss, D. Krüger, J. Sann and J. Janek, *Solid State Ionics*, 2015, **278**, 98.
99. K. P. C. Yao, D. G. Kwabi, R. A. Quinlan, A. N. Mansour, A. Grimaud, Y.-L. Lee, Y.-C. Lu and Y. Shao-Horn, *J. Electrochem. Soc.*, 2013, **160**, A824.
100. M. I. Nandasiri, L. E. Camacho-Forero, A. M. Schwarz, V. Shutthanandan, S. Thevuthasan, P. B. Balbuena, K. T. Mueller and V. Murugesan, *Chem. Mater.*, 2017, **29**, 4728.
101. A. Schwöbel, R. Hausbrand and W. Jaegermann, *Solid State Ionics*, 2015, **273**, 51.
102. A. Etxebarria, D.-J. Yun, M. Blum, Y. Ye, M. Sun, K.-J. Lee, H. Su, M. Á. Muñoz-Márquez, P. N. Ross and E. J. Crumlin, *ACS Appl. Mater. Interfaces*, 2020, **12**, 26607.
103. J. Maibach, J. Rizell, A. Matic and N. Mozhzhukhina, *ACS Mater. Lett.*, 2023, **5**, 2431.
104. X. Wu, C. Villevieille, P. Novák and M. El Kazzi, *Phys. Chem. Chem. Phys.*, 2018, **20**, 11123.
105. D.-J. Yun, S. Kim, S. Heo, H.-s. Choi, J. Baik, J. Chung, S. Park, D. Yu, J. Lee, S. Kang, C. Jung and D.-S. Ko, *Adv. Energy Mater.*, 2024, **14**, 2401473.
106. F. G. Capone, J. Sottmann, V. Meunier, L. Pérez Ramírez, A. Grimaud, A. Iadecola, M. Scardamaglia, J. P. Rueff and R. Dedryvère, *Energy Environ. Sci.*, 2024, **17**, 1509.
107. K. Hikima, K. Shimizu, H. Kiuchi, Y. Hinuma, K. Suzuki, M. Hirayama, E. Matsubara and R. Kanno, *J. Am. Chem. Soc.*, 2022, **144**, 236.
108. C. Guhl, P. Kehne, Q. Ma, F. Tietz, L. Alff, P. Komissinskiy, W. Jaegermann and R. Hausbrand, *Rev. Sci. Instrum.*, 2018, **89**, 073104.

109. H. Kiuchi, K. Hikima, K. Shimizu, R. Kanno, F. Toshiharu and E. Matsubara, *Electrochem. Commun.*, 2020, **118**, 106790.
110. R. Endo, T. Ohnishi, K. Takada and T. Masuda, *J. Phys. Commun.*, 2021, **5**, 015001.
111. X. He, J. M. Larson, H. A. Bechtel and R. Kostecki, *Nat. Commun.*, 2022, **13**, 1398.
112. R. S. Weatherup, B. Eren, Y. Hao, H. Bluhm and M. B. Salmeron, *J. Phys. Chem. Lett.*, 2016, **7**, 1622.
113. R. Koerver, F. Walther, I. Aygün, J. Sann, C. Dietrich, W. G. Zeier and J. Janek, *J. Mater. Chem. A*, 2017, **5**, 22750.
114. S. Saxena, Y. Xing, D. Kwon and M. Pecht, *Int. J. Electr. Power Energy Syst.*, 2019, **107**, 438.
115. P. M. Dietrich, L. Gehrlein, J. Maibach and A. Thissen, *Crystals*, 2020, **10**, 1056.
116. C. H. Wu, R. S. Weatherup and M. B. Salmeron, *Phys. Chem. Chem. Phys.*, 2015, **17**, 30229.
117. M. Favaro, B. Jeong, P. N. Ross, J. Yano, Z. Hussain, Z. Liu and E. J. Crumlin, *Nat. Commun.*, 2016, **7**, 12695.
118. J. Maibach, I. Källquist, M. Andersson, S. Urpelainen, K. Edström, H. Rensmo, H. Siegbahn and M. Hahlin, *Nat. Commun.*, 2019, **10**, 3080.
119. R. Lee, T. S. Nunney, M. Isaacs, R. G. Palgrave and A. Dey, *ACS Appl. Mater. Interfaces*, 2024, **16**, 35675.
120. J. Qian, A. Baskin, Z. Liu, D. Prendergast and E. J. Crumlin, *J. Chem. Phys.*, 2020, **153**, 044709.
121. J. Maibach, C. Xu, S. K. Eriksson, J. Åhlund, T. Gustafsson, H. Siegbahn, H. Rensmo, K. Edström and M. Hahlin, *Rev. Sci. Instrum.*, 2015, **86**, 044101.
122. R. S. Weatherup, *Top. Catal.*, 2018, **61**, 2085.
123. Y.-H. Lu, C. Morales, X. Zhao, M. A. van Spronsen, A. Baskin, D. Prendergast, P. Yang, H. A. Bechtel, E. S. Barnard, D. F. Ogletree, V. Altoe, L. Soriano, A. M. Schwartzberg and M. Salmeron, *Nano Lett.*, 2020, **20**, 6364.
124. J. S. Bunch, S. S. Verbridge, J. S. Alden, A. M. van der Zande, J. M. Parpia, H. G. Craighead and P. L. McEuen, *Nano Lett.*, 2008, **8**, 2458.
125. F. Roth, M. Borgwardt, L. Wenthaus, J. Mahl, S. Palutke, G. Brenner, G. Mercurio, S. Molodtsov, W. Wurth, O. Gessner and W. Eberhardt, *Nat. Commun.*, 2021, **12**, 1196.

CHAPTER 5

XPS as a Tool for the Investigation of Heterogeneous Catalysts

ANNA MARIA VENEZIA

Institute of Nanostructured Materials, ISMN-CNR, via Ugo la Malfa 153, Palermo 90146, Italy
E-mail: annamaria.venezia@cnr.it

5.1 Introduction

A catalyst is a chemical substance that affects the rate and selectivity of a chemical reaction without being part of the end products. The primary reasons for using catalysts are the enhancement of the reaction rate and the increase of selectivity toward the desired products. Particularly in the case of heterogeneous catalysis, where the catalyst is a solid material and the reaction medium is liquid or gaseous, the catalytic process occurs at the catalyst surface and involves a sequence of elementary steps including adsorption, surface diffusion, chemical rearrangements of the adsorbed reaction intermediates and desorption of the product.[1] Therefore, for the understanding of the catalytic process, a detailed characterisation of the catalyst surface and of the interface between the solid and the adsorbed reagents is crucial. The X-ray photoelectron spectroscopy technique, based on the detection of photoelectrons with smaller mean free paths in solid materials as compared

Inorganic Materials Series No. 19
Photoelectron Spectroscopy in Inorganic Materials Science
Edited by Robert G. Palgrave and Philip R. Davies

Published by the Royal Society of Chemistry, www.rsc.org

to techniques based on the detection of photons like UV and IR spectroscopies and X-ray Diffraction (XRD), is a surface-sensitive tool and XPS has become one of the most valuable techniques for catalyst investigation. In the past century, catalysis has contributed enormously to the advancement of society, with the development of important industrial processes such as oil refining, production of key commodity chemicals produced in large volume and in dedicated plants, such as ammonia and sulfuric acid, and the production of fine chemicals.[2–4] The synthesis of ammonia, used as a fertiliser, was developed by Fritz Haber and Carl Bosch who won the Nobel prize in 1918 and 1931 respectively. The process, fundamental to avoid human starvation, consisted of the reaction between N_2 and H_2 under high pressure and high temperature and catalysed by fused iron catalysts. Along with the production of the large-scale chemicals cited above, where the heterogeneous catalysis is well established, the modern pharmaceutical industry dealing with smaller volume chemicals of greater added value, aims to design very highly selective heterogeneous catalysts to replace the less environmentally friendly homogeneous ones. The research in this field focuses on the design of innovative catalytic materials with specific structure, chemistry and morphology, able to drive the catalytic process in the desired direction, with affordable cost and time reduction.[5] Since catalysis, as said above, involves the most external layer of a solid material, the active species is generally supported over an inert and inexpensive carrier in order to decrease the final costs. Generally, the larger the number of active sites deposited at the catalyst surface the larger the catalytic activity. For this reason, innovative synthesis procedures are continuously being developed to maximise the surface concentration of the supported active species over an inert carrier.

Moreover, in the present day, catalysis is dealing with two even more challenging tasks: easing environmental pollution and helping the full exploitation of alternative energy sources. The first task requires the development of materials such as catalytic converters for automotive applications, to abate CH_x and NO_x compounds and particulate matter (PM) arising from engine fuel combustion, and catalysts for the removal of volatile organic compounds (VOC) present in domestic and industrial environments. The second task, dealing with the so-called carbon neutral or net-zero carbon emission actions to combat climate change, focuses on the development of new materials for application in hydrogen fuel cells (FC), or for the transformation of sequestered CO_2 or of carbon-containing waste materials into carbon neutral alternative fuels such as e-fuel or biofuel respectively.[4] For many decades, catalysis has been treated as an empirical science with catalysts being selected on the bases of a trial-and-error approach. Thanks to surface science and to different techniques able to probe the outermost layers of a material, it is now becoming possible to understand how the catalysts work on the atomic scale. Therefore, through the selection of supports, synthesis procedure and promoters, it is possible to rationally design and modulate the surface properties of a catalyst for a specific chemical reaction. XPS is one of the most complete analytical methods in this area, which alone or in combination with other surface analysis techniques provides compositional, electronic and even morphological information of a catalyst surface.[3–5] Before using any analytical technique, it is mandatory to ask what types of information are needed to tackle a particular issue. In the case of catalyst investigations, it is important to know: (i) the surface elemental composition; (ii) the active element dispersion corresponding to the number of exposed atoms of the active element over the total number of atoms present in the material; (iii) the chemical state of the different elements composing the catalyst material.[5] These quantitative and qualitative properties may change from the material as just synthesised to the material after an activation procedure, generally needed before the catalytic reaction, and after being used in a particular process. Therefore, it is important to investigate the changes of the surface at different moments of a catalytic trial. The comparison between the outcomes of such analyses carried out at the different stages of the catalyst life would provide information on the catalyst transformation and on the origin of its ultimate deactivation. Understanding the causes of deactivation is indeed important in order to adopt appropriate strategies for catalyst regeneration.[6,7] Thanks to the technological advancement, an enormous contribution to the elucidation of the catalytic processes and materials has been recently provided by the use of near ambient pressure X-ray photoelectron spectroscopy (NAP-XPS), able to analyse the catalyst surface in *operando* conditions, *i.e.*, while the reaction is running.[8–16] Due to the heterogeneity of most catalytic materials, which are generally composed of small aggregates of powder, some advanced features of the XPS technique such as 'XPS imaging' and 'XPS depth profiling' are more difficult to apply to catalysts and will not be discussed here.[6,17]

In catalysis technology a large variety of materials are used, depending on the chemical process to be activated. The present chapter will focus on three main categories of catalysts: (i) supported metal catalysts; (ii) supported oxide catalysts; (iii) sulfur-based catalysts. The application of XPS in the case of heterogenisation or immobilisation of homogeneous catalysts over a solid oxide will also be highlighted.[18–21] In the following we will describe the main features of each catalyst group, as they have emerged from the XPS surface investigation, with a detailed description of selected case studies.

5.2 Catalytic Materials

5.2.1 Supported Metal Catalysts

Supported metal catalysts represent a large group of materials being used for several catalytic processes. Their main advantage resides in the fact that their active element, which can vary from noble metals to transition metals, is highly dispersed over the surface of a carrier, therefore maximising the number of active sites.[5,22–26] As mentioned above, catalytic reactions involve the outermost layer of a material, whereas XPS monitors up to a few layers below. The appropriate selection of photoelectron peaks with different surface sensitivity and the use of specific models allows the attainment of dispersion information and the detection of the active chemical species.[5,27] When dealing with catalysts formed by supported metal particles, the binding energy (BE) of the metal element may undergo negative or positive shift with respect to the corresponding bulk metal, depending on the chemical environment of the element and on the particle size. Indeed, the loss of the metallic character, due to a reduced particle size, alloying with another metal, or electron transfer towards or from the carrier, altering the charge density of the metal particle, would influence the kinetic energy of the photoelectron and therefore its binding energy.[28,29] Because of its surface enhanced sensitivity, XPS is also a valid tool for particle size determination. Indeed, XPS intensity ratios of the metal particle related peaks, I_p, and support related peaks, I_s, are strongly dependent on the catalyst dispersion; catalysts with the same chemical composition, *i.e.*, same loading of metal over the same support, exhibit higher I_p/I_s values in the case of small metal particles as compared to large particles.[3] It is important to point out that a high surface area of the supports, although favouring the dispersion of metal particles, may produce a lower I_p/I_s intensity ratio by virtue of the large contribution of the support elements. Based on morphology, structure and density properties of the catalyst, the dispersion of supported metal particles is generally estimated using the Kerkhof–Moulijn and the Davis models for particles homogenously distributed on high surface area supports.[27,30] The deviation from a monolayer prediction indicated particle growth, multilayer formation or metal segregation on the external surface of the supports. In catalysis, as supports, high surface area and inert oxides such as alumina and silica or other oxides with different degrees of chemical activity, are used. With appropriate preparation procedures, very efficient catalysts in terms of activity and selectivity are produced. The main properties of supported metal catalysts, regardless of the chemical process they are being used for, are the surface concentration of the metal and its oxidation state. Other instrumental techniques, such as IR, Raman spectroscopy, TEM, and XRD, generally yield these features separately. XPS is a unique technique able to provide this information at the same time.

5.2.1.1 Gold Supported Catalysts

Among the supported metals, noble metals, despite their high cost, are the most active for many reactions and in terms of cost/benefits are still the preferred metals especially for smaller scale production of chemicals. An interesting case is given by gold, which, at variance with platinum and palladium, has been for a long time regarded as a quite inert material, maintaining its metallic state in harsh oxidative environments and therefore of no interest for catalytic applications.[31] However, since the pioneering work of Haruta, in the late eighties, reporting an unusually high catalytic activity of supported gold nanoparticles in CO oxidation, there has been a tremendous industrial interest for this element with application in several catalytic reactions such as catalytic combustion of hydrocarbons, water gas shift reaction and hydrochlorination of acetylene, to cite a few.[32,33] Nowadays, due to their high efficiency at low temperature, gold-based catalysts are considered suitable as gas sensors and for pollution abatement devices. The factors responsible for the peculiarly high activity of gold at lower temperature have been debated from both experimental and theoretical points of view. A direct correlation between the activity of gold and the size of the gold particles was attributed to changes in the electronic properties of the gold.[34] The increase of the catalytic activity with the decrease of the particle size was attributed to the so-called quantum size effect consisting of a loss of metallic character for particles below a certain size.[35] Indeed, the loss of metallic character implied a decrease of the electron density at the gold atoms with the formation of slightly positive gold clusters, more reactive toward the substrates.[36] The variations in the electronic properties with the particle sizes were demonstrated by photoelectron spectroscopy; both UV photoelectron spectroscopy (UPS) and XPS.[37] When gold was supported on oxides as nanoparticles (NPs) of sizes of a few nanometres, it became quite active in the oxidation of the CO molecule.[38,39] Besides particle size, modification of the electronic properties of the gold particles could also be attained by an appropriate choice of the support oxide and by maximising the interaction between gold and support and enlarging their interface.[37] In Figure 5.1, the experimental and the curve fitted spectra of the Au 4f core level for different gold catalysts are shown.[40] The original work from our group aimed to investigate the effect of the preparation method and of the type of oxide on the electronic properties of gold. The effect of the synthesis procedure was clearly demonstrated by the comparison between the two compositionally similar samples of Au/SiO_2, one prepared by deposition precipitation (DP) and the other one by solvated metal atom deposition (SMAD). As previously described, the DP method consisted in adding the gold precursor, $HAuCl_4$, to an aqueous suspension of SiO_2 at pH = 8, stirring at 353 K for 24 h, then filtering, washing, and drying at 373 K.[40] The SMAD technique originally developed by Klabunde was based on the metal thermal evaporation from high purity Au foil followed by low temperature (liquid N_2) acetone trapping on the walls of the metal vapour reactor.[41] Upon raising the temperature to room temperature, the solvated metal atom solution was used to impregnate the selected support. Excess solvent (acetone) was removed under vacuum and the SMAD-prepared samples were dried at room temperature. Such a mild procedure would have ensured the deposition of nanogold metal particles without any chemical interaction with the support and, due to the low operating temperature procedure, without being sintered.[41] As seen in Figure 5.1 and as reported in Table 5.1, summarising the XPS results and the temperature T_{50} (K) at which 50% conversion of CO was obtained, the two Au/SiO_2 catalysts' spectra were different. The Au 4f region was fitted using the two spin orbit components Au $4f_{7/2}$ and Au $4f_{5/2}$ separated by 3.7 eV and in a fixed intensity ratio of 4 : 3, in accord with the atomic core level spin orbit coupling theory.[35]

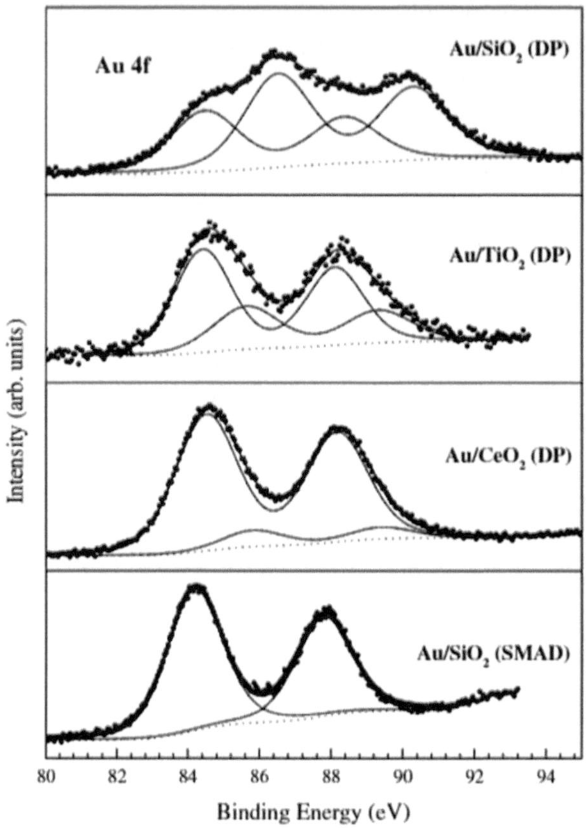

Figure 5.1 XPS curve fitting of the Au 4f photoelectron peak in supported gold catalysts prepared by different methods. Reproduced from ref. 40 with permission from John Wiley & Sons, Copyright 2006.

The full widths at half maximum (FWHM), sometimes indicative of multiple components, were also reported in Table 5.1. Concerning the silica supported gold, the SMAD procedure yielded purely metallic gold, characterised by Au $4f_{7/2}$ binding energy of 84.2 eV, whereas through the DP procedure, metallic gold and Au^{3+} species, characterised by Au $4f_{7/2}$ binding energy of 86.5 eV, in a ratio of 2 : 3, were produced. In the case of the silica supported gold catalysts, better catalytic activity was registered for the SMAD sample. The presence of metallic gold was not detected by XRD due to particle sizes smaller than 2–5 nm, below the detection limit of the diffraction technique. Similar XP spectral differences were observed between the SMAD and the DP catalysts with the other two supports. The SMAD technique always produced metallic species, whereas the DP method produced both metallic and oxidised ones. However, in the case of the two 'reducible oxides', TiO_2 and CeO_2, Au^+ species were obtained instead of Au^{3+}.[40] The difference with the silica support was probably due to the oxygen mobility of the two oxides which

Table 5.1 Au $4f_{7/2}$ binding energy with FWHM in parentheses, percentage of surface Au species and catalytic T_{50} reaction temperature for CO oxidation. Reproduced from ref. 40 with permission from John Wiley & Sons, Copyright 2006.

Sample	Au $4f_{7/2}$ BE (eV)	Surface Au species	T_{50} (K)
Au/SiO_2 (SMAD)	84.2 (2.0)	Au^0	580
Au/CeO_2 (SMAD)	84.4 (2.0)	Au^0	380
Au/TiO_2 (SMAD)	83.9 (1.9)	Au^0	450
Au/SiO_2 (DP)	84.4 (2.3)	Au^0 (40%)	—
	86.5 (2.2)	Au^{3+} (60%)	630
Au/TiO_2 (DP)	84.4 (1.9)	Au^0 (67%)	—
	85.6 (2.2)	Au^{1+} (33%)	276
Au/CeO_2 (DP)	84.5 (2.0)	Au^0 (90%)	250
	85.8 (2.0)	Au^{1+} (10%)	—

helped to compensate for the positive charge formation yielding a final oxidation state close to Au^+. Correspondingly, the temperatures of the 50% conversion were considerably lower. As seen in Table 5.1, the surface oxidised Au^+ stabilised as $(AuO)^-$ by the cations of the cerium oxide or titanium oxide contributed to the enhancement of the catalytic activity. The study showed that, more than a size effect, it was the formation of slightly oxidised gold particles, stabilised by the interaction with the supports, that contributed to the CO oxidation activity of the gold catalysts at low temperature. In this case, the size and the type of carrier, both contributing to the extension of the interface between the gold and carrier, played an important role.[37]

Gold-based catalysts have been successfully used for the reverse water gas shift (RWGS) process at low temperature, described by the equation $CO_2 + H_2 \rightleftharpoons CO + H_2O$. With this reaction, CO_2, arising from fuel combustion and considered responsible for global warming, is transformed into the more valuable CO which is used as a feedstock for the synthesis of fine chemicals and fuels through the Fischer–Tropsch process. In a recent study, a small amount of gold (0.1 wt%) supported on TiO_2 and on ZrO_2 was tested in the RWGS reaction to investigate the role of defects such as oxygen vacancies on the gold activity.[42] To get a closer insight into the nature of the gold species, the Au 4f XP spectra of the catalysts were recorded and are displayed here in Figure 5.2. Both samples prepared by the deposition–precipitation method were characterised by the Au $4f_{7/2}$ binding energies in Au/TiO_2 and Au/ZrO_2 of 83.4 eV and 83.9 eV, respectively (note the strong background signal from titanium in the former case). These values are close to the binding energy of Au^0, thus indicating the metallic character of the supported Au species. The shift of about 0.5 eV to lower binding energies, observed for Au/TiO_2 with respect to Au/ZrO_2, was correlated with a charge transfer from the oxygen vacancies of the TiO_2 support to Au, leading to a partially negatively charged Au species ($Au^{\delta-}$) compared to the ZrO_2 support case. The XPS results suggested the formation of electron-enriched Au species ($Au^{\delta-}$) on the surface of TiO_2. Such species have been suggested to play an important role

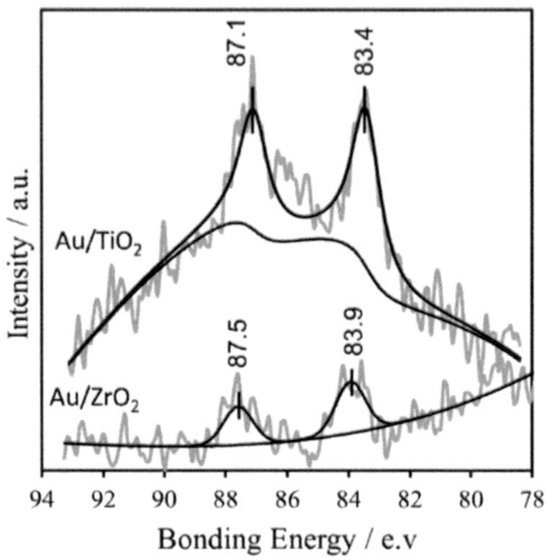

Figure 5.2 Au 4f XP spectra of Au/TiO_2 and Au/ZrO_2. Reproduced from ref. 42 with permission from Elsevier, Copyright 2023.

in favouring the adsorption of CO_2 and its activation.[43] *In situ* electron paramagnetic resonance (EPR) measurements confirmed the presence of Ti^{3+} species and oxygen vacancies O_v. These results suggested that O_v–Ti^{3+} sites participate in the reaction mechanism and served as active sites for RWGS. In summary, through the combination of two different analytical techniques, XPS and EPR, it was demonstrated that both the surface defects (oxygen vacancies) and the gold NPs play critical roles in the activation of CO_2 and H_2, validating the superiority of the gold supported on TiO_2 as compared to gold supported on ZrO_2.

5.2.1.2 Bimetallic AuPt, AuPd Supported Catalysts

Supported bimetallic catalysts have been used for different reactions.[44] Pt and Pd, characterised by elevated hydrogenation activity, are used in deep hydrodesulfurisation (HDS) reactions.[45,46] Differently from the industrial CoMo HDS catalysts, activated by a sulfidation treatment, Pt and Pd are active in the metallic state and allow operation at lower temperatures. Moreover, a second metal, *e.g.*, Au, is added to increase the tolerance to sulfur.[46] In our lab, SiO_2-supported monometallic and bimetallic platinum–gold catalysts with 1 wt% loading of each metal, were prepared by deposition of metal nanoparticles stabilised by mercaptopropyltriethoxysilane (MPTES) after different aging times of the solution containing metal ions and MPTES. After calcination at 400 °C the catalysts were tested in the thiophene hydrodesulfurisation (HDS) reaction.[47] In order to discriminate the effect of the ligand–metal interaction, different aging times of the metal–MPTES solution before the addition of silica were considered. As seen in Figure 5.3, the Pt 4f and Au

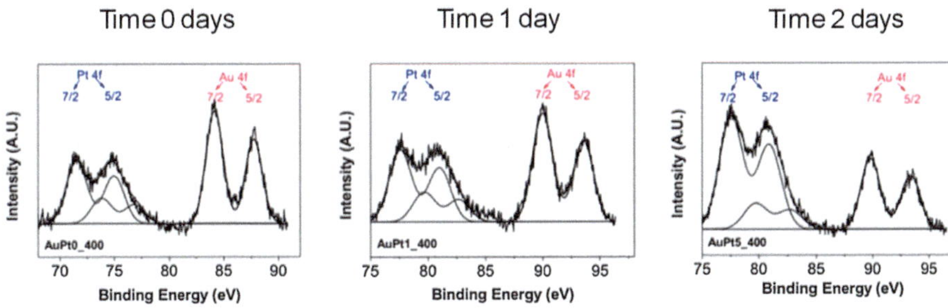

Figure 5.3 Pt 4f and Au 4f XP spectral region of the AuPt/SiO_2 catalysts after different aging times. Reproduced from ref. 44 with permission from Elsevier, Copyright 2017.

4f XP spectra of the catalysts obtained with different aging times were different in terms of quantitative information. The same chemical species, Pt, Au and PtO_2, were present in all cases. However, drastic changes occurred concerning their relative intensity ratios. Indeed, as observed from the relative changes of the Pt 4f and Au 4f peak intensities, the gold enriched particles detected at time zero, *i.e.*, without aging of the solution, transformed into platinum enriched particles with increasing the aging time up to five days. The results were in accord with a structural change from the less active Pt@AuPt core–shell like structure to the more active Au@AuPt core–shell like structure with an increased surface exposure of platinum. The change took place by simply extending the time the two metal ions had spent in solution with the MPTES, before adding the SiO_2 support.[44,47] The XPS analyses were also fundamental to explain the XRD and the IR results.

The positive effect of the –SH functionalisation of silica was also demonstrated in a study of the HDS activity of Pd–Au catalysts.[48] The XPS analyses of the main Pd 3d and Au 4f core levels of the catalysts prepared with and without –SH mesoporous HMS silica, clearly demonstrated the positive effect of the –SH functionalisation on the increased metal dispersion. As shown in Figure 5.4, the intensity of the gold peaks in the energy range 84–88 eV with respect to the intensity of the silicon satellite peak at 93 eV directly probed the surface compositional changes at different stages of the catalyst life.

From the displayed spectra it was evident how in the absence of functionalisation, the surface gold concentration in PdAuHMS changed with respect to silica, differently from the PdAuHMS-4SH exhibiting stable gold concentration after the different catalyst stages. The chemical stability along with a structural stability detected by XRD analyses were crucial for the improvement of the catalytic activity of the bimetallic catalyst prepared through the –SH functionalisation of the silica support.[48]

In another study of some years ago, the effect of gold and of the metal deposition procedure on the HDS activity and stability of palladium catalysts was reported.[49] The study demonstrated very well the role of the XPS

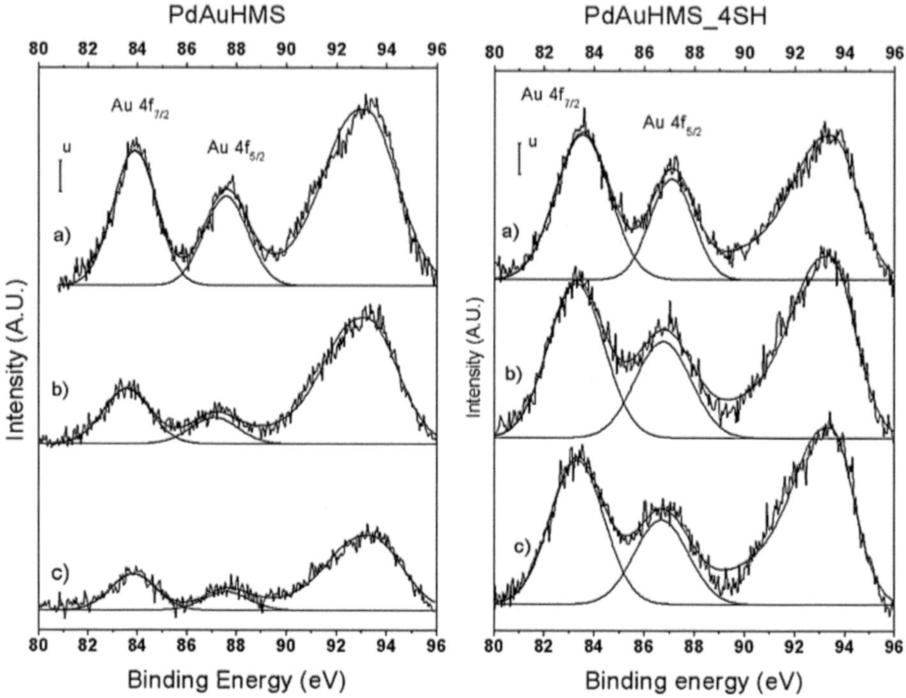

Figure 5.4 Au 4f XP spectra of bimetallic catalysts as: (a) fresh; (b) reduced; (c) aged. Reproduced from ref. 48 with permission from Elsevier, Copyright 2012.

technique in detecting the structural and electronic effect of gold acting as a Pd promoter. If on one side the XRD analyses indicated the formation of bulk alloyed Pd_xAu_y particles, on the other side the XPS detected surface segregation of gold. The presence of gold at the surface particle, acting as a diluting agent, would decrease the size of the HDS active Pd ensemble, contributing to the enhanced activity of the surface sensitive HDS reaction. Moreover, the increase of the intensity ratio between Pd 3d attributed to metal and Pd 3d due to PdO with the increase of the gold amount confirmed the alloying effect of gold, inducing the reduction of palladium.[49] The negative ~0.5 eV binding energy shift of Au $4f_{7/2}$ of the monometallic Au with respect to the bimetallic Pd–Au catalyst, indicated an electron transfer from palladium to gold on the grounds of the bulk electronegativity differences. In Figure 5.5, the modification of the Pd 3d spectra as a function of gold loading in the catalysts is displayed. The Pd 3d spectra are characterised by the spin orbit components Pd $3d_{5/2}$ and Pd $3d_{3/2}$ about 5.4 eV apart. As shown in Figure 5.5, two palladium species components are present in both monometallic and bimetallic catalysts. The monometallic Pd catalysts was characterised by two palladium oxide Pd 3d components, typical of PdO and PdO_2.

The bimetallic $Au_{25}Pd_{75}$ with excess of palladium contained PdO_2 and PdO, with less PdO_2 as compared to the monometallic catalyst. The bimetallic

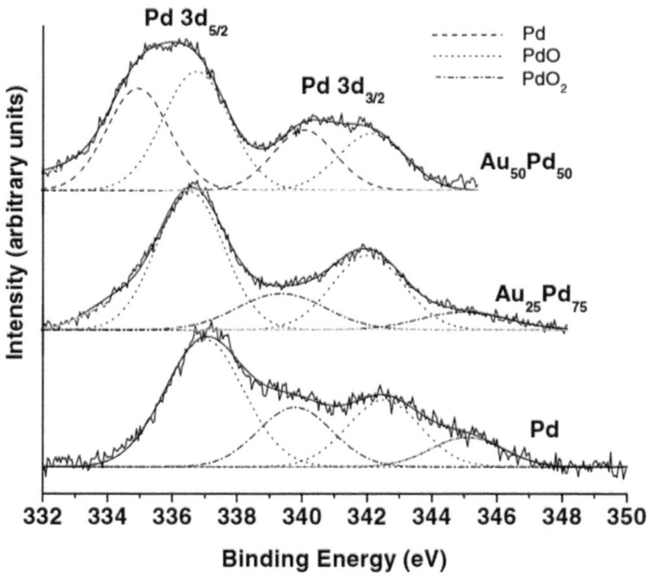

Figure 5.5 Experimental and fitted Pd 3d photoelectron spectra of monometallic palladium and bimetallic catalysts after calcination at 673 K. Reproduced from ref. 49 with permission from Elsevier, Copyright 2002.

$Au_{50}Pd_{50}$ contained metallic Pd(0) and PdO. The Pd(0) component was found to increase as a function of the Au loading. The similarity of Pd 3d and Au 4f spectra of the catalysts in the precursor state, *i.e.*, calcined state, with those of the activated catalyst (*i.e.*, after reduction), demonstrated the good structural stability of the catalytic system in the presence of gold.

The synergetic effect in bimetallic particles was recently investigated in TiO_2 supported Au–Pd catalysts for the selective base-free oxidation of glucose.[23] In this study, Au–Pd bimetallic catalysts supported on TiO_2 were found to be more active and selective in the oxidation of glucose to gluconic acid in correspondence to a specific Au/Pd atomic ratio equivalent to the composition of 0.5 wt% Au and 0.5 wt% Pd. Given the low metal loading, the XRD technique was not able to provide any useful information. In contrast, chemical and structural information were obtained from XPS analyses. As seen in Figure 5.6, the Au 4f XP spectrum increased its intensity with the increase of the wt% Au indicating a rather uniform gold deposition. Moreover, the asymmetry on the high energy side of the Au $4f_{7/2}$ and Au $4f_{5/2}$ spin–orbit components was attributed to an additional chemical component of gold. With appropriate curve fitting, metallic Au^0 with Au $4f_{7/2}$ binding energy equal to 83.3 eV and a small fraction of oxidised $Au^{\delta+}$ with Au $4f_{7/2}$ at 85.9 eV were ascertained. On the palladium side, the authors claimed the presence of two Pd $3d_{5/2}$ components, one at ~335.0 eV, typical of metallic Pd^0, and another one at 336.1 eV typical of Pd^{2+}. An increase of the Pd 3d intensity with the increase of the wt% Au and the total absence of the Pd 3d signal

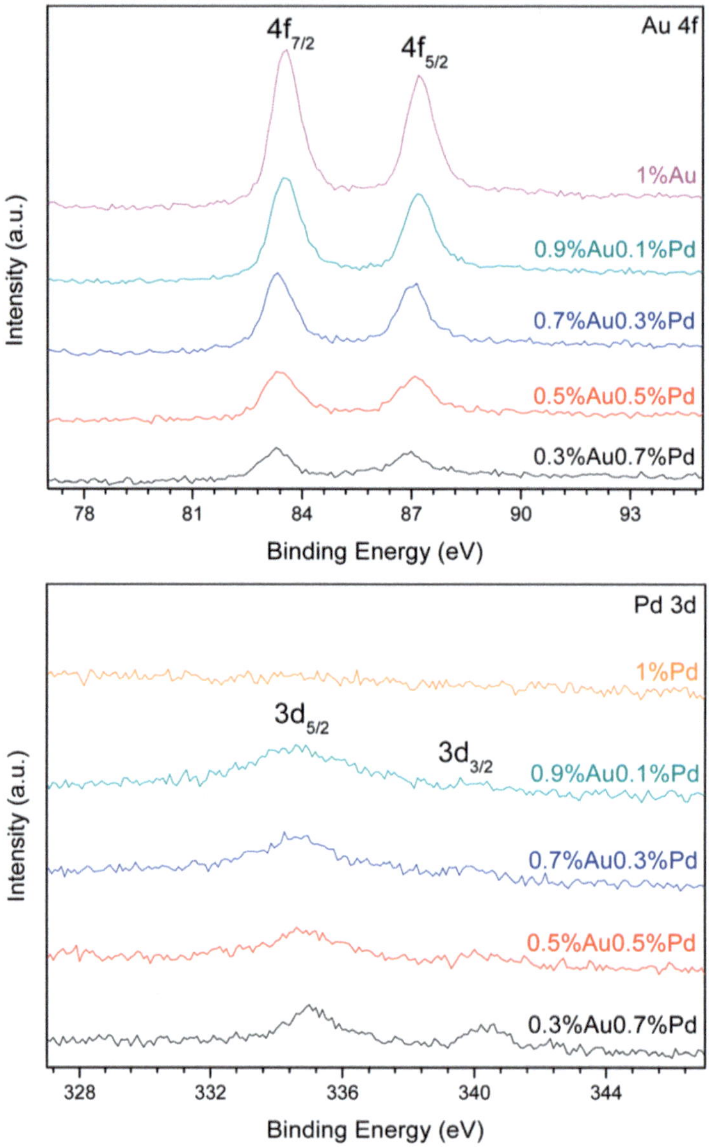

Figure 5.6 XP spectra of the Au 4f and Pd 3d photoelectron peaks in the series of catalysts prepared by the sol-immobilisation method. Reproduced from ref. 23 with permission from the Royal Society of Chemistry.

when no gold was present is notable. Therefore, the effect of gold on driving the palladium to the catalyst surface was inferred. The volcano-type plot of glucose conversion *versus* Au content in the catalyst, was indicative of the synergetic effect between palladium and gold. From the XPS data it could be determined that gold which was the electronic promoter of palladium, was mostly reduced, and its metallic character decreased with the increase of the Pd content in the samples. At the same time the oxidation state of Pd progressively moved from Pd^0 to Pd^{2+} with the increase of gold content in the sample. It was concluded that there was an optimum ratio between Pd^0 and Pd^{2+} species that would maintain a high activity as well as a high selectivity.

5.2.1.3 *Silver*

Another important noble element for catalytic applications is silver which is used as a catalyst component in several oxidation and dehydrogenation reactions.[50–52] The effect of the interaction between silver nanoparticles dispersed on a high surface area silica, and sulfur compounds during the adsorptive desulfurisation process is here briefly described with particular attention to the XPS results. The deep adsorptive desulfurisation process of real diesel fuel is indeed a promising alternative to the more classical hydrodesulfurisation method.[51,53] In the study reported here, in addition to techniques such as HRTEM and XRD, indicating the presence of highly dispersed Ag particles, the XPS analyses contributed to the understanding of the chemical interaction occurring as a consequence of the catalytic reaction.[50,54,55] The spectra in Figure 5.7 refer to silver catalysts, obtained by wet

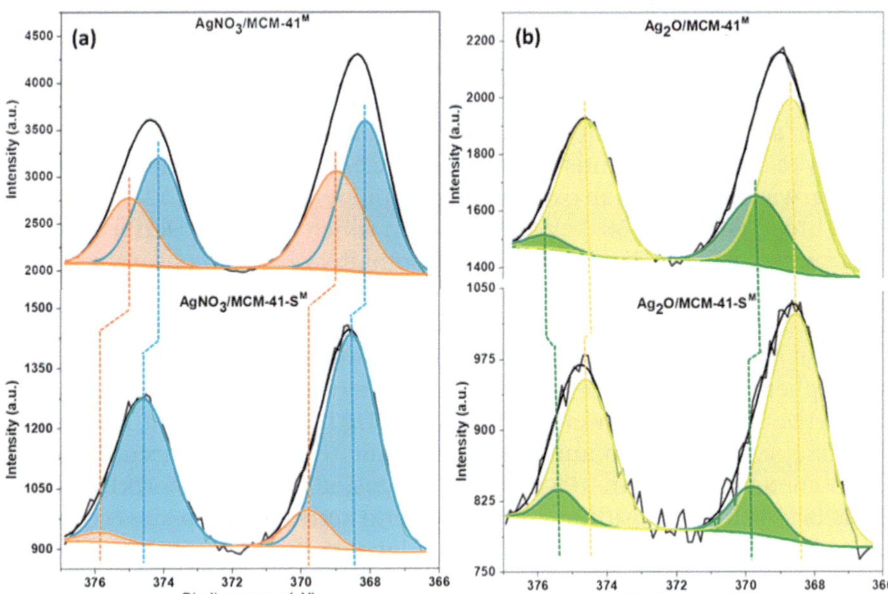

Figure 5.7 Ag 3d XP spectra for (a) $AgNO_3$/MCM-41^M and (b) Ag_2O/MCM-41^M before and after desulfurisation. Reproduced from ref. 51, https://doi.org/10.3390/catal9080651, under the terms of the CC BY 4.0 license, https://creativecommons.org/licenses/by/4.0/.

impregnation of a home-made mesoporous silica MCM-41 with an ethanolic solution of $AgNO_3$, before and after the sulfur adsorption test. The spectra on the left side (a) refer to the 393 K dried sample and the spectra on the right side refer to the 773 K calcined sample. The spectra are typical of the Ag 3d core–shell characterised by the Ag $3d_{5/2}$ and Ag $3d_{3/2}$ spin–orbit components approximately 6 eV apart. Considering the spectra of Figure 5.7a, the spin–orbit components at 368.1 eV and 374.1 eV are attributed to AgO and the weaker components at 369.0 eV and 375.0 eV are assigned to Ag(0) nanoparticles highly dispersed in the support.[50] The presence of the metallic state of silver was related to the light-induced photolysis of the $AgNO_3$ molecule. Furthermore, as reported in the literature, some oxygenated metal-adsorbents could undergo auto-reduction processes in the presence of lattice oxygen and water. Ag(0) could also interact with oxygen as Ag–O_{ads} and Ag–O_{lat}. All these oxygenated species (Ag–O) are highly active for oxidation of aromatic pollutants and even for weak π-complexation interactions with aromatic organosulfur.[51] It is worth noting that it is impossible to distinguish different silver oxides on the basis of Ag 3d binding energies, because there is no chemical shift.

On the contrary the Auger line M_4VV is quite sensitive to the Ag oxidation state, although it is more difficult to identify.[50] After the sulfur adsorption test, the increase in BEs of oxygenated Ag species (368.4 eV and 374.5 eV) was assigned to oxygen loss after the oxidation reactions, while the broad peaks which emerged at 369.4 eV and 375.6 eV were ascribed to the formation of an Ag–S–R bridging configuration *via* π-complexation (S–R = DBT) performed by Ag(0) and also by active Ag–O species.[55] According to the authors, Ag(0) ([Kr]$4d^{10}5s^1$) and Ag(I) ([Kr]$4d^{10}5s^0$) species are able to form bonds with their empty or semi-filled s orbitals, and their d orbitals can retro-donate electronic density to the π anti-ligands (π^*) of aromatic organosulfur rings. For the sample that was calcined in an oxidising atmosphere, the XP spectra (Figure 5.7b) show broad peaks at 368.5 eV and 374.6 eV, which could be ascribed to Ag_2O/AgO species. Highly dispersed Ag nanodomains were also observed at 369.4 eV and 375.9 eV. The spectra recorded after adsorptive desulfurisation show that the Ag_2O/AgO peak binding energy remained almost unchanged; however, the positive shift in Ag(0) BE (369.8 eV and 375.4 eV) demonstrated the same phenomenon observed for the non-calcined sample, where metallic and active silver species were complexed by DBT molecules. The results from the kinetic, equilibrium and thermodynamic tests were in accord with the phenomena observed in XPS data and suggested that chemisorptive interactions between organosulfur compounds and active Ag nano-sites controlled the sulfur adsorption. According to the higher sulfur concentration detected by XPS, the uncalcined $AgNO_3$/MCM-41 material synthesised from mesoporous silica presented significant efficiency for adsorptive desulfurisation; quite promising as a complementary process to the expensive HDS method for its low-cost, sustainability and efficient removal of recalcitrant organosulfur compounds.

It should be noticed that in contrast to most metals exhibiting positive core level binding energy shift from the metal to the oxide state, with the extent of shift depending on the oxidation state, the behaviour of the Ag 3d binding energy, as for the case described above, has been largely reported to follow an opposite trend.[56] In a recent work performed with synchrotron radiation on an *in situ* created beam of free Ag particles, without any fabrication related contamination and charge accumulation, the authors reconsidered such a statement and established a positive 1.2 eV shift between the oxidised Ag and the metallic Ag in accord with other metal behaviours. They explained the contrast to previous experimental results with a lack of a proper charging compensation procedure. In any case, a peculiarity of the silver atoms in real catalysts, which respond differently to the charging effect, exists and would deserve further investigation.

5.2.1.4 Heterogenisation (Immobilisation) of Pd Catalysts

Palladium (II) complexes are often used as homogeneous catalysts for fine chemicals and pharmaceutical products. To avoid the poor thermal stability of the active complex and the difficulties in separating the catalysts from the products, many attempts of heterogenisation of metal complexes have been reported.[57,58] Recently, new heterogeneous catalytic systems have been developed by supporting the catalytically active $[PdI_4]^{2-}$ anion on an imidazolium network grown on multi-walled carbon nanotubes (MWCNTs) through a simple two-stage procedure represented by Scheme 5.1.

The obtained catalysts, PdI_4@MWCNT-imi-X (with X = I, Br) have proven to be efficient heterogeneous catalysts in the important oxidative carbonylation process, namely, the oxidative amino-carbonylation of terminal alkynes to give high added value 2-ynamides.[19] According to the XPS analyses, the immobilisation procedure of the $[PdI_4]^{2-}$ on the MWCNT support was successful, in terms of the functionalisation of the CNT with a poly-ionic liquid network

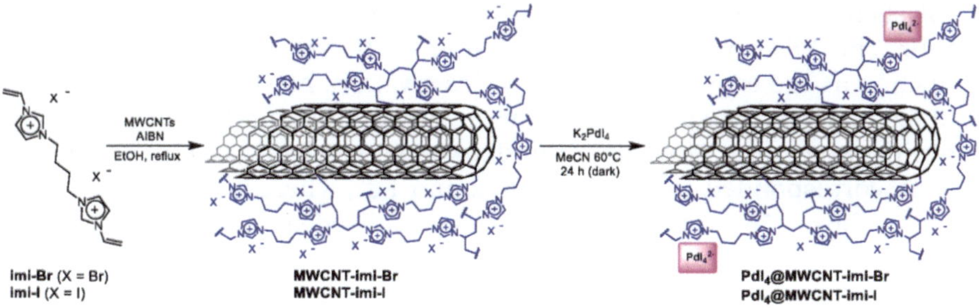

Scheme 5.1 Preparation of PdI_4@MWCNT-imi-X (X = Br, I) heterogeneous materials. Reproduced from ref. 19 with permission from Elsevier, Copyright 2022.

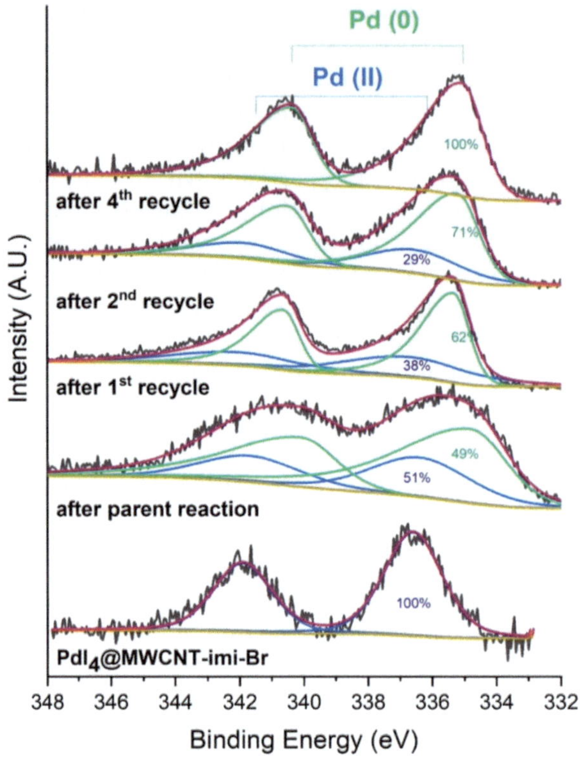

Figure 5.8 Pd 3d spectra of pristine PdI_4@MWCNT-imi-Br and of the catalyst after the parent experiment and the first, second, and fourth recycle under CO/air = 16/4 atm. Reproduced from ref. 19 with permission from Elsevier, Copyright 2022.

and the subsequent anchoring of the complex $[PdI_4]^{2-}$ without modification of the palladium oxidation state. In consecutive tests, after each cycle the catalysts lose part of their activity, to a major extent for the PdI_4@MWCNT-imi-Br. The XPS analyses performed after each cycle indicated palladium reduction as the cause of the catalyst deactivation. As shown in Figures 5.8 and 5.9, each of the spin–orbit components, Pd $3d_{5/2}$ and Pd $3d_{3/2}$, was fitted with two curves. The spin–orbit doublet with Pd $3d_{5/2}$ binding energy at 335 eV corresponded to metallic palladium Pd^0, whereas the doublet with Pd $3d_{5/2}$ above 336 eV corresponded to the oxidised state Pd^{2+}. The progressive reduction of palladium with each reaction cycle was evident. As clearly seen in Figure 5.9, the carbonylation conditions also had an important effect on the recyclability of the PdI_4@MWCNT-imi-I catalyst. The higher amount of oxygen when using CO/air = 32/8 atm in the reaction mixture, ensured a more efficient Pd^0 reoxidation.

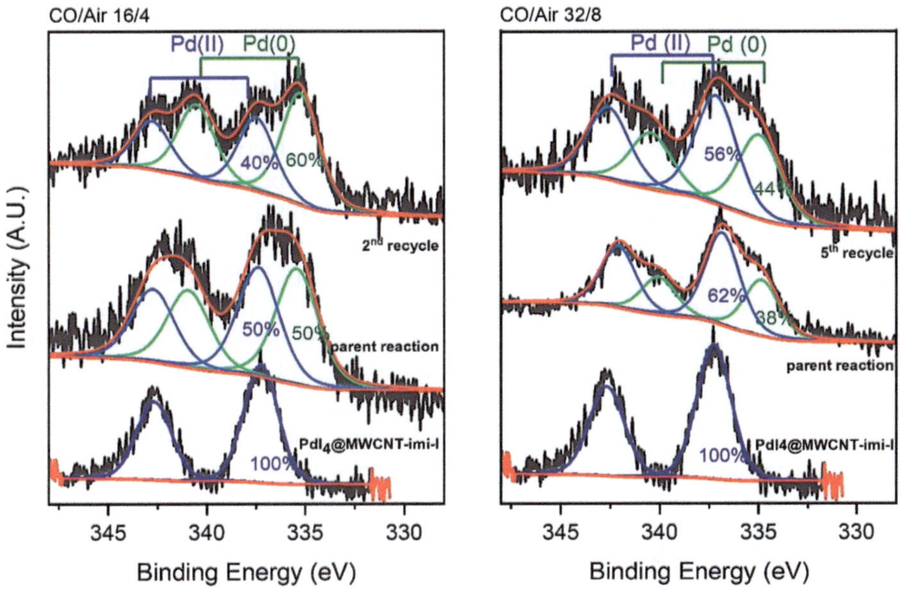

Figure 5.9 Pd 3d spectra of pristine PdI_4@MWCNT-imi-I and of the catalyst after the parent experiment and catalyst recycles under CO/air = 16/4 atm (left) and 32/8 atm (right). Reproduced from ref. 19 with permission from Elsevier, Copyright 2022.

5.3 Oxide Catalysts

Metal oxides play an important role in heterogeneous catalysis.[59,60] Besides being used as supports of active species as in the above-described cases, they can be used as catalytically active components. XPS analysis of metal oxide catalysts involves a careful evaluation of the binding energies of the metal centre, which often exists in different valence states, complicating the spectral component assessment. Accurate fitting of the O 1s photoelectron region would provide additional information on the bulk and surface species.[61] Moreover, some difficulty in obtaining reliable O 1s data may arise from hydroxylation and/or carbonation upon prolonged exposure to air.[62] The binding energy of the bulk contribution to the O 1s spectrum is controlled by the valence state mixing of oxygen and metal centres, whereas the energy of the surface species depends on the experimental environment the oxide has been exposed to. Generally, a large binding energy range (~3 eV) characterises the O 1s peaks associated with a formal charge of −2. The electronegative oxygen tends to extract electrons from the metal centres, resulting in high local electron densities at O atoms and consequent low O 1s binding energies of about 528.5 eV for alkali and alkaline-earth oxides, and around 530.5 eV for the transition metal systems. Main group elements, being less effective charge donating partners, give rise to higher O 1s binding energy positions, up to 533 eV for water and silicon oxides.[59] However, the simple relationship between charge variation on the ionised atom, q_a, and binding energy shift, DE, as derived from the charge potential model, is altered by two other important factors: the charge distribution of the atoms surrounding the ionised atoms, the so-called initial state effect, and the relaxation effect or final state effect related to the redistribution of the electronic charge during the photoionisation process.[3] Quite often the O 1s peak exhibits a weaker component on the high binding energy side of the main peak, between 530.5 eV and 531.1 eV due to –OH or to *per*-oxo O_2^{2-}. Other components between 531.1 eV and 532 eV are attributed to oxygen vacancies in the surfaces characterised by a lower electron density charge with respect to the O^{2-}.[59] However, some recent density functional calculations have reconsidered the chemical shift effect attributed to the oxygen vacancy.[63] As will be shown later the O 1s chemical shift attributed to oxygen vacancies helps to explain the reaction mechanism of important catalytic reactions. In the following, some selected XPS investigations of the most used oxides in catalysis are reported.

5.3.1 CeO_2-based Catalysts

In this section, catalytic systems based on CeO_2 as a carrier or as an additive will be illustrated from the XPS analysis point of view. Ce oxide with its high oxygen storage capacity (OSC), due to the rapid interchange within the Ce^{3+} and Ce^{4+} redox couple, provides oxygen vacancies, which are very important for certain environmental reactions involving substrate oxidation.[64–68] Our group has explored the role of CeO_2 as a support of nickel catalysts for the catalytic partial oxidation of methane (CPOM). In particular, the effect of the catalyst preparation method was investigated. Different syntheses allowed the tuning of the interaction between nickel and ceria by achieving small CeO_2 and NiO crystallites with different degrees of metal–support interaction.[69–71] A comparison between CeO_2 and SiO_2 supports, representative of active and inert oxides respectively, was also considered.[72,73] The whole regions containing Ni 2p and Ce 3d photoelectron spectra of two differently prepared nickel catalysts are shown Figure 5.10.

The Ni/CeO_2*MW*, was prepared by co-precipitation assisted by microwave. The Ni/CeO_2*NTA* was prepared by wet impregnating a previously prepared CeO_2 with nickel nitrate in the presence of the tridentate chelating agent, nitrilotriacetic acid (NTA).[71] The corresponding spectra of the aged samples, after the CPOM, are also shown. Using the CasaXPS software, the experimental spectra were fitted with many components of the Ni 2p and Ce 3d photoelectron transitions. The Ni 2p spectra of both samples, before and after the CPO reaction, were characterised by the two spin–orbit components of Ni $2p_{3/2}$ and Ni $2p_{1/2}$ with shakeup features 6 eV above the binding energy of the main peaks. The binding energy of Ni $2p_{3/2}$, located at around 854–855 eV, and the presence of the shakeup were indicative of NiO.[74] The Ce 3d spectra are always rather complex, due to multiplet splitting of the final states. All the possible electronic states arising from Ce(III) and Ce(IV) positive ions are

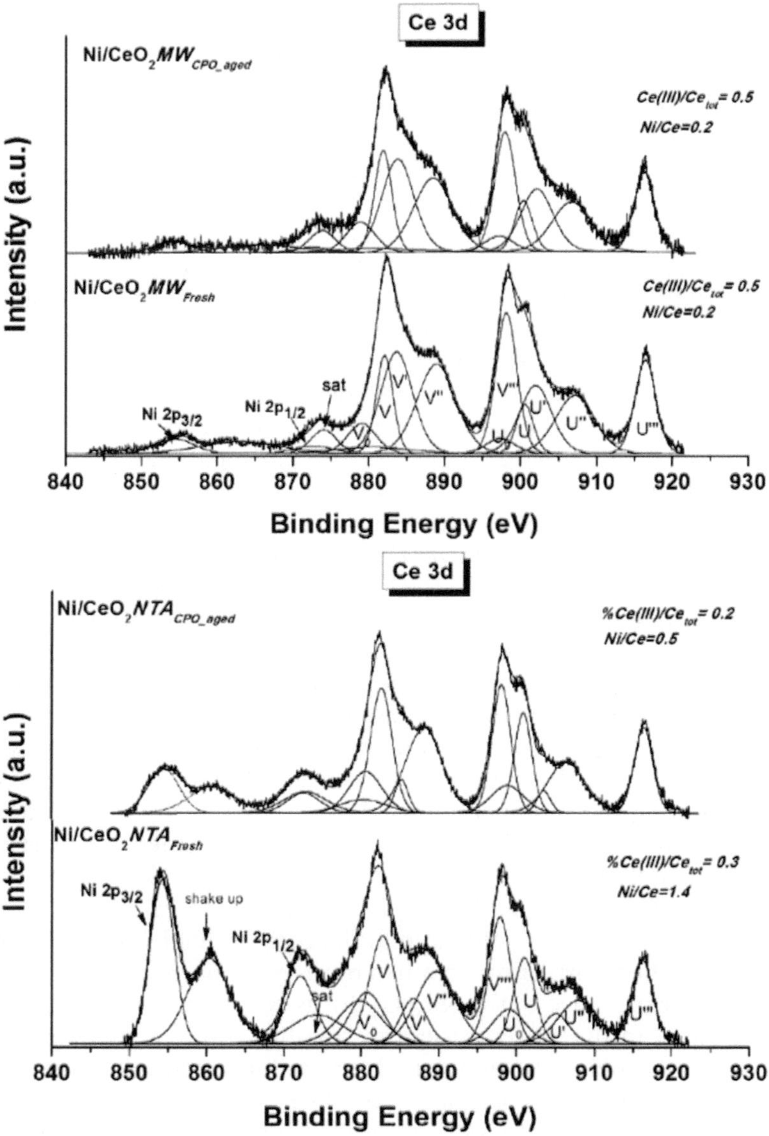

Figure 5.10 Ni 2p and Ce 3d photoelectron spectra of the catalysts as fresh and after the methane catalytic partial oxidation test. Reproduced from ref. 71, https://doi.org/10.3390/catal8050220, under the terms of the CC BY 4.0 license, https://creativecommons.org/licenses/by/4.0/.

labelled according to the conventional notation.[75] Curve fitting allowed the identification of ten peaks, six of which arise from Ce(IV) species and four from Ce(III) species. The two sets of spin–orbit multiplets, corresponding to the $3d_{3/2}$ and $3d_{5/2}$ states are labelled with the letters U and V respectively. U‴ and V‴ originate from the Ce $3d^9$ O $2p^6$ Ce $4f^0$ final states of Ce(IV). The additional states of Ce(IV), U, V, U″ and V″ result from a mixture of the Ce $3d^9$ O $2p^5$ Ce $4f^1$ and Ce $3d^9$ O $2p^4$ Ce $4f^2$ final states. The contribution of Ce(III) to the Ce 3d spectrum consists of two doublet pairs: U′, V′ and U_0, V_0. These doublets correspond to a mixture of the Ce $3d^9$ O $2p^5$ Ce $4f^2$ and Ce $3d^9$ O $2p^6$ Ce $4f^1$ final states. As the sum of the areas of the peaks corresponding to the Ce(IV) and Ce(III) ions is proportional to the concentration of ions, the concentration of the two oxidation states in the oxide sample could be obtained. For each set of data, the relative amount of Ce(III) with respect to the total Ce is indicated in Figure 5.10. Due to its high lattice and surface oxygen mobility, as reported in the literature, CeO_2 under the X-ray exposure inside the XPS apparatus may undergo reduction, giving rise to Ce(III) ions even in the fresh (calcined) samples.[76] Therefore, to have reliable information from spectra comparison, the same experimental conditions for the XPS analyses, in terms of X-ray power and exposure, were used. The Ni/Ce atomic ratios, as given from XPS quantitative analyses, are also indicated in each spectrum of Figure 5.10. As expected from the two different preparation procedures, the Ni/Ce ratio of the Ni/CeO_2NTA sample was much larger than the Ni/Ce ratio of the Ni/CeO_2*MW* sample and larger than the analytical ratio. Since the analytical composition was the same, the larger surface segregation of nickel on the former sample was evident, intrinsic to the wet-impregnation procedure. At the same time, the ratio Ce(III)/Ce_{tot} that was obtained for the NTA sample was about half that of the corresponding ratio obtained for the Ni/CeO_2*MW* catalyst. Since the exposure to the X-rays of the XPS measurements was the same, the larger ratio reflected a larger number of oxygen vacancies, which is very important for the catalytic promoting effect of CeO_2.[64] Moreover, it is worth noting the modification undergone by the Ni/Ce atomic ratios upon catalytic reaction. While a stable value was maintained in Ni/CeO_2*MW*, a decrease by more than 60% occurred in Ni/CeO_2*NTA*. The difference was attributable to the superior structural stability of the microwave-assisted co-impregnated sample, as compared to the complexing ligand wet-impregnated sample. The information obtained by the XPS analysis enabled the better catalytic performance of the NiCeO_2*MW* as compared to the NiCeO_2*NTA* to be attributed to the better structural stability and to the enhancement of the oxygen vacancies caused by the increased interaction between nickel and ceria driven by the preparation procedure. It is worth noticing that in this case, opposite to what was discussed about the supported metal, the better surface dispersion of the Ni as NiO at the surface of the carrier did not favour the catalytic performance. As confirmed by temperature-programmed reduction (TPR) analyses, the interaction between the two oxides was such to enhance the reducibility of the ceria and at the same time, to inhibit the reducibility of the Ni(II) species. In accord with the XPS results, despite this inhibited reducibility, the catalytic performance of this sample was improved by the increased number of oxygen vacancies accelerating the redox mechanism involved in the catalytic oxidation of methane.

A recent study of the role of the oxygen vacancies in CeO_2-catalysed styrene degradation will be described here with particular attention to the role of photoelectron spectroscopy in identifying the principal factors contributing to the oxidation activity.[77] CeO_2 catalysts with four different morphologies (sphere CeO_2–S, octahedral CeO_2–O, rod CeO_2–R and cube CeO_2–C) were synthesised and their catalytic oxidation performance for styrene oxidation was evaluated. According to the characterisation results, lattice distortion was the key factor to form oxygen vacancies on CeO_2 catalysts with the four different morphologies. The oxygen vacancies may adsorb and activate more oxygen to produce surface adsorbed oxygen, and also adsorb more styrene to be activated. The Ce 3d and O 1s spectra are shown in Figure 5.11. As seen in Figure 5.11a, the Ce 3d was fitted with 8 peaks corresponding to some of the multiplet splitting of the Ce $3d_{5/2}$ and Ce $3d_{3/2}$ labelled by the authors with the notation γ and λ respectively. The peaks corresponding to the Ce(III) species were marked as γ' and λ'. In Figure 5.11b the O 1s spectra were divided into four peaks. The four peaks were assigned to lattice oxygen (O_{latt}, O^{2-}) at 529.2 eV, surface and adsorbed oxygen species (O_{ads}, O_2^- or O_2^{2-}) at 529.8 eV and 531.8 eV, respectively, and surface hydroxyl oxygen (O_{ads}O–H) at 533.3 eV. The $O_{ads}/(O_{latt} + O_{ads} + O_{ads}$O–H) ratio provided a measure of the concentration of surface oxygen vacancies. According to the Ce 3d and O 1s XPS results, the concentration of surface oxygen vacancies was found to be in the order of CeO_2–C < CeO_2–O < CeO_2–R < CeO_2–S. A linear relationship, confirmed also by the UV-Raman spectra, was found between the oxygen vacancy concentration and the amount of surface adsorbed oxygen species, and between

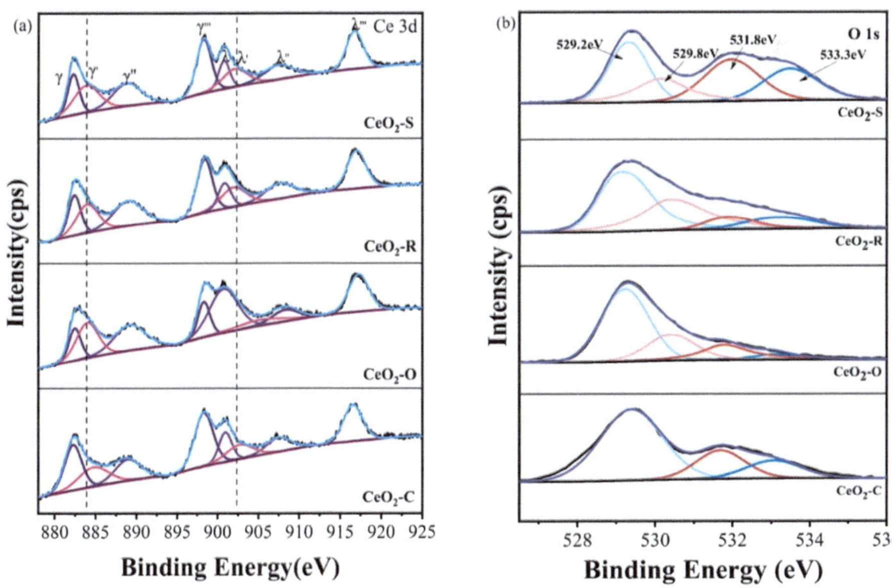

Figure 5.11 XP spectra of fresh CeO_2–S, CeO_2–R, CeO_2–O and CeO_2–C samples. (a) Ce 3d XP spectra of samples; (b) O 1s XP spectra of samples. Reproduced from ref. 77 with permission from Elsevier, Copyright 2022.

the amount of surface adsorbed oxygen species and the styrene oxidation rate. Among all samples, CeO_2–S showed the highest catalytic activity (T_{50} = 133 °C and T_{90} = 184 °C, with T_{50} and T_{90} representing the temperatures at which 50% and 90% of conversion were registered) due to its highest oxygen vacancy concentration.

The experimental results were supported by DFT theoretical calculations demonstrating the role of defects and crystal planes in the catalytic oxidation of styrene and revealing the intermediates of the catalytic oxidation of styrene. From the XP spectra of the used catalysts, given in Figure 5.12, the decrease of the Ce^{3+} component after catalytic reaction was clear, demonstrating the involvement of the Ce^{4+} in the reaction mechanism. Combining the XPS analysis with other techniques, the study provided the understanding of the relationship between lattice distortion and oxygen vacancy formation and identified the role of oxygen vacancies in the mechanism of styrene degradation through sequential oxidation to CO_2.[77]

Thanks to the technological advancement of the last decade, it has become possible to use near ambient pressure X-ray photoelectron spectroscopy (NAP-XPS) to investigate heterogeneous catalysts under conditions closer to the operative ones, *i.e.*, in *operando* conditions.[16] A quite exhaustive review of the application of NAP-XPS to ceria-based catalysts has been recently published.[78] An interesting case study deals with the important issue of the oxidation of soot particles. Such particles would consist of an amorphous

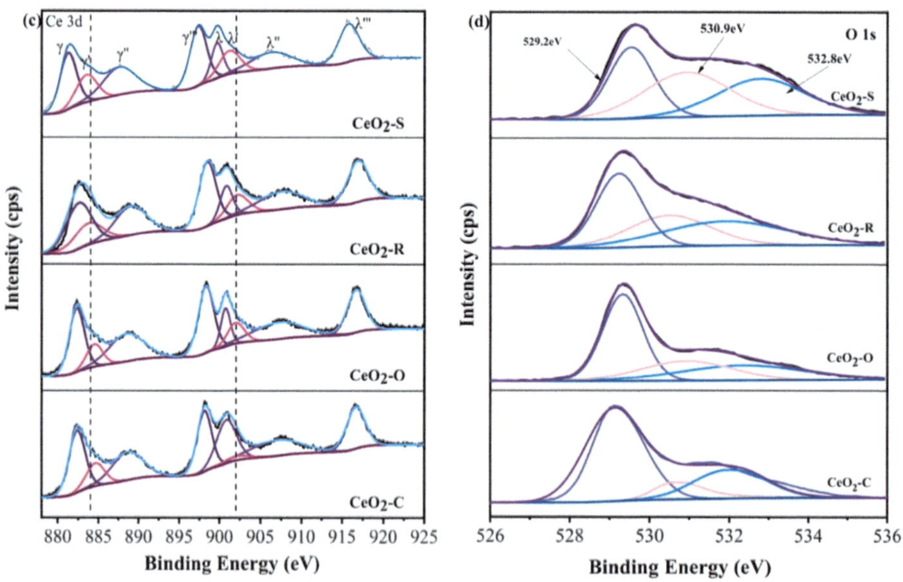

Figure 5.12 XP spectra of CeO_2–S, CeO_2–R, CeO_2–O and CeO_2–C samples after catalytic oxidation of styrene. (c) Ce 3d XP spectra of samples; (d) O 1s XP spectra of samples. Reproduced from ref. 77 with permission from Elsevier, Copyright 2022.

carbon core of few nanometres surrounded by a graphitic shell. To achieve total soot combustion in a diesel engine it is necessary to develop catalysts able to lower the ignition temperature. In this respect ceria-based catalysts were found to be quite promising.[79,80] It is generally assumed that the soot combustion reaction follows the Mars–van Krevelen mechanism, in which lattice oxygen atoms of the outermost layers of ceria are transferred onto soot, and the subsequent exposure to gaseous O_2 fills up the vacancies created on the oxide.[80] It is claimed that paramagnetic O_2^- superoxide and diamagnetic O_2^{2-} peroxide species are formed when reduced CeO_{2-x} is exposed to molecular O_2. The spillover of these species onto the soot surface would be responsible for the soot oxidation.[81] By means of NAP-XPS, Soler *et al.* provided direct evidence of the redox chemistry and the influence of the reaction conditions in the oxidation of carbon soot over ceria-based catalysts.[79] They investigated a sample of pure ceria and a sample of $Ce_{0.8}Zr_{0.2}O_2$, mixed with soot, which were subjected to increasing temperatures (from 300 to 823 K) under 1 mbar argon atmosphere in a first set of experiments (with a final replacement of Ar with O_2 at 823 K), and under 1 mbar oxygen atmosphere in a second set of measurements. Heating ceria under an inert atmosphere, such as Ar, resulted in the formation of oxygen vacancies and the concurrent reduction of Ce^{4+} to Ce^{3+}, increasing with the temperature. However, when samples were treated with O_2, the amount of Ce^{3+} species remained low over the entire temperature range, since molecular oxygen rapidly reacted with the oxygen vacancies created at the ceria–soot interface. Interesting results were obtained when comparing the behaviour of CeO_2–soot and $Ce_{0.8}Zr_{0.2}O_2$–soot systems, since the amount of Ce^{3+} species upon heating under an Ar atmosphere was higher for the Zr-doped sample. Figure 5.13 shows Ce 3d (a, b) and O 1s spectra (c, d) of the $Ce_{0.8}Zr_{0.2}O_2$–soot sample recorded at 823 K under an argon environment (a, c) and after replacing Ar with O_2 at the same temperature (b, d). As expected, the replacement of the atmosphere caused an immediate and huge decrease of the Ce^{3+} amount (Figure 5.13a and b), but this phenomenon was also detected through O 1s spectra, which showed three different components. The fitted peaks at binding energies of 528.7 eV and 530.2 eV (Figure 5.13c and d) were attributed to ceria lattice oxygen and surface oxygen atoms with low coordination due to vacancy formation, respectively, and the peak at higher binding energy (532.2 eV) corresponded to superoxide species O_2^-. When the argon gas phase was replaced by O_2, cerium(III) oxide re-oxidised and the relative O 1s signal intensities changed noticeably. The O 1s peak of the surface oxygen associated with Ce^{3+}, located at 530.2 eV, decreased in intensity (Figure 5.13d). Concurrently, the signal at 532.2 eV attributed to superoxide species became more intense, demonstrating that these active oxygen species resulted from the reaction between molecular O_2 and oxygen vacancies. Therefore, the NAP-XPS technique allowed the authors to demonstrate that soot oxidation over ceria-based materials involved a two-way cooperative mechanism: on one side, the formation of oxygen vacancies and Ce^{3+} species at the ceria–soot interface caused by the reducing action by soot, and, on the other side, the oxidation

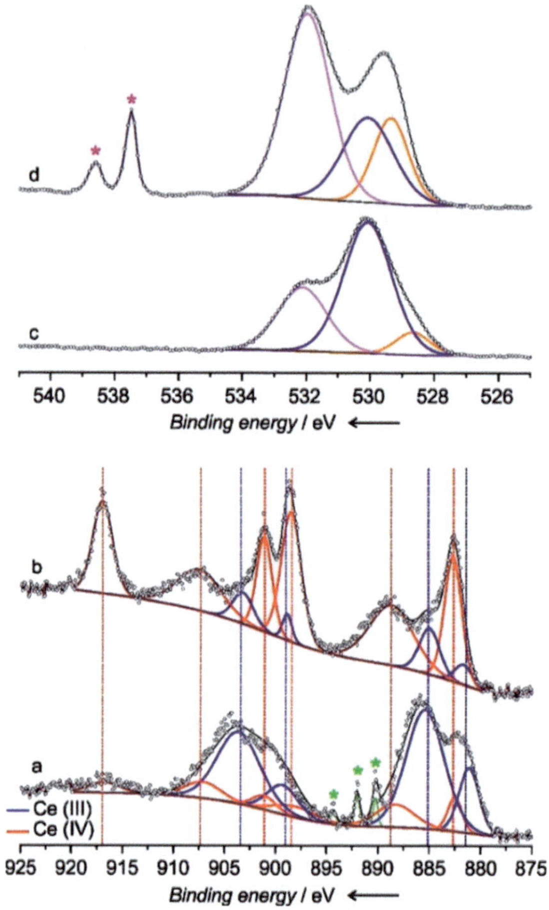

Figure 5.13 Ce 3d (a) and (b) and O 1s spectra (c) and (d) recorded over $Ce_{0.8}Zr_{0.2}O_2$–soot at 823 K under an argon atmosphere (a) and (c) and by replacing Ar with O_2 (b) and (d). Green *, Argon gas phase; pink *, oxygen gas phase. Reproduced from ref. 79 with permission from John Wiley & Sons, Copyright 2016.

of the soot surface by active superoxide species, generated from the reaction between the oxygen vacancies and gaseous O_2. Both routes occurred simultaneously and reinforced each other.

Another important process where CeO_2 is used as a promising support is the dry reforming of methane (DRM) consisting of the following reaction:[78,82]

$$CH_4 + CO_2 \rightleftharpoons 2CO + 2H_2 \left(\Delta H_{298K} = +247\ \text{kJ mol}^{-1}\right)$$

Along with the partial oxidation (CPO) and the steam reforming of methane (SRM), DRM is one of the processes producing synthesis gas, or *syngas* (CO/H_2), an important precursor for several industrial processes. As compared to the other two, besides the benefit of converting two greenhouse gases, the DRM reaction has the advantage of yielding syngas in a 1 : 1 molecular ratio, ideal for the Fischer–Tropsch synthesis of fuels and for the synthesis of oxygenates.[83] However, the high stability of the C–H bond and the large activation energy of the CO_2 require efficient catalysts that would not sinter at elevated temperatures and would not accumulate carbon on their surfaces. Although noble metals are the most active, they easily sinter at the high temperature of the reaction. Nickel, on the other hand, due to its low cost, even if less active, represents a good alternative. By adjusting the supports, and with different syntheses and addition of promoters, it has been possible to obtain active and stable nickel catalysts.[84–86] To understand the chemistry of the DRM process, the interaction of CH_4, CO_2 and CH_4/CO_2 mixture with a model catalyst of Ni nanoparticles dispersed on a CeO_{2-x} (111) was studied by NAP-XPS.[87]

The relevant XP spectra obtained at the Synchrotron Facility at Brookhaven National Laboratory are shown in Figure 5.14. The spectra refer to a model catalyst $Ni/CeO_2(111)$, with weight composition 20%NiO/80%CeO_2, formed by small Ni particles dispersed on a solid solution of $Ce_{1-x}Ni_xO_{2-y}$, under a mixture of methane and carbon dioxide in a range of temperatures between 300 and 700 K.[88] It is worth noticing that in this study, carried out with synchrotron radiation, the Ce 4d and Ni 3p levels were analysed instead of the more generally used Ce 3d and Ni 2p core levels. At 300 K, the C 1s region exhibits peaks of the gaseous reactants, CO_2 at 294 eV and CH_4 at

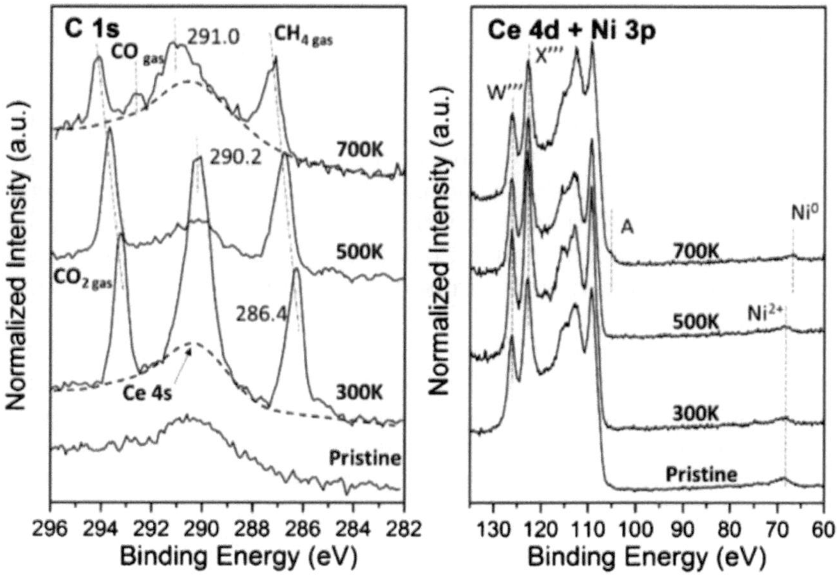

Figure 5.14 C 1s and Ce 4d + Ni 3p spectra of the $Ni/CeO_2(111)$ catalyst (Θ_{Ni} ~ 0.1 ML) surface under 100 mTorr CH_4 and 100 mTorr CO_2 at different temperatures. Reproduced from ref. 87 with permission from John Wiley & Sons, Copyright 2016.

287 eV, and features at 285–286 eV of adsorbed CH_x species and at 290.2 eV of adsorbed CO_x. The presence of the adsorbed species indicated that the catalyst was able to dissociate methane at room temperature. Increasing the temperature, the adsorbed CO_x and CH_x species disappeared and up to 500 K no catalytic activity was observed. At 700 K only adsorbed gas species of CH_4, CO and CO_2 remained. At this temperature the system became catalytically active, with the production of gaseous H_2 and CO and reduction of Ni^{2+} to Ni^0 and of Ce^{4+} to Ce^{3+}. The comparison of these results with those obtained under the catalyst exposure to only methane indicated that the carbon arising from the decomposition of methane would react with the O centres of the CeO_2, explaining the XPS derived larger ratio of Ce^{3+}/Ce^{4+} obtained in the case of the methane-only exposure.[89] It is worth noting that no C 1s peak below 284 eV attributable to elemental carbon or nickel carbide was observed, confirming that contrary to other supported nickel catalysts, the strong interaction between nickel and ceria favoured methane decomposition and carbon removal by the reaction of the CH_4 decomposition carbon with the O atoms of the catalysts. The formation of the solid solution $Ce_{1-x}Ni_xO_{2-y}$ with adjacent Ni and O atoms as adsorption sites favoured the catalytic process.

5.3.2 VO_x-based Catalysts

Vanadium oxide, generally written as VO_x, is an important material for many environmental and industrial processes such as selective catalytic reduction of NO_x with ammonia, or non-oxidative or oxidative dehydrogenation of propane to cite a few.[88–93] A crucial point for the understanding and optimisation of its catalytic activity is the identification of the vanadium oxidation state. The principal oxidation states in its oxides are V^{2+} (VO), V^{3+} (V_2O_3), V^{4+} (VO_2) and V^{5+} (V_2O_5).[94,95] The XP spectra of the V 2p core level, characterised by the two spin orbit components, V $2p_{3/2}$ and V $2p_{1/2}$, provide information on the vanadium species present at the oxide surface. As mentioned for the metal catalysts, to increase the surface area available for the catalytic reaction, the VO_x is generally supported on an *inert* oxide. The loading of the vanadium oxides and the support nature and morphology are tuned to optimise the catalytic behaviour.

In a recent study, vanadium supported by a wet impregnation method on a hexagonal mesoporous SiO_2 was used as a catalyst for non-oxidative propane dehydrogenation to propylene (PDH).[91] The catalytic activity of the catalysts denoted as *n*V-HMS with *n* representing the wt% of vanadium, was evaluated as a function of different VO_x loadings. The V 2p experimental spectra with the fitted curves are shown in Figure 5.15 and the results are summarised in Table 5.2 for the as-prepared samples. Three binding energies arising from the fitted curves, at around 518.5 eV, 517.3 eV and 516.5 eV, were assigned respectively to VO_x species with low polymerisation degree such as *mono* or *low* polyvanadate structures with highly dispersed V^{5+} species interacting with SiO_2, to bulk V^{5+} species including clusters of V_2O_5, and to V^{4+} and/or V^{3+}

species.[96] The photoelectron spectra indicated the coexistence of vanadium species in different chemical oxidation states. According to the normalised proportion of VO_x species, obtained from the intensity ratios of the V $2p_{3/2}$ components, listed in Table 5.2, a larger percentage of highly dispersed V^{5+} species was maintained up to a vanadium loading of 8 wt% (Figure 5.15c). The percentage of the bulk V^{5+} species with binding energy at around 517.3 eV rose substantially when the vanadium loading increased further up to 15 wt% (Figure 5.15e). In accord with the other characterisation results such as

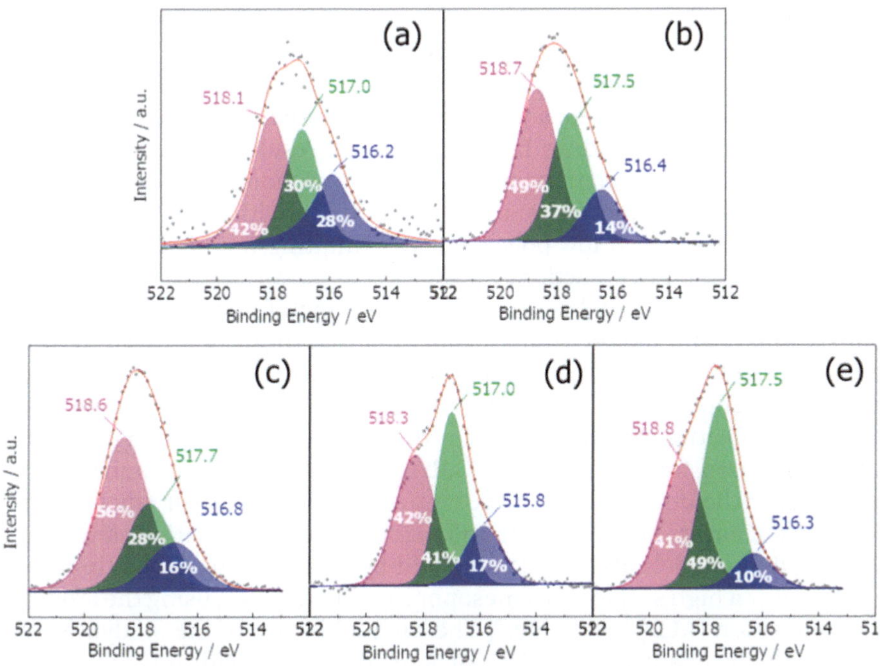

Figure 5.15 Deconvolution of the V $2p_{3/2}$ signal of as-prepared (a) 3V-HMS, (b) 5V-HMS, (c) 8V-HMS, (d) 10V-HMS and (e) 15V-HMS. Reproduced from ref. 91 with permission from Elsevier, Copyright 2022.

Table 5.2 V $2p_{3/2}$ signal deconvolution results of as-prepared V-HMS.

	V/Si mole ratio		V $2p_{3/2}$ binding energy (eV) (VO_x species proportion)		
Catalyst	By ICP	By XPS	V^{5+}	V_2O_5	V^{4+}, V^{3+}
1V-HMS	0.01	0.008	—	—	—
3V-HMS	0.03	0.02	518.1 (42%)	517.0 (30%)	516.2 (28%)
5V-HMS	0.06	0.03	518.7 (49%)	517.5 (37%)	516.4 (14%)
8V-HMS	0.10	0.06	518.6 (56%)	517.7 (28%)	516.8 (16%)
10V-MS	0.10	0.07	518.3 (42%)	517.0 (41%)	515.8 (17%)
15V-MS	0.20	0.10	518.8 (41%)	517.5 (49%)	516.3 (10%)

UV-vis spectroscopy and H_2-TPR, the XPS analysis proved that V_2O_5 clusters, characterised by the V $2p_{3/2}$ component at 517.3 eV, were formed on the surface of 15V-HMS.

Moreover, as reported in Table 5.2, the XPS-derived V/Si mole ratios are lower than those derived from ICP analysis in the as-prepared catalysts, suggesting that VO_x species were dispersed mainly inside HMS and for this reason not accessible to the XPS information depth.[97] When V loading increased from 3 wt% to 15 wt%, the fraction of V^{4+}/V^{3+} species decreased from 28% to 10%, while that of V^{5+} species as clusters of V_2O_5 increased from 30% to 49%. According to the XPS analysis, monovanadate and low-polyvanadate coexisted at 3 wt% and 5 wt% V loading. 8V-HMS possessed the highest amount of low-polyvanadate compared with the other samples. With further increase of the amount of vanadium, the polymerisation degree of vanadium oxide species continually increased, and V_2O_5 clusters, as the main species, existed at 15 wt% of vanadium loading. In conclusion, the results from the XPS analyses confirmed a direct relationship between the vanadium loading, the dispersion of VO_x species and the catalytic performance, with the 8V-HMS in the form of highly dispersed low-polyvanadate exhibiting the highest catalytic performance in the PDH reaction.

Regarding environmental concerns, the emission of nitrogen oxide (NO_x) from stationary and mobile sources should be significantly reduced to mitigate potential health risks to humans and other living organisms. Selective catalytic reduction of NO_x with NH_3 (NH_3-SCR) is nowadays becoming one of the most effective catalytic routes for this objective.[88,98,99] Among the catalysts being used for this process, VO_x-based ones are quite promising. As discussed for the other catalytic process described above, their efficiency is related to the nature of the support and to the vanadium oxidation state. In a recent study, a model catalyst was prepared by consecutively depositing VO_x and TiO_2 over a high surface area mesoporous silica SBA-15, using the atomic layer deposition (ALD) technique.[90] The main objective of the study was to elucidate the mechanism of the NH_3-SCR, involving the acid sites of the catalyst, Brønsted or Lewis, where the NH_3 would adsorb, and the reaction of the NO with the adsorbed ammonia, accompanied by a redox process. In all these processes the role of the VO_x surface structure was fundamental. The authors used several techniques such as Raman, *in situ* diffuse reflectance infrared Fourier transform spectroscopy (DRIFTS), XRD and XPS to explore the nature of the surface sites. The contribution given by the XPS surface analysis is shown in Figure 5.16. The spectra refer to the sample obtained from 3 ALD cycles of titania deposited over SBA-15 followed by VO_x deposition from 3 cycles leading to a V:Ti loading of 5.1:5.9. The top spectrum of Figure 5.16 contains the binding energy region of the O 1s, V $2p_{1/2}$, and V $2p_{3/2}$ photoemission peaks located at 532.9 eV, 524.9 eV, and 517.4 eV, respectively. At lower energy the spectrum of the Ti 2p core level is present. As expected, the spectral region of Figure 5.16 is dominated by the O 1s contribution at 532.9 eV that, by comparison with the literature, was attributed to SiO_2.[68] Detailed analysis with least-squares fitting of the V $2p_{3/2}$ region, based on

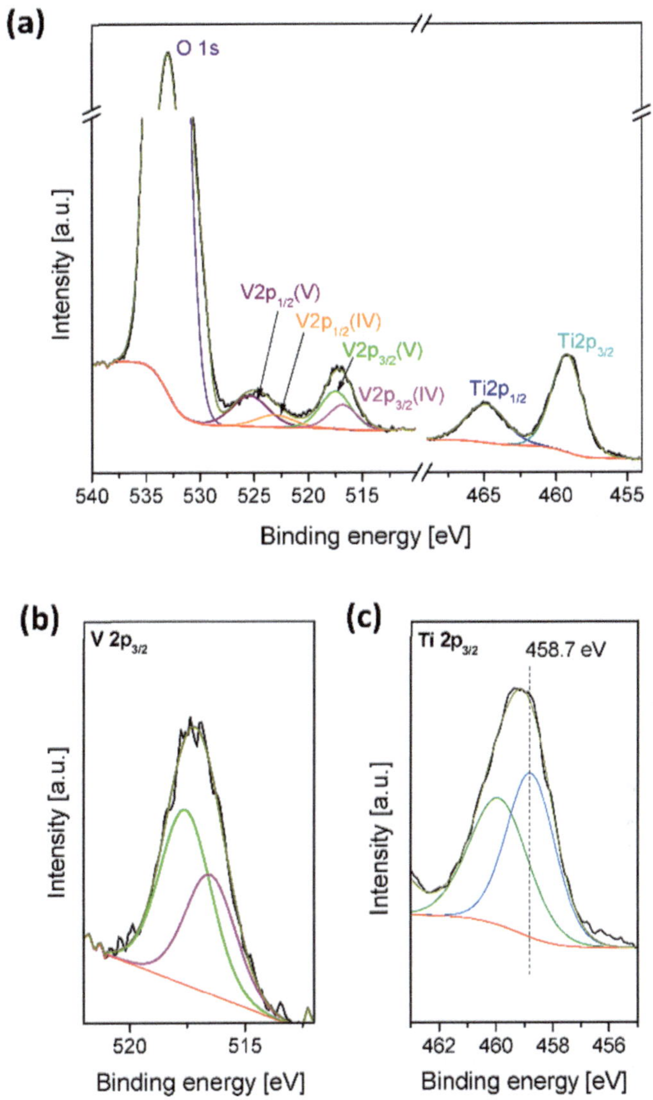

Figure 5.16 (a) O 1s, V 2p, and Ti 2p photoemission of TiO_2/SBA-15+3$x$$VO_x$. Detailed view of (b) the V $2p_{3/2}$ region and (c) the Ti $2p_{3/2}$ region. The coloured lines represent the result of a least-squares fit analysis. Reproduced from ref. 89, https://doi.org/10.3390/catal10121386, under the terms of the CC BY 4.0 license, http://creativecommons.org/licenses/by/4.0/.

literature data for binary vanadium oxides (V_2O_5, V_2O_4, V_2O_3), revealed the presence of V^{5+} at 517.6 eV (60%) and V^{4+} at 516.3 eV (40%), while a small contribution from V^{3+} at 515.7 eV could not be ruled out.[59–69]

The Ti $2p_{3/2}$ peak was deconvoluted into two contributions located at 458.7 eV and 459.9 eV, which were attributed to TiO_2 domains and to

Ti–O–Si bonds, respectively.[100,101] The NO_x conversion followed a volcano shape with an increase from 100 °C to 250 °C, a maximum conversion level at 250–350 °C, and a decrease above 350 °C. At high temperatures a decrease of both NO_x conversion and N_2 selectivity was observed, which was attributed to increased NH_3 oxidation leading to the formation of N_2O. *Operando* DRIFT experiments enabled the detection of the main role of the Lewis acid sites where the NH_3 would adsorb followed by extraction of hydrogen and subsequent interaction with NO leading to the intermediate NH_2NO followed by N_2 and H_2O formation.[89] The high NO_x conversion was attributed to the presence of highly dispersed VO_x species and to the moderate acidity of VO_x supported on TiO_2/SBA-15. The presence of amorphous domains such as Ti–O–Si and V–O–Ti would improve the good N_2 selectivity at the expense of the N_2O formation favoured by bulk like V_2O_5.[102]

Moreover, the good dispersion of the active species on similar types of catalysts prepared by the ALD technique was measured by XPS of adsorbed NH_3.[103] Using NH_3 as a basic probe molecule, only strong and medium strength surface acid sites could be qualified and quantified, because weak acid sites would not retain ammonia under the ultra-high vacuum (UHV) conditions of the XPS apparatus. The N 1s spectra of TiSi, ALD A-VTiSi and impregnated I-VTiSi are presented in Figure 5.17.[104] From previous literature, only Lewis

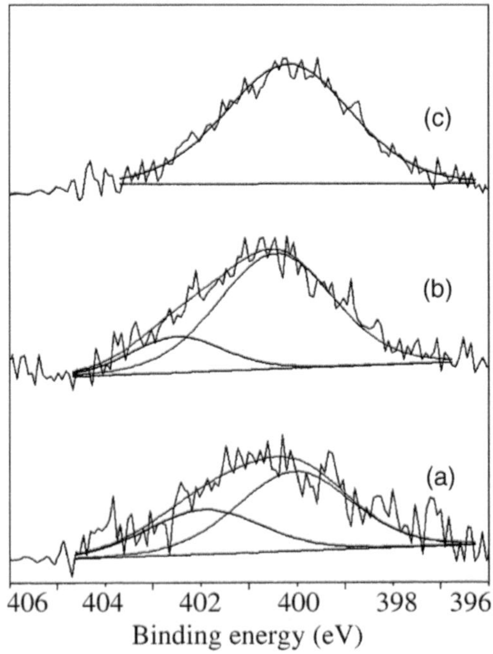

Figure 5.17 N 1s XP spectra obtained after ammonia adsorption at 353 K on titania/silica and vanadia catalysts supported on TiO_2/SiO_2: (a) TiSi; (b) A-VTiSi; (c) I-VTiSi. Reproduced from ref. 104 with permission from Elsevier, Copyright 2003.

acidity was detected.[105] As shown in Figure 5.17, the N 1s XP spectra could be decomposed into two peaks related to the nature of the acid sites; the Brønsted acid sites were associated with the high binding energy peak of the N 1s envelope at ~402 eV and the Lewis acid sites with the low binding energy peak at ~400 eV. According to the spectra in Figure 5.17, a larger number of Brønsted acid sites were present on the ADL prepared catalyst as compared to the impregnated one.

The titania/silica support, as seen in Figure 5.17a exhibited rather strong Brønsted acid sites. Upon addition of vanadia by ALD, the relative contribution of strong Lewis acid sites increased from 65% to 80%. However, when vanadia was deposited by impregnation, only one peak due to the Lewis acid sites was detected. The absence of strong Brønsted acidity was due to the overall less acidic character of the I-VTiSi catalyst, as confirmed also by microcalorimetry, and arose from the change in the support nature toward a more 'titania-like' behaviour by agglomeration.

5.4 Sulfide-based Catalysts

Due to legal restrictions aiming to improve environmental protection, industry is nowadays forced to replace fossil fuel energy sources with more green alternatives such as solar, wind and green hydrogen-derived energy sources. However, it is foreseeable that for the next 30–50 years to come, fossil fuels will continue to be used at least for long distance transportation. Within this scenario the development of highly active and selective hydrodesulfurisation (HDS) catalysts for the removal of sulfur from heavy oil will continue to be an important and quite challenging issue for the petroleum industry.[106–110] Supported molybdenum sulfide catalysts promoted by cobalt and nickel have been widely used in industrial HDS processes.[111] The active species in the process are represented by sulfides and for the investigation of these chemical species, XP spectroscopy has played an important role. A study from our group is here reported to show the main information derived from the XPS analysis.[112] The aim of the study was to investigate the effect of the support composition in terms of alumina and silica and then the effect of addition of sodium on the HDS of thiophene.[113] Modification of the support acidity by changing the Al/Si ratio produced changes in Brønsted/Lewis acidity with consequent variation in the metal–support interaction and on the catalytic activity. The best activity was obtained in correspondence to Al/Si = 2.3, affording a good compromise between the Brønsted acidity of the pure silica leading to weaker interaction between metal and support, and the Lewis acidity of the pure alumina leading to strong metal–support interaction with consequent decrease of the metal reducibility. Mo–O–X links, with X = Al or Si, better dispersed on the support surface as compared to the pure silica case and more reducible species during the sulfiding treatment with respect to the pure alumina case were obtained. XPS characterisation identified

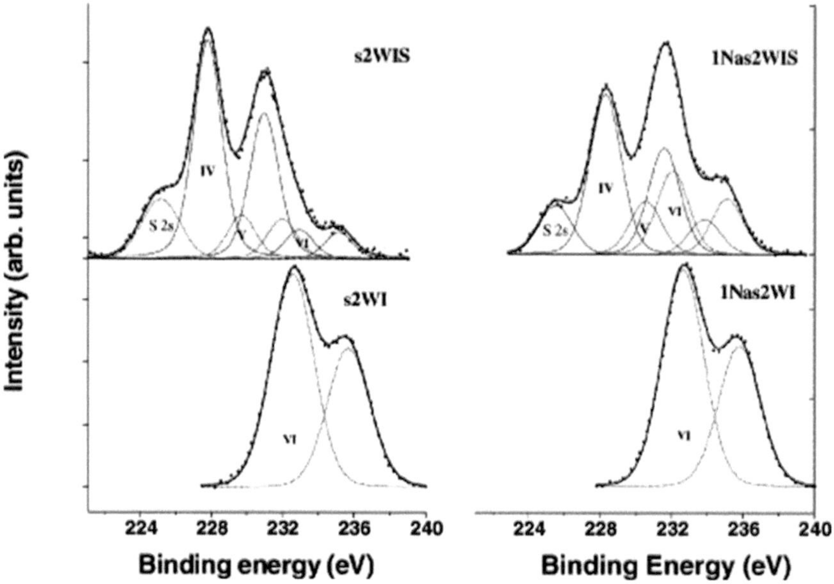

Figure 5.18 Mo 3d and S 2s photoelectron spectra of fresh and aged s2WI and 1Nas2WI samples. Reproduced from ref. 112 with permission from Elsevier, Copyright 2004.

the chemical species and the changes occurring during the HDS reaction. In Figure 5.18, spectra of Mo 3d of fresh and aged samples are shown. The Mo 3d spectra of the fresh samples were characterised by the two spin–orbit components, $3d_{5/2}$ and $3d_{3/2}$, with an energy separation of 3.1 eV and intensity ratio of 1 : 5.[110] No significant changes of the Mo 3d binding energy with the different support composition were found. Binding energies were typical of Mo(VI). However, the broadness of the peaks was an indication of several Mo(VI) oxo-species. As seen in Figure 5.18, the presence of 1 wt% Na did not affect the Mo 3d spectra, except for a narrowing of the peaks in the case of the silica-rich samples attributed to the sodium-induced transformation of polymolybdate clusters into monomeric units. All the catalysts were surface enriched in Mo and Co with respect to the nominal concentration. The sodium surface concentration as detected by XPS was in accord with its analytical content, indicating a rather uniform distribution through the samples. As seen in Figure 5.18, the spectra of the samples after being used in the HDS reaction changed significantly. Concerning the Mo 3d region, the spectra of the aged samples became quite complex. The region also included the S 2s photoelectron peak centred at 225.6 eV and three Mo 3d doublets.[114] The Mo 3d doublet characterised by the spin orbit coupling component Mo $3d_{5/2}$ at the highest energy ~232 eV, was attributed to unreduced Mo(VI).

The Mo 3d doublet with a Mo $3d_{5/2}$ component at ~230 eV was indicative of an intermediate oxidation state which was not clearly identified. According to the literature it could have been attributed to Mo(V). The presence of mixed oxy-sulfide, –S–Mo^{6+}–O– could not be excluded. Sulfiding at low temperature might proceed through an exchange between oxygen and sulfur. The replacement of the oxygen atoms by the less negative sulfur atoms produced a decrease in the Mo(VI) binding energy.[116] This replacement explained also the small down energy shift of the binding energy value of the high energy peak of the samples after HDS. The most intense peak at ~228 eV was attributed to Mo(IV) possibly as an oxide or as a sulfide.[114] The corresponding Co 2p spectra of the fresh samples were characterised by two spin orbit components, Co $2p_{3/2}$ and Co $2p_{1/2}$, about 16 eV apart. Each component was accompanied by satellite peaks (shake up peaks), 6 eV above each main peak, typical of Co(II) species.[74] After HDS reaction, a new spin orbit doublet, characterised by the Co $2p_{3/2}$ at 777.6 eV appeared. Such a value could have indicated Co(0) or Co_9S_8.[115] However, considering that the XPS-derived intensity ratio S/Mo(IV) was in accord with the presence of MoS_2, the Co 2p low energy peak was attributed to metallic cobalt Co(0). In another study a correlation between the S/Mo(IV) ratio and the hydrodesulfurisation activity of a silica supported CoMo catalyst was demonstrated.[114]

The importance of the extent of sulfidation in the HDS activity of CoMo catalysts was recently investigated using XPS.[116] In their work the authors demonstrated that by using appropriate background removal models it was possible to estimate with great accuracy the extent of MoS_2 crucial for the catalytic performance of the HDS catalysts.

5.5 Final Remarks

This chapter intended to demonstrate the valuable contribution XP spectroscopy has made to the investigation of solid catalysts and to the understanding of heterogeneous catalytic processes. A large variety of catalytic materials are being developed to make chemical processes more efficient and selective. Examples from three main categories of catalysts have been considered here: (i) supported metals; (ii) supported oxides; (iii) sulfide-based compounds. For the supported metals, studies of noble metals such as Au, Pd and Pt, have been discussed, showing the importance of the surface distribution of the active metal and the modification of the oxidation state upon alloying with another noble metal and after the catalytic reaction. For the supported oxides, case studies on CeO_2 and V_xO_y, used in oxidation reactions and in the selective catalytic reduction of ammonia (SCR-NH_3) respectively, were reported. In particular, the capability of XPS analysis to detect oxygen vacancies, important for the oxidation activity of reducible oxides like CeO_2 and the capability to discriminate Brønsted and Lewis acid sites for V_xO_y was outlined. Finally, a case study on the application of XPS to sulfide-based catalysts such as CoMoS was described. The great potential of near ambient pressure XPS (NAP-XPS) as an *operando* technique in catalysis has also been highlighted.

References

1. G. A. Somorjai, *Introduction to Surface Chemistry and Catalysis*, John Wiley & Sons Inc., New York, 1994.
2. C. Lucarelli and A. Vaccari, *Green Chem.*, 2011, **13**, 1941.
3. A. M. Venezia, *Catal. Today*, 2003, **77**, 359.
4. M. A. Isaacs, J. Davies-Jones, P. R. Davies, S. Guan, R. Lee, D. J. Morgan and R. Palgrave, *Mater. Chem. Front.*, 2021, **22**, 7931.
5. A. M. Tarditi, M. F. Mori and L. M. Cornaglia, *Top. Catal.*, 2019, **62**, 822.
6. B. Kobe, M. Badley, J. D. Henderson, S. Anderson, M. C. Biesinger and D. Shoesmith, *Surf. Interface Anal.*, 2017, **49**, 1345.
7. J. Sobczak, I. Wysocka, S. Murgrabia and A. Rogala, *Energies*, 2022, **15**, 5420.
8. L. Nguyen, F. F. Tao, Y. Tang, J. Dou and X. J. Bao, *Chem. Rev.*, 2019, **119**, 6822.
9. A. Bergmann and B. Roldan Cuenya, *ACS Catal.*, 2019, **9**, 10020.
10. L. Lukashuk, N. Yigit, R. Rameshan, E. Kolar, D. Teschner, M. Hävecker, A. Knop-Gericke, R. Schlögl, K. Föttinger and G. Rupprechter, *ACS Catal.*, 2018, **8**, 8630.
11. S. Blomberg, U. Hejral, M. Shipilin, S. Albertin, H. Karlsson, C. Hulteberg, P. Lomker, C. Goodwin, D. Degerman, J. Gustafson, C. Schlueter, A. Nilsson, E. Lundgren and P. Amman, *ACS Catal.*, 2021, **11**, 9128.
12. L. Zhong, D. Chen and S. Zafeiratos, *Catal. Sci. Technol.*, 2019, **9**, 3851.
13. L. Nguyen, F. F. Tao, Y. Tang, J. Dou and X. J. Bao, *Chem. Rev.*, 2019, **119**, 6822.
14. M. A. Newton, *Catalysts*, 2017, **7**, 58.
15. Y. Han, H. Zhang, Y. Yu and Z. Liu, *ACS Catal.*, 2021, **11**, 1464.
16. X. Lian, J. Gao, Y. Ding, Y. Liu and W. Chen, *J. Phys. Chem. Lett.*, 2022, **13**, 8264.
17. D. J. Morgan, *J. Electron Spectrosc. Relat. Phenom.*, 2019, **231**, 109.
18. Y. Wang, Y. Tian, S. Y. Pan and S. W. Snyder, *ChemSusChem*, 2022, **15**, e202201290.
19. I. Ziccarelli, R. Mancuso, F. Giacalone, C. Calabrese, V. La Parola, A. De Salvo, N. Della Ca', M. Gruttadauria and B. Gabriele, *J. Catal.*, 2022, **413**, 1098.
20. X. Rao, M. Tatoulian, C. Guyon, S. Ognier, C. Chu and A. A. Hassan, *Nanomaterials*, 2019, **9**, 1034.
21. A. Mercadante, V. Campisciano, A. Morena, L. Valentino, V. La Parola, C. Aprile, M. Gruttadauria and F. Giacalone, *Eur. J. Org Chem.*, 2022, e202200497.
22. I. Lucentini, A. Casanovas and J. Llorca, *Int. J. Hydrogen Energy*, 2019, **44**, 12693.
23. J. Sha, S. Paul, F. Dumeignil and R. Wojcieszak, *RSC Adv.*, 2019, **51**, 29888.
24. L. Xu, S. Hua, J. Zhou, Y. Xu, C. Lu, F. Feng, J. Zhao, X. Xu, Q. Wang, Q. Zhang and X. Li, *Appl. Catal., A*, 2022, **642**, 118690.

25. J. N. Díaz de León, S. Loera-Serna, T. A. Zepeda, D. Domínguez, B. Pawelec, A. M. Venezia and S. Fuentes-Moyado, *Fuel*, 2020, **266**, 117031.
26. M. Y. Smirnov, A. V. Kalinkin, E. I. Vovk, P. A. Simonov, E. Y. Gerasimov, A. M. Sorokin and V. I. Bukhtiyarov, *Appl. Surf. Sci.*, 2018, **428**, 972.
27. S. M. Davis, *J. Catal.*, 1989, **117**, 432.
28. R. J. Isaifan, S. Ntais and E. A. Baranova, *Appl. Catal., A*, 2013, **464–465**, 87.
29. Z. Jiang, W. Zhang, L. Jin, X. Yang, F. Xu, J. Zhu and W. Huang, *J. Phys. Chem. C*, 2007, **111**, 12434.
30. F. P. I. M. Kerkhof and J. A. Moulijn, *J. Phys. Chem.*, 1979, **83**, 1612.
31. M. Okumura and M. Haruta, *Catal. Today*, 2016, **259**, 81.
32. M. Gabrovska, I. Ivanov, D. Nikolova, J. Krstić, A. M. Venezia, D. Crişan, K. Tenchev, V. Idakiev and T. Tabakova, *Nanomaterials*, 2021, **11**, 366.
33. M. Haruta, *Catal. Today*, 1997, **36**, 153.
34. F. Vindigni, M. Manzoli, A. Chiorino and F. Boccuzzi, *Gold Bull.*, 2009, **42**, 106.
35. S. P. Chenakin and N. Kruse, *Phys. Chem. Chem. Phys.*, 2016, **18**, 22778.
36. G. C. Bond, *Molecules*, 2012, **17**, 1716.
37. L. Guczi, G. Petö, A. Beck, K. Frey, O. Geszti, G. Molnár and C. Daróczi, *J. Am. Chem. Soc.*, 2003, **125**, 4332.
38. A. M. Venezia, G. Pantaleo, A. Longo, G. Di Carlo, M. P. Casaletto, F. L. Liotta and G. Deganello, *J. Phys. Chem. B*, 2005, **109**, 2821.
39. M. P. Casaletto, A. Longo, A. Martorana and A. Prestianni, *Appl. Catal., A*, 2006, **302**, 309.
40. M. P. Casaletto, A. Longo, A. Martorana, A. Prestianni and A. M. Venezia, *Surf. Interface Anal.*, 2006, **38**, 215.
41. K. J. Klabunde, Y. X. Li and B. J. Tan, *Chem. Mater.*, 1991, **3**, 30.
42. A. I. M. Rabee, D. Zhao, S. Cisneros, C. R. Kreyenschulte, V. Kondratenko, S. Bartling, C. Kubis, E. Kontratenko, A. Brüchner and J. Rabeat, *Appl. Catal., B*, 2023, **321**, 122083.
43. C. Wu, P. Zhang, Z. Zhang, L. Zhang, G. Yang and B. Han, *ChemCatChem*, 2017, **9**, 3691.
44. A. M. Venezia, V. La Parola and L. F. Liotta, *Catal. Today*, 2017, **285**, 114.
45. A. M. Venezia, R. Murania, G. Pantaleo, V. La Parola, S. Scirè and G. Deganello, *Appl. Catal., A*, 2009, **352**, 296.
46. A. M. Venezia, V. La Parola, G. Deganello, B. Pawelec and J. L. G. Fierro, *J. Catal.*, 2003, **215**, 317.
47. V. La Parola, M. Kantcheva, M. Milanova and A. M. Venezia, *J. Catal.*, 2013, **298**, 170.
48. V. La Parola, M. L. Testa and A. M. Venezia, *Appl. Catal., B*, 2012, **119**, 248.
49. A. M. Venezia, V. La Parola, V. Nicolì and G. Deganello, *J. Catal.*, 2002, **212**, 56.
50. P. Hu, Z. Amghouz, Z. Huang, F. Xu, Y. Chen and X. Tang, *Environ. Sci. Technol.*, 2015, **49**, 2384.

51. R. V. Sales, H. O. M. A. Moura, A. B. F. Câmara, E. Rodríguez-Castellón, J. A. B. Silva, S. B. C. Pergher, L. M. A. Campos, M. M. Urbina, T. C. Bicudo and L. S. de Carvalho, *Catalysts*, 2019, **9**, 651.
52. T. Pu, A. Setiawan, B. M. Lis, M. Zhu, M. E. Ford, S. Rangarajan and I. E. Wachs, *ACS Catal.*, 2022, **12**, 4375.
53. M. Umar, A. Tanimu, S. A. Ganiyu, H. Dafalla, I. Kozhevnikov and K. Alhooshani, *Energy Fuels*, 2020, **34**, 9948.
54. M. Losurdo, I. Bergmair, B. Dastmalchi, T. H. Kim, M. M. Giangregroio, W. Jiao, G. Bianco, A. Brown, K. Hingerl and G. Bruno, *Adv. Funct. Mater.*, 2014, **24**, 1864.
55. C. Battocchio, C. Meneghini, I. Fratoddi, I. Venditti, M. V. Russo, G. Aquilanti, C. Maurizio, F. Bondino, R. Matassa, M. Rossi, S. Mobilio and G. Polzonetti, *J. Phys. Chem. C*, 2012, **116**, 19571.
56. E. Tzomos, M. H. Mikkelä, G. Öhrwall, O. Björneholm and M. Tchaplyguine, *Surf. Sci.*, 2023, **733**, 122307.
57. M. O. Ivanytsya, S. V. Ryabukhin, D. M. Volochnyuk and S. V. Kolotilov, *Theor. Exp. Chem.*, 2020, **56**, 283.
58. F. Giacalone, V. Campisciano, C. Calabrese, V. La Parola, Z. Syrgiannis, M. Prato and M. Grattadauria, *ACS Nano*, 2016, **10**, 4627.
59. D. Teschner, E. M. Vass and R. Schlögl, in *Photoelectron Spectroscopy of Catalytic Oxide Materials in Metal Oxide Catalysis*, ed. S. D. Jackson and J. S. J. Hargreaves, Wiley-VCH, Weinheim, 2009, pp. 243–298.
60. M. I. Ariëns, L. G. A. van de Water, A. I. Dugulan, E. Brück and E. J. M. Hensen, *J. Catal.*, 2022, **405**, 391.
61. A. C. Dupuis, M. Abu Haija, B. Richter, H. Kuhlenbeck and H. J. Freund, *Surf. Sci.*, 2003, **539**, 99.
62. J. C. Dupin, D. Gonbeau, P. Vinatier and A. Levasseur, *Phys. Chem. Chem. Phys.*, 2000, **2**, 1319.
63. A. Posada-Borbón, N. Bosio and H. Grönbeck, *Surf. Sci.*, 2021, **705**, 12761.
64. T. Wu, R. T. Guo, C. F. Li and W. G. Pan, *J. Environ. Chem. Eng.*, 2023, **11**, 109136.
65. T. C. Grinter and G. Thornton, *J. Phys.: Condens. Matter*, 2022, **34**, 253001.
66. A. M. Venezia, L. F. Liotta, G. Pantaleo and A. Longo, in *Catalysis by Ceria and Related Materials*, ed. A. Trovarelli and P. Fornasiero, Imperial College Press, Singapore, 2nd edn, 2013, pp. 813–879.
67. A. S. Al-Fatesh, R. Kumar, S. O. Kasim, A. A. Ibrahim, A. H. Fakeeha and A. E. Abasaeed, *Ind. Eng. Chem. Res.*, 2022, **61**, 164.
68. J. Liu, X. Wu, Y. Chen, Y. Zhang, T. Zhang, H. Ai and Q. Liu, *Int. J. Hydrogen Energy*, 2022, **47**, 6089.
69. G. Pantaleo, V. La Parola, F. Deganello, P. Calatozzo, R. Bal and A. M. Venezia, *Appl. Catal., B*, 2015, **164**, 135.
70. G. Pantaleo, V. La Parola, F. Deganello, R. K. Singha, R. Bal and A. M. Venezia, *Appl. Catal., B*, 2016, **189**, 233.
71. V. La Parola, G. Pantaleo and A. M. Venezia, *Catalysts*, 2018, **8**, 220.

72. A. Emamdoust, V. La Parola, G. Pantaleo, M. L. Testa, S. Farjami and A. M. Venezia, *J. Energy Chem.*, 2020, **47**, 1.
73. V. La Parola, G. Pantaleo and A. M. Venezia, *Inorg. Chim. Acta*, 2022, **531**, 120710.
74. M. C. Biesinger, B. P. Payn, A. P. Grosvenor, L. W. M. Lau, A. R. Gerson and R. S. C. Smart, *Appl. Surf. Sci.*, 2011, **257**, 2717.
75. N. K. Renuka, N. Harsha and T. Divya, *RSC Adv.*, 2015, **5**, 38837.
76. E. Paparazzo, G. M. Ingo and N. Zacchetti, *J. Vac. Sci. Technol. A*, 1991, **9**, 1416.
77. Y. Zhang, J. Lu, L. Zhang, T. Fu, J. Zhang, X. Zhu, X. Gao, D. He, Y. Luo, D. D. Dionysiou and W. Zhu, *Appl. Catal., B*, 2022, **309**, 121249.
78. X. Garcia, L. Soler, N. J. Divins, X. Vendrell, I. Serrano, I. Lucentini, J. Prat, E. Solano, M. Tallarida, C. Escudero and J. Llorca, *Catalysts*, 2020, **10**, 286.
79. L. Soler, A. Casanovas, C. Escudero, V. Pérez-Dieste, E. Aneggi, A. Trovarelli and J. Llorca, *ChemCatChem*, 2016, **8**, 2748.
80. E. Aneggi, C. De Leiterburg and A. Trovarelli, in *Catalysis by Ceria and Related Materials*, ed. A. Trovarelli and P. Fornasiero, Imperial College Press, Singapore, 2nd edn, 2013, pp. 565–621.
81. G. Preda, A. Migani, K. M. Neyman, S. T. Bromley, F. Illas and G. Pacchioni, *J. Phys. Chem. C*, 2011, **115**, 5817.
82. R. Singh, A. Dhir, S. Kumar Mohapatra and S. Kumar Mahla, *Biomass Convers. Biorefin.*, 2020, **10**, 567.
83. V. La Parola, G. Pantaleo, L. F. Liotta, A. M. Venezia, M. Gabrovska, D. Nikolova and T. Tabakova, *Catalysts*, 2023, **13**, 606.
84. A. I. Tsiotsias, N. D. Charisiou, V. Sebastian, S. Gaber, S. J. Hinder, M. A. Baker, K. Polychronopoulou and M. A. Goula, *Int. J. Hydrogen Energy*, 2022, **47**, 5337.
85. Q. L. M. Ha, U. Armbruster, H. Atia, M. Schneider, H. Lund, G. Agostini, J. Radnik, H. T. Vuong and A. Martin, *Catalysts*, 2017, **7**, 157.
86. T. Fujita, X. Peng, A. Yamaguchi, Y. Cho, Y. Zhang, K. Higuchi, Y. Yamamoto, T. Tokunaga, S. Arai, M. Miyauchi and H. Abe, *ACS Omega*, 2018, **3**, 16651.
87. Z. Liu, D. C. Grinter, P. G. Lustemberg, T. Nguyen-Phan, Y. Zhou, S. Luo, I. Waluyo, E. I. Crumlin, D. J. Stacchiola, J. Zhou, J. Carrasco, H. F. Busnengo, M. V. Ganduglia-Pirovano, S. D. Senanayake and J. A. Rodriguez, *Angew. Chem., Int. Ed.*, 2016, **55**, 7455.
88. J. K. Lai and I. E. Wachs, *ACS Catal.*, 2018, **8**, 6537.
89. J. Shen and C. Hess, *Catalysts*, 2020, **10**, 1386.
90. A. Mohammadi, C. Praty, A. Farzi, H. Soleimanzadeh, S. Schwarz, M. Stöger-Pollach, J. Bernardi, S. Penner and A. Niaei, *Catal. Lett.*, 2023, **153**, 2176.
91. Y. Li, X. Yu, Q. Zhang, V. A. Kondratenko, Y. Wang, G. Cui, M. Zhou, C. Xu, E. V. Kontratenko and G. Jiang, *J. Catal.*, 2022, **413**, 658.
92. E. V. Kondratenko, O. V. Buyevskaya and M. Baerns, *Top. Catal.*, 2001, **15**, 175.

93. E. Santacesaria, M. Cozzolino, M. Di Serio, A. M. Venezia and R. Tesser, *Appl. Catal., A*, 2004, **270**, 177.
94. I. V. Bacherikova and V. O. Zazhigalov, *Theor. Exp. Chem.*, 2021, **57**, 126.
95. D. Kim, M. Kim, J. Yi, S.-H. Nam, J.-H. Boo, Y. S. Park and J. Lee, *Sci. Adv. Mater.*, 2017, **9**, 1415.
96. L. Shi, X.-Q. Zhu, Y. Su, W.-Z. Weng, H. Feng, X.-D. Yi, Z.-X. Liu and H.-L. Wan, *J. Catal.*, 2013, **307**, 316.
97. P. Hu, Y. Chen, X. Yan, W.-Z. Lang and Y.-J. Guo, *Ind. Eng. Chem. Res.*, 2019, **58**, 4065.
98. A. Białas, K. Rugała, C. Czosnek, G. Mordarski and J. Gurgul, *Catalysts*, 2020, **10**, 1388.
99. L. Zhang, L. Li, Y. Cao, Y. Xiong, S. Wu, J. Sun, C. Tang, F. Gao and L. Dong, *Catal. Sci. Technol.*, 2015, **5**, 2188.
100. M. A. Reiche, E. Ortelli and A. Baiker, *Appl. Catal., B*, 1999, **23**, 187.
101. Y. Pan, W. Zhao, Q. Zhong, W. Cai and H. Li, *J. Environ. Sci.*, 2013, **25**, 1703.
102. I. Song, S. Youn, H. Lee, S. G. Lee, S. J. Cho and D. H. Kim, *Appl. Catal., B*, 2017, **210**, 421.
103. J. Keränen, C. Guimon, A. Auroux, E. Iiskola and L. Niinistö, *Phys. Chem. Chem. Phys.*, 2003, **5**, 5333.
104. J. Keränen, C. Guimon, E. Iiskola, A. Auroux and L. Niinistö, *Catal. Today*, 2003, **78**, 149.
105. I. E. Wachs, G. Deo, B. M. Weckhuysen, A. Andreini, M. A. Vuurman, M. De Boer and M. Amiridis, *J. Catal.*, 1996, **161**, 211.
106. R. Palcheva, L. Kaluža, L. Dimitrov, G. Tyuliev, G. Avdeev, K. Jirátová and A. Spojakina, *Appl. Catal., A*, 2016, **520**, 24.
107. R. Obeso-Estrella, J. L. G. Fierro, J. N. Diaz de León, S. Fuentes, G. Alonso-Nuñez, E. Lugo-Medina, B. Pawelec and T. A. Zepeda, *Fuel*, 2018, **233**, 644.
108. F. Jin, J. Fan, C. Guo, H. Liu, J. Zang, H. Yu, Y. Wang and X. Wang, *Pet. Process. Petrochem.*, 2020, **51**, 62.
109. R. Nava, R. A. Ortega, G. Alonso, C. Ornelas, B. Pawelec and J. L. G. Fierro, *Catal. Today*, 2007, **127**, 70.
110. V. La Parola, G. Deganello, C. R. Tewell and A. M. Venezia, *Appl. Catal., A*, 2002, **235**, 171.
111. C. Song, *Catal. Today*, 2003, **86**, 211.
112. V. La Parola, G. Deganello and A. M. Venezia, *Appl. Catal., A*, 2004, **260**, 237.
113. A. M. Venezia, F. Raimondi, V. La Parola and G. Deganello, *J. Catal.*, 2000, **194**, 393.
114. A. M. Venezia, V. La Parola, G. Deganello, D. Cauzzi, G. Leonardi and G. Predieri, *Appl. Catal., A*, 2002, **229**, 261.
115. P. Dufresne, N. Brahma, F. Labruyère, M. Lacroix and M. Breysse, *Catal. Today*, 1996, **29**, 251.
116. M. Bravo-Sanchez, A. Romero-Galarza, J. Ramírez, A. Gutiérrez-Alejandre and D. A. Solís-Casados, *Appl. Surf. Sci.*, 2019, **493**, 587.

CHAPTER 6

Photoemission Spectroscopy of Solar Cell Materials

VIN DHANAK AND TIM VEAL*

Stephenson Institute for Renewable Energy and Department of Physics, University of Liverpool, UK
*E-mail: timveal@liverpool.ac.uk

6.1 Introduction

Photoelectron spectroscopy (PES) is a powerful analytical technique widely used in the field of energy materials research. This method provides valuable insights in the characterisation of materials used in energy conversion and storage processes, from the elemental composition and chemical state to the electronic structure of materials used in energy storage devices, fuel cells and photovoltaics, contributing significantly to the advancement of various energy-related technologies. From catalysts to energy storage devices, fuel cells and photovoltaics, as energy demands continue to grow, the insights provided by PES will undoubtedly remain integral to the evolution of energy materials science. It should be noted that in this chapter, the term PES is used as a broad and encompassing term, which includes techniques such as X-ray photoelectron spectroscopy (XPS), ultraviolet photoelectron spectroscopy

Inorganic Materials Series No. 19
Photoelectron Spectroscopy in Inorganic Materials Science
Edited by Robert G. Palgrave and Philip R. Davies

Published by the Royal Society of Chemistry, www.rsc.org

(UPS) and hard X-ray photoelectron spectroscopy (HAXPES). While these techniques may vary in terms of the photon energy utilised, it is important to note that the underlying photoemission process remains consistent across all of them. Whether employing XPS, UPS, or HAXPES, the fundamental principles of photoelectron spectroscopy are retained.

In the context of energy storage materials (see Chapter 4), XPS plays a pivotal role in several aspects. By analysing the binding energies of core-level electrons, it can identify the oxidation state of elements present in the materials, which is crucial for understanding their reactivity and performance.[1,2] For instance, it can reveal the changes in chemical composition and electronic structure that occur during charging and discharging cycles, as well as investigating side reactions that can lead to capacity fading and electrode degradation, enabling a deeper understanding of the degradation mechanisms and leading to the development of more efficient and durable energy storage technologies. For example, in lithium–sulfur batteries, XPS is crucial in understanding the chemistry of the sulfur cathode and the formation of various sulfur species during cycling. From the XP spectra, it is possible to identify polysulfide intermediates and the formation of solid-state lithium sulfides on the electrode surface. This knowledge aids in developing strategies to mitigate the dissolution of polysulfides, which is a major challenge in these batteries.[3,4] In solid-state batteries, which promise enhanced safety and energy density compared to traditional Li-ion batteries, XPS is used to study the chemical interaction and electronic structure changes at the interface between solid electrolytes and electrode materials, helping to design stable and efficient solid-state battery architectures.[2,5]

Photoemission spectroscopy also plays a crucial role in the analysis of photovoltaic materials which is the subject and focus of this chapter, offering insights into the electronic structure, chemical composition, interface properties and heterointerface band alignments that are vital for optimising solar cell efficiency. For example, PES is used to analyse the surface composition and electronic structure of thin-film solar cell materials like CIGS (copper indium gallium selenide) and corresponding CdS/CIGS interface.[6] Measuring the chemical states of copper, indium, gallium, and selenium aids in understanding their oxidation states and how they affect the band alignment and charge transport within the material. PES also aids in characterising interfaces between different layers, such as the absorber layer and the buffer layer, helping to optimise the overall device performance.[7,8] In perovskite solar cells, which have gained significant attention in recent years, by analysing the binding energies of elements like lead, iodine, and caesium (if used), XPS can provide insights into the formation of grain boundaries, the presence of surface defects, and the influence of compositional variations on the electronic structure. This information guides efforts to improve the stability and efficiency of perovskite solar cells.[9–11] In a typical perovskite device configuration, there is an interface between the perovskite layer and a layer of titanium dioxide (TiO_2). This TiO_2 layer is often situated on top of fluorine-doped tin oxide (FTO) coated glass. By conducting measurements of the binding energies of core-level electrons and the valence bands within these materials, it becomes feasible to ascertain the energy levels of their valence and conduction bands and determine the band alignment between the layers. This data is of paramount importance for fine-tuning the device structure to improve the efficiency of charge separation and collection efficiency.[12–14]

PES is also used to investigate the electronic structure of organic photovoltaic materials, including polymer donors and fullerene acceptors. It makes it possible to measure the energy levels of these materials and how they align at the donor–acceptor interface, affecting charge separation and transport.[15–17] By comparing the valance band maximum (VBM) of the polymer with the lowest unoccupied molecular orbital (LUMO) of the fullerene acceptor, it is possible to determine the energy level alignment at the donor–acceptor interface. If the LUMO of the fullerene is higher in energy than the VBM of the polymer, a favourable energy offset exists for efficient electron transfer from the polymer to the fullerene. The energy offset obtained from PES measurements indicates that photoexcited electrons in the polymer have sufficient energy to overcome the barrier and transfer to the acceptor material. This facilitates effective charge separation, as electrons are transferred from the polymer to the fullerene, leaving behind positively charged holes.[16] By utilising PES to study band alignment in photovoltaic materials, it is possible to fine-tune the properties of organic solar cells and other types of solar devices to achieve better charge extraction, reduced recombination, and improved energy conversion efficiency.

PES is also valuable for probing the chemical interaction between different layers and identifying degradation products that can impact device performance over time. In solar cell devices, transparent conducting oxides (TCOs) are commonly used as transparent electrodes. PES is often used to investigate the electronic properties and surface chemistry of these materials, understanding how doping and surface treatments affect their conductivity, transparency, and work function. This information is essential for optimising the electrical and optical properties of the transparent electrodes.[18] Moreover, in a typical thin-film solar cell, the interface between the absorber layer (*e.g.*, CdTe, CIGS or perovskite) and the TCO layer (*e.g.* FTO) plays a critical role in electron–hole separation. PES can be employed to precisely measure the energy levels of the valence and conduction bands at this interface and investigate their alignment.[19]

Photoemission is clearly a valuable tool for characterising PV materials, providing insights into their electronic structure, band alignment, surface chemistry, and contaminants. This information is critical for designing and improving the performance of photovoltaic devices. In the following sections, we will focus on the methodology for determining band alignments in PV material interfaces and illustrate this with examples.

6.2 Measuring Work Function and Vacuum Level by XPS

Measuring the Fermi level position and work function in XPS is a valuable technique with broad applications in materials science, surface science, and device engineering.[20] It provides critical insights into the electronic properties of materials, aids in material identification, helps optimise device interfaces, and plays a key role in various technological advancements. The work function is closely related to the energy band diagram near the surface of a material and corresponds to the difference between the Fermi energy and the vacuum level (Figure 6.1). The local vacuum level E_{VAC} holds significant importance in defining and determining all parameters outlined in Figure 6.1. The work function, denoted as WF, is the energy required to extract an electron initially positioned at the Fermi level (E_F) within the material and relocate it to a point in free space just outside the surface, precisely at E_{VAC}. In metals, E_F signifies the demarcation between occupied and unoccupied states in a continuous range of states, making ionisation potential (IP) and electron affinity (E_A) equivalent to the material's work function. Conversely, non-degenerate semiconductors lack states, and thus electrons, at the Fermi level, E_F. Consequently, WF in these materials adopts a statistical value, residing somewhere between IP and E_A, in the band gap E_g (refer to Figure 6.1). In semiconductors, WF is intricately tied to the positioning of E_{VAC} and E_F, with E_F's position

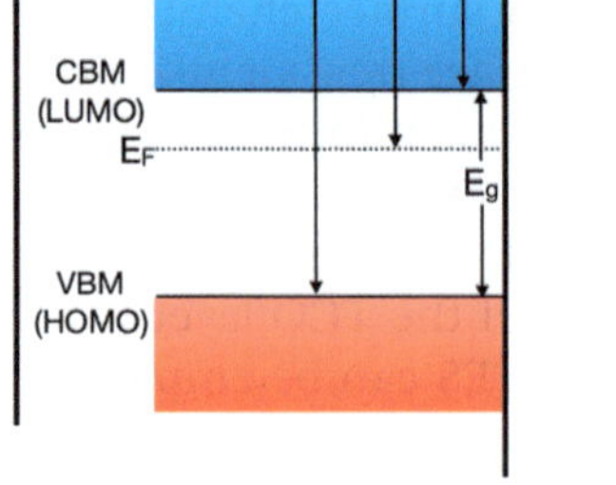

Figure 6.1 Flat band energy diagram of a semiconductor. CBM and VBM refer to conduction and valence band edges. Their equivalents in an organic semiconductor are the lowest unoccupied and highest occupied molecular orbitals (LUMO, HOMO), respectively. Valence band and conduction band are separated by the band gap (E_g), and the Fermi level (E_F) is located within the band gap. The work function (WF), ionisation potential (IP) and electron affinity (E_A) are defined with respect to the vacuum level (E_{VAC}).

being contingent upon factors like density of states, temperature, carrier density, and doping concentration within the material. The quantification of WF is achieved through photoemission spectroscopy, a technique which provides precise measurements of both E_F and E_{VAC} positions. In electronic and optoelectronic devices, achieving proper band alignment at interfaces is essential for efficient charge transport and device performance, so measuring the work function of materials used in these devices helps in optimising the design of device interfaces to ensure desirable energy level alignments.

Photoemission spectra are commonly presented with respect to binding energies relative to the Fermi energy of the sample. However, what is measured in the spectrometer are the kinetic energies of emitted electrons resulting from photoexcitation (see Chapter 2, Section 2.3.1.1 for a full discussion). These spectra essentially consist of two principal components: the first comprises primary electrons, unaffected by inelastic collisions, displaying distinctive spectral features such as core-levels and valence bands, closely mirroring the sample's density of states (DOS). The second component includes secondary electrons, which are primary electrons that have lost varying amounts of energy. The secondary electrons exhibit a continuous energy spectrum extending all the way down to zero kinetic energy and overlapping with the primary electron spectrum. These dual aspects of the spectra provide diverse types of information: the primary peaks facilitate the determination of binding energies of electronic states within the sample, whereas secondary electrons are primarily utilised for measuring the sample's work function.

Figure 6.2 provides a schematic representation of the photoemission process.[21] Here, electrons are excited out of a metallic sample with a specific kinetic energy, determined by a range of factors including the photon energy ($h\nu$), the sample's work function (Φ_s), the binding energy (E_b) of the excited electron, and any inelastic scattering that occurs within the sample. The resulting spectrum, outside the sample surface shown in the upper part of Figure 6.2a, comprises the two main components: primary electrons, characterised by a kinetic energy $E_{kin} = h\nu - E_b - \Phi_s$, and a secondary electron continuum. Among these electrons, the fastest are the primary electrons emitted directly from the Fermi edge (where $E_b = 0$), possessing an E_{kin} value of $h\nu - \Phi_s$. In contrast, the slowest electrons are the secondary electrons that just manage to exit the sample surface but have previously lost energy due to scattering processes, resulting in a kinetic energy of $E_{kin} = 0$ eV.

To measure and display the kinetic energy spectrum of the photoemitted electrons, a spectrometer consisting of an energy analyser and detector is employed. Since the spectrometer analyser has a work function Φ_a, which is usually less than the sample work function, and since the sample and analyser are electrically connected so that their Fermi energies are aligned, all the photoemitted electrons gain in kinetic energy by the contact potential $\Phi_s - \Phi_a$. Consequently, the kinetic energy in the analyser is $E_{kin} = h\nu - E_b - \Phi_s + \Phi_s - \Phi_a = h\nu - E_b - \Phi_a$. This spectrum, which is what is measured at the detector, is shown in Figure 6.2b. Note that usually the work function of the analyser is

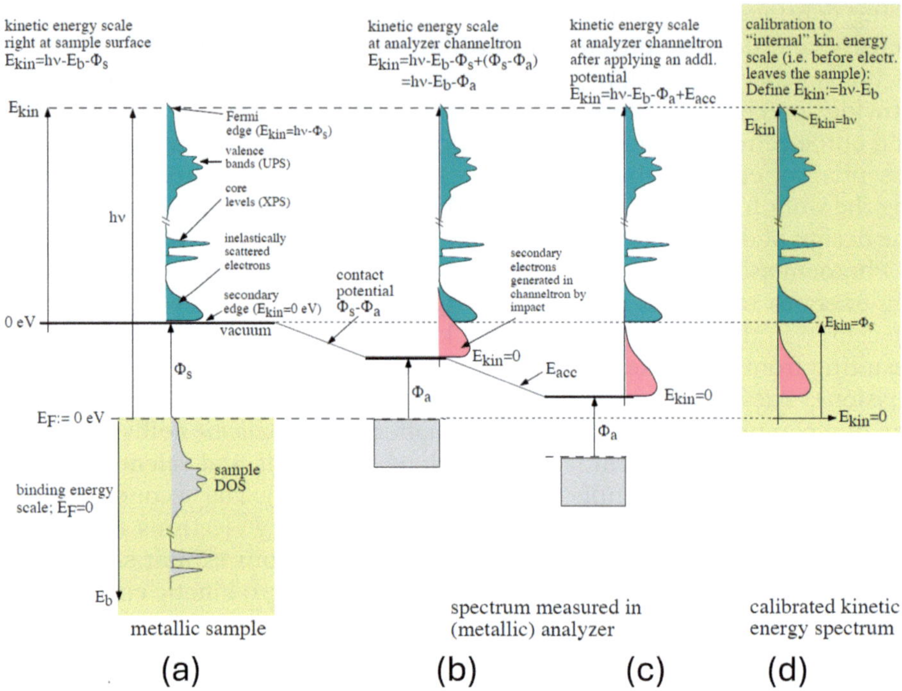

Figure 6.2 (a) Schematic of the photoemission process in a metallic sample, the lower figure shows the spectra in binding energy in the initial state inside the sample, the upper figure shows the spectra in kinetic energy in the final state outside the sample surface, including the superimposed secondary electrons onto the primary electrons. (b) Spectra measured in the spectrometer, where additional secondary electrons originating from the spectrometer surfaces are superimposed on the spectrum from the sample. (c) Spectra after a negative bias is applied on the sample, separating the secondary edges from the sample and the spectrometer. (d) The spectrometer is calibrated so that the Fermi level of the sample appears at binding energy $E_b = 0$ eV, and the work function of the sample is then determined by subtracting the entire spectrum width from the photon energy. Reproduced from ref. 21 with permission from R. Schlaf.

unknown, so the kinetic energy scale is calibrated using the Fermi edge from a metal sample such as gold or silver, and defining the Fermi energy $E_b = 0$, and the kinetic energy $E_{kin} = h\nu$. Also shown in the figure is the continuum of secondary electrons that is generated in the analyser, arising from stray photoelectrons that strike the surfaces of the spectrometer, and which is superimposed on the spectrum from the sample. The presence of the secondary electrons from the spectrometer makes it difficult to determine the secondary edge from the sample, which is needed to measure the work function of the sample. In order to separate the secondary edge of the sample from the secondary electrons due to the analyser, the sample is negatively biased

(E_{acc}) relative to the analyser. This only affects the electrons from the sample, which are accelerated by the bias potential, yielding $E_{kin} = h\nu - E_b - \Phi_a + E_{acc}$ (Figure 6.2c), separating the secondary electrons from the analyser. Since the kinetic energy scale is calibrated so that the Fermi edge from a metal sample is at $E_{kin} = h\nu$, the secondary electron edge kinetic energy corresponds to the work function of the sample (Figure 6.2d). By calibrating the spectrometer so that the Fermi level binding energy is $E_b = 0$ eV, the secondary edge from the sample appears at $E_b = h\nu - \Phi_s$, so the work function is then determined by subtracting the photon energy from the binding energy of the secondary edge of the sample.

This process works also for semiconductors even though the Fermi level here lies in the band gap, where there is very low DOS. This is because the secondary electrons originate from inelastic processes from all primary electrons in the semiconductor (Figure 6.3). As indicated in the figure, an important additional parameter that is obtained is the ionisation potential (IP or E_{ion}) which is the difference in energy between the VBM and the vacuum level.

In practice it is important to realise that the count rate at the secondary electron edge is extremely high and can saturate the spectrometer detector, so this measurement should be made at low photon power and with the spectrometer pass energy and slits set at the lowest possible setting.

Numerous techniques are available for investigating different aspects of a material's electronic structure, yet some are particularly well-suited for photovoltaic (PV) absorbers due to their relevance to crucial properties in this context. Key attributes for solar absorbers are the band gap, the position of band edges, and optical absorption characteristics. Furthermore, structural and chemical stability hold significance for materials in a broader context. Generally, spectroscopies such as X-ray absorption, or inverse photoemission techniques are employed to ascertain the conduction band minimum (CBM) and the vacant states above it. Conversely, valence band maximum (VBM) and occupied states beneath it are most effectively probed through XPS. These methods are complementary to optical absorption which probes the joint density of states of the occupied and unoccupied band structure. Analysing the band structures of materials enables the establishment of relationships between energy levels, elucidating orbital hybridisation and the origins of electronic band positions, including the band gap. Moreover, understanding the characteristics of states within the conduction and valence bands facilitates the determination of a material's transport properties and optical behaviour.

In solar cells, losses and inefficiencies arise due to mismatched bandgaps or misaligned bands within the cell structure. Consequently, investigating band alignments in heterojunctions becomes critically important. When two semiconductors come into contact, several changes occur: charge equilibration, alignment of Fermi levels, and bending of energy bands. Nevertheless, anticipating these changes is possible by understanding the inherent band positions of a material in isolation. Hence, measurements of the parameters

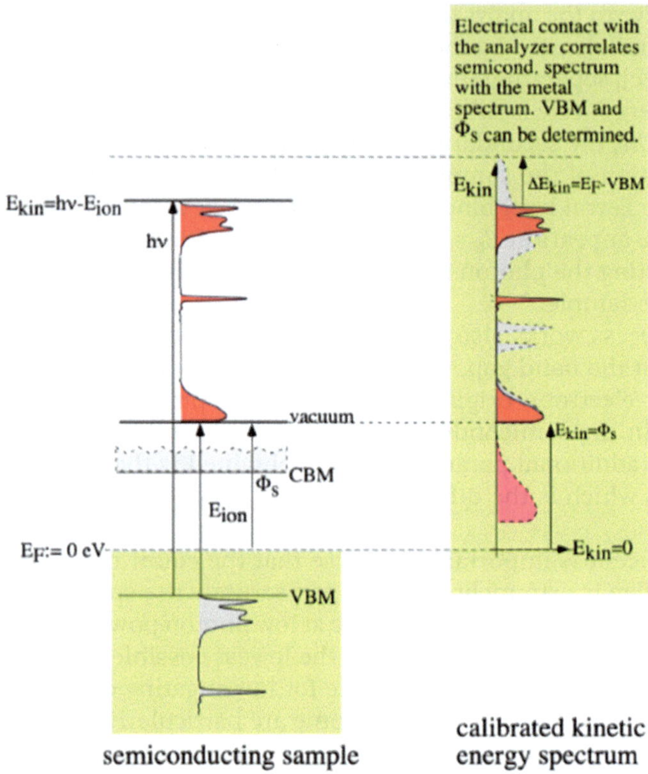

Figure 6.3 Photoemission spectra from a semiconducting sample where steps (b) and (c) shown in Figure 6.2 have been omitted. Once the spectrometer has been calibrated using a metallic sample as described, and the semiconductor sample is in electrical contact with the spectrometer, the work function can be determined as for a metallic sample. An additional important parameter for a semiconductor that can be determined is the electron ionisation potential, IP, which is the difference between the vacuum level and the VBM. Reproduced from ref. 21 with permission from R. Schlaf.

depicted in Figure 6.1 become crucial for assessing and developing novel solar absorber materials. These ideas are illustrated in the next section by considering the electronic structure of the tin sulfides (SnS, SnS_2 and Sn_2S_3) for use in PV.

6.3 The Electronic Structures of the Binary Chalcogenides for Use in PV

The binary tin sulfides, denoted as Sn_xS_y, represent a class of materials that has gained recent prominence in photovoltaic (PV) applications, despite their long-known existence. Their relevance extends beyond PV and holds significance in the development of an environmentally friendly and earth abundant PV absorber CZTS (copper zinc tin sulfide). Tin exhibits two oxidation states: Sn^{2+} ([Kr]$4d^{10}5s^{2}5p^{0}$) and Sn^{4+} ([Kr]$4d^{10}5s^{0}5p^{0}$). Within the Sn–S system, three primary stable phases of tin sulfide have been identified: tin monosulfide (SnS), tin disulfide (SnS_2), and the mixed oxidation state tin sesquisulfide (Sn_2S_3). These compounds are all characterised as layered structures. While the concept of using SnS as a solar absorber was initially proposed in the late 1980s, progress in this area has been somewhat limited, with recently developed devices consistently yielding conversion efficiencies below 4%. It has been postulated that the poor alignment of energy bands between SnS and the commonly employed CdS window layer in PV device configurations accounts for these subpar efficiencies. Previous studies, both theoretical and experimental, have supported this notion by determining the band positions of SnS. Consequently, alternative combinations of window layer materials have been explored, yielding improved efficiencies when materials other than CdS are used. Nevertheless, challenges related to band alignments persist in these structures.

Whittles *et al.*[22,23] in their studies used a combination of laboratory based XPS and IPES measurements of the clean crystals of the three phases of tin sulfides to determine the values of IP, WF, EA and E_g. They also show that the oxidation states of tin, and therefore the different phases of tin sulfide, can be easily distinguished using properly energy referenced XPS of a sufficiently high resolution. The positions of the Fermi levels in the region probed by the PES measurements and their proximity to either the VBM or CBM allows determination of the type of semiconductor and here, confirms the expected p-type nature of SnS and n-type natures of SnS_2 and Sn_2S_3. It is worth noting that Whittles[22] showcased the utility of XPS in examining absorber materials. XPS was employed to uncover the presence, impact, and mechanisms of material contamination. It revealed how the existence of such contaminants could negatively affect cell performance, especially in terms of oxidation. Additionally, it shed light on how these factors had been overlooked in previous research. When combined with complementary techniques, XPS was used to assist in identifying the phases present in a grown material and detecting undesirable phases that can harm the device's performance. To effectively utilise XPS, line-shape fitting procedures, analytical considerations and characterisation methods were developed to support the exploration of novel approaches to cell design.

By following vacuum-alignment procedures, natural band alignment diagrams were constructed between the tin sulfides and other appropriate materials for band alignment comparisons. In Figure 6.4, we observe the alignment of the p-type SnS with other prevalent p-type absorber materials like CIGS, CZTS, and CdTe. Additionally, the diagram showcases the alignment of the n-type SnS_2 alongside the commonly used n-type window layer CdS. This comprehensive analysis not only sheds light on the potential utility of SnS and SnS_2 as standalone materials in solar cells, but also provides

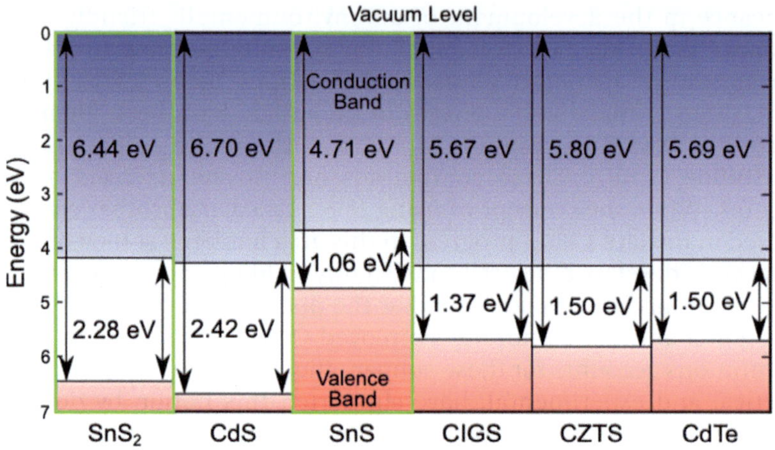

Figure 6.4 Vacuum-aligned band diagram for SnS and SnS_2 in comparison with other common solar cell materials. IP and E_g values for SnS and SnS_2 were determined from combined XPS/IPES measurements and values for the common n-type window layer material CdS, and the common absorbers: CIGS, CZTS, and CdTe are shown for comparison.

insights into the potential consequences should these materials be incorporated as unintentional impurity phases, for instance, within CZTS.

When compared to other commonly employed thin-film PV absorber materials, it becomes evident that SnS exhibits a relatively low ionisation potential. This characteristic can be attributed to the presence of additional states at the uppermost part of its valence band, as detailed below. Consequently, this disparity in energy levels leads to a mismatch when SnS is coupled with conventional partner materials in a solar cell, such as CdS. This mismatch offers some insight into the observed relatively modest solar cell efficiencies thus far. Additionally, the substantial negative conduction band offset (CBO) of −0.63 eV in the CdS/SnS junction contributes to heightened interface recombination. In a solar cell, band alignments between the window layer and the absorber layer are typically described in terms of the CBO between these materials (Figure 6.5). A 'spike-like' band offset occurs when the CBM of the window layer is positioned above that of the absorber, while a 'cliff-like' offset is observed when the CBM of the window layer is situated below that of the absorber. An excessively positive CBO (as depicted in Figure 6.5a) results in electrons excited within the absorber encountering a potential barrier that hinders their movement into the window layer, thereby reducing the short-circuit current (J_{sc}) of the solar cell. Conversely, an overly negative CBO (as illustrated in Figure 6.5b) creates a region for carrier recombination at the interface between the conduction band of the window layer and the valence band of the absorber. Additionally, it constrains the built-in voltage (V_{bi}) of the junction, ultimately leading to a lower open-circuit voltage (V_{oc}).

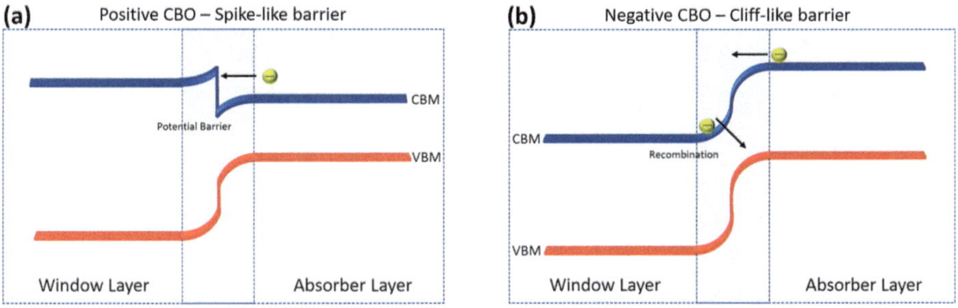

Figure 6.5 Schematic diagram showing (a) a very positive CBO leading to a potential barrier in the conduction band and (b) a very negative CBO leading to a recombination centre and low built-in voltage. Reproduced from ref. 27, https://doi.org/10.1021/acsaem.0c01477, under the terms of the CC BY 4.0 license, https://creativecommons.org/licenses/by/4.0/.

From the perspectives of both band alignment and environmental safety, it thus becomes apparent that CdS is not an optimal choice as a window layer for SnS-based solar cells. In response to this challenge, alternative window layer materials with more favourable conduction band offsets (CBOs), including ZnO, $ZnIn_2Se_4$, In_2S_3, and Zn(O,S), have been proposed. It is worth noting that, given SnS's low ionisation potential, a revised architecture for SnS-based solar cells could involve substituting the commonly used Mo back contact with a contact metal possessing a lower work function.

As depicted in Figure 6.4, it becomes evident that the band positions determined for SnS_2 closely align with those of CdS. Consequently, at least from an electronic perspective, the n-type SnS_2 presents itself as a viable, earth-abundant alternative to CdS in solar cells where CdS is already appropriately matched in terms of energy bands. Additionally, the alignment showcased in Figure 6.4 suggests that the introduction of SnS or SnS_2 within a CZTS cell could have detrimental effects on its performance. This is due to the low ionisation potential (IP) of SnS, which might act as an electron barrier, and the substantial band gap of SnS_2, which could impede charge transport by acting as an insulator. Furthermore, given that SnS_2 (and even Sn_2S_3) exhibits n-type behaviour, it could potentially lead to the formation of a second diode, thereby mitigating the desired rectifying behaviour of the cell.

Figure 6.6a illustrates the determined natural band alignments among the three phases of tin sulfide, facilitating an assessment of their applicability within solar cell devices and the impact of secondary tin sulfide phase formation within SnS cells. It is apparent that SnS and Sn_2S_3 exhibit similar band-level positions, while SnS_2 noticeably deviates to lower energy levels on the band-alignment diagram. This discrepancy was resolved through an examination of the density of states in the valence band, as elaborated below. The band alignment indicates that the emergence of SnS_2 within a SnS solar cell would detrimentally affect its performance, a phenomenon observed during SnS growth at the substrate interface. SnS_2 possesses a substantial negative

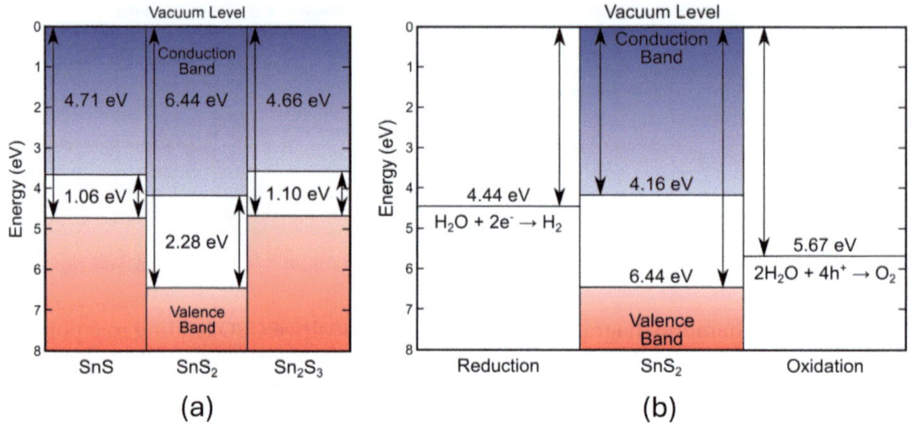

Figure 6.6 (a) Vacuum-aligned band diagram between the three phases of tin sulfide and (b) vacuum-aligned band diagram between SnS_2 and the reduction and oxidation potentials of water. Values of IP and E_g taken from combined XPS/IPES measurements. Reproduced from ref. 22 with permission from T. Whittles.

CBO of −0.51 eV with respect to SnS, thus functioning as a recombination centre. Moreover, the formation of n-type SnS_2 or Sn_2S_3 could potentially introduce second diodes within a SnS cell.

This study also explored the prospect of constructing viable photovoltaic devices exclusively from tin sulfide materials, aiming to streamline production and sustain the earth-abundant nature of these materials throughout the device. Figure 6.6 reveals that SnS/SnS_2 and SnS/Sn_2S_3 would create type-II heterojunctions. However, establishing a solar cell based on a SnS/SnS_2 junction proves unfeasible due to the unfavourable CBO between them. Conversely, a cell based on a SnS/Sn_2S_3 junction seems electronically advantageous, boasting a slight positive CBO (0.09 eV). Nevertheless, the consistent production of high-quality Sn_2S_3 films has been shown to pose challenges, and a comparison with other absorber materials suggests that a SnS cell would probably favour a type-I heterojunction architecture. Nonetheless, the results presented here for the relatively underexplored Sn_2S_3 hold promise for its future development.

Incidentally, beyond its potential use in photovoltaic devices, SnS_2 has displayed potential for photocatalytic water splitting. Figure 6.6b shows the band edges of SnS_2 and the potentials required for water reduction and oxidation. Notably, SnS_2 possesses a sufficiently wide bandgap to encompass both potentials, and its band edges are suitably positioned, eliminating the need for external bias to drive the reaction. Additionally, SnS_2's layered structure provides a substantially larger surface area for light absorption. Consequently, this study highlights SnS_2 as a promising candidate for water splitting, surpassing some other materials currently under investigation in this area, including CdS.[24]

The observed positions of the band edges in tin sulfides and their implications for devices can be correlated with the intrinsic electronic structure of these materials. Analysing the DOS for these materials and comparing them to DFT-calculated DOS in the valence band provides valuable insights into the bonding characteristics. This, in turn, sheds light on the crystal structure and electronic properties, including the underlying reasons for the low ionisation potential in phases containing the Sn^{2+} oxidation state. The XPS measurements showed that in these phases, the raising of the VBM is primarily driven by the presence of occupied Sn 5s orbitals. The bonding of these electrons, as described by the revised lone pair model, accounts for the additional states found at the upper end of the valence band. This understanding, coupled with the band misalignment, offers a more comprehensive explanation for the suboptimal efficiencies observed in tin sulfide solar cells.

Figure 6.7a displays the calculated total and partial density of states (pDOS) curves for the three tin sulfide phases, alongside the experimentally acquired valence band (VB) spectra derived from XPS measurements. The adjusted DOS curves are in good agreement with the XPS data across all three phases, showcasing the presence of all spectral features. Minor disparities are observed in the positions and relative intensities of these features. The primary deviation is noticeable in the high binding energy feature, which is a shared characteristic in all spectra. This divergence has been attributed to final-state relaxation effects inherent in XPS, known to shift features located near the lower edge of the valence band closer to the valence band edge. These relaxation effects were not considered in the calculations. Furthermore, it is worth noting that lifetime broadening increases with binding energy, whereas the calculated DOS adjustments employed a constant broadening factor, which may not fully capture this effect.

All three phases show a valence band consisting of majority S p states which extend from the VBM at 0 eV to −9 eV. The lack of S 5s contribution to the valence band shows that this orbital (shown around −14 eV in Figure 6.7b(ii) for SnS) is somewhat localised.

The top VB of SnS is comprised of three main features. The first, at −6.5 to −10 eV binding energy, is dominated by Sn s states hybridised slightly with S p states. The second, at −2.5 to −6.5 eV, is comprised mainly of S p states, with a slight contribution from Sn p states. The top of the valence band is formed from a hybridisation of S p, Sn s, and Sn p states. The first CB of SnS (Figure 6.7b(ii)) extends from 1 to 6 eV and has majority Sn p states hybridised with S p states and a slight contribution from S s and also Sn s states near the bottom of the CB around 2 eV.

The valence band of SnS exhibits three primary characteristics. The initial feature, spanning from −6.5 to −10 eV in binding energy, is primarily a result of Sn s states with a slight hybridisation involving S p states. The second feature, ranging from −2.5 to −6.5 eV, is predominantly composed of S p states, with a minor contribution from Sn p states. At the top of the valence band, a hybridisation involving S p, Sn s, and Sn p states is responsible for shaping its structure. In the case of SnS's first conduction band (CB), depicted in

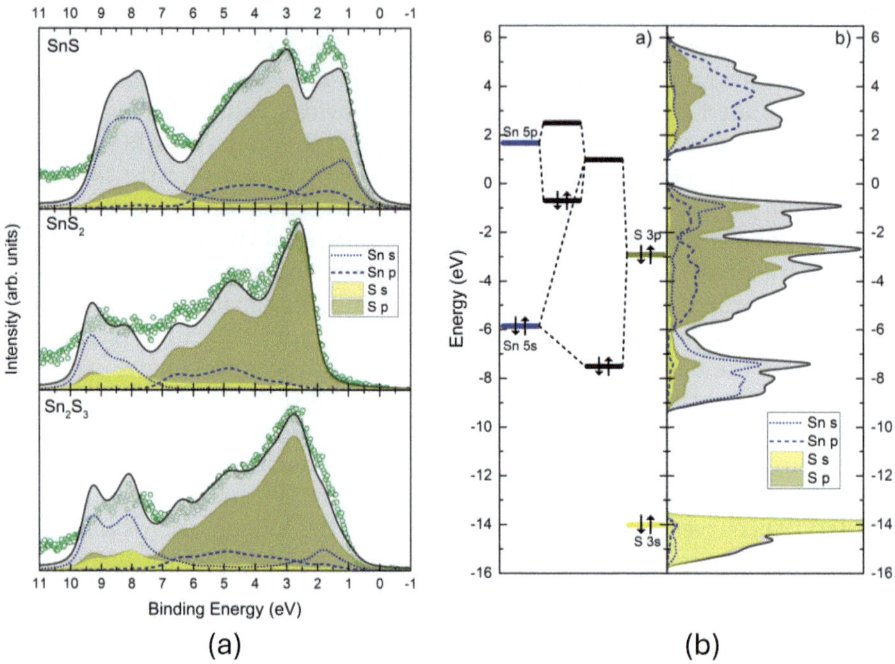

Figure 6.7 (a) Comparison between calculated and experimentally measured valence-band spectra for SnS, SnS_2, and Sn_2S_3, referenced to the Fermi level at 0 eV binding energy. The experimental data were obtained using XPS and have been background corrected. The calculated total and partial density of states (DOS) curves have been adjusted for cross-section considerations and broadened to facilitate a more accurate comparison with the XP spectra, illustrated as green open circles. (b) (i) Visualisation of configuration energies (CE) in relation to the valence orbitals of SnS, showcasing the proposed lone pair bonding mechanism discussed in the text. It is essential to note that these CE values are presented solely as a schematic guide and do not account for ionisation, multi-electron occupancy, or hybridisation. (ii) Computed DOS curves (before cross-section corrections and experimental broadening have been applied) for SnS, intended to assess the contributions of orbital hybridisation.[22,23] Reproduced from ref. 22 with permission from T. Whittles.

Figure 6.6b(ii), it extends from 1 to 6 eV and is primarily comprised of Sn p states with a hybridisation involving S p states, complemented by a slight contribution from S s and Sn s states, particularly around the lower region of the CB at approximately 2 eV.

The top valence band of SnS_2 manifests two key characteristics. The first, spanning from −6 to −8 eV, primarily results from Sn s states with a slight hybridisation involving S s and p states. The second feature, situated between 0 and −5.5 eV, consists of three distinct peaks. Among these, the two lower-energy peaks are predominantly composed of S p states with a hybridisation involving Sn p states. Meanwhile, the largest peak at the pinnacle of the valence band is primarily attributed to S p states, with minimal contributions from other states.

The valence band of Sn_2S_3 likewise exhibits two primary features. The first, covering a range from −6.5 to −9 eV, is chiefly shaped by Sn s states with a slight hybridisation involving S s and p states. The second, spanning from 0 to −6.5 eV, closely resembles the corresponding feature in SnS_2, albeit with less distinct peaks. In the range of −2 to −6 eV, S p states dominate, accompanied by a minor contribution from Sn p states. However, unlike SnS_2, an additional peak appears at the top of the valence band (−1 eV), attributed to a modest hybridisation involving Sn s states. Notably, upon comparing the VB spectra of the three phases, it becomes evident that the VB density of states (VBDOS) for Sn_2S_3 seems to be an equal combination of the VBs of SnS and SnS_2. This observation supports the idea of a 1:1 mixing of Sn^{2+} and Sn^{4+} cations within this material.

In SnS, an interesting feature emerges in the form of Sn s states at the upper end of the valence band, a characteristic that is absent in SnS_2. This phenomenon can be elucidated by considering interactions within the revised lone pair model.[25] Conventionally, the Sn 5s orbital is deemed too energetically distant to directly interact with the Sn 5p orbital. However, a mixing of these states becomes apparent at the top of the valence band in SnS. In the Sn^{2+} oxidation state, the Sn 5s orbital is formally occupied and initially forms a hybridisation with the fully occupied S 3p orbital. Subsequently, the higher-energy antibonding states originating from this interaction further combine with the Sn 5p orbital, giving rise to fully occupied states positioned prominently at the upper part of the valence band. These states exhibit Sn s, Sn p, and S p characteristics. This bonding mechanism is conceptually illustrated in Figure 6.6b concerning the core electrons (CE) for SnS. In contrast, SnS_2 does not display Sn s states at the top of its valence band due to the formal unoccupancy of the Sn 5s orbital in the Sn^{4+} oxidation state. Consequently, it primarily forms bonds with the entirely occupied S 3p orbital, leading to fully occupied bonding states located at lower energies within the valence band, along with corresponding empty antibonding states in the conduction band (not depicted here). As for Sn_2S_3, which comprises a blend of Sn oxidation states, it does exhibit some Sn s states at the top of the valence band, owing to the revised lone-pair mechanism expounded earlier. However, these states are not as conspicuous as in SnS, as they stem exclusively from the Sn^{2+} cations within the material and not from the Sn^{4+} cations. Similar XPS studies have been reported for the germanium and antimony-based chalcogenide PV materials.[26,27]

In related studies of the ternary chalcogenide materials $CuSbS_2$ (CAS)[28] and Cu_3BiS_3 (CBS),[29] which also exhibit favourable optical characteristics as photovoltaic solar absorber materials, yet have displayed relatively lacklustre performance in solar cells, XPS has unravelled this puzzling discrepancy. These studies conducted an in-depth analysis using XPS to scrutinise the core levels of its constituent elements, surface contaminants, ionisation potential, and valence band spectra. Notably, the ionisation potential and electron affinity for CAS stand at 4.98 eV and 3.43 eV, respectively, which are lower than those of other common absorbers such as CIGS. Supported by DFT calculations and validated through experiments, they show that the VBM in CAS is elevated due to the presence of lone pair electrons from the antimony cations, which contribute additional states compared to indium or gallium cations in CIGS. This elevation in the VBM results in a misalignment of the conduction band when CAS is combined with CdS. This misalignment is a key factor contributing to the subpar performance of solar cells incorporating a CAS/CdS heterojunction. In the case of CBS, XP spectra of the valence band and model calculations reveal that the conduction band structure exhibits resemblances to CIGS and CZTS materials at a superficial level. However, the presence of Bi^{3+} cations, along with the occupied Bi 6s orbital, exerts a profound influence. These factors result in a notably low ionisation potential, measured at 4.98 eV. This suggests that the commonly used CdS window layer, which is favoured for CIGS and CZTS, leads to unfavourable band alignment in CBS devices. Consequently, it is advisable to consider alternative materials with better-aligned band structures for CBS devices to advance effectively. These findings further emphasised the critical importance of considering the electronic structure when designing solar cells to optimise open-circuit voltages and overall cell efficiencies.

These XPS investigations, coupled with calculated DOS from DFT shed light on how the hybridisation of the Sn (or Sb cations in CAS or in Sb_2Se_3) lone pair results in the introduction of additional states situated higher within the valence band for SnS. It also elucidates how the valence band of the mixed-valency Sn_2S_3 is a composite of spectral features originating from the single-valency materials, a phenomenon that had not been directly explored previously. These characteristics within the DOS profiles provide a rationale for the previously discussed low IP value and the resultant band misalignments. Both SnS and Sn_2S_3 exhibit low IP values, which can be attributed to the contribution of extra lone pair states situated at the uppermost portion of the VB, leading to a shift in the VBM and consequently a reduction in the ionisation potential. Conversely, the absence of a contribution from these states to the VB in SnS_2 results in a lower VBM compared to the other two phases.

Lone pairs of electrons are revealed to play a significant role in our understanding of band alignment in a number of photovoltaic materials where Ge, Sn, Sb, Bi and Pb cations are present and their outermost s orbitals are occupied. They provide a framework for designing more efficient and sustainable PV technology, not only in chalcogenide-based materials but also in perovskite solar cells. For example, in halide perovskite $MAPbI_3$ the revised lone pair model provides insights into the role of the lone pair electrons of the Pb 6p orbitals in enhancing charge separation, exciton formation, and overall device performance.[30] Photoemission spectroscopic techniques, such as UPS and XPS, provide critical experimental data for understanding the role of lone pair electrons in PV materials. By analysing the electronic structure in the valence band and band alignment, valuable insights into how lone pair electrons influence the photovoltaic behaviour of materials is gained, ultimately contributing to the development of more efficient solar cells.

6.4 Energy Band Alignment at the Heterointerface in PV

The tin sulfide system demonstrated that using the valence band and secondary electron cut-off in an XP spectrum, the ionisation potential and work function of a material can be measured relative to the vacuum level. Together with DFT calculations, this offers a valuable tool for the evaluation and advancement of innovative solar absorber materials. Nonetheless, these parameters alone do not provide a comprehensive description of the interface between an absorber and window layers. Some insight can be gained using the natural band alignment, as discussed in the preceding section, which is related to what is called Anderson's rule. Anderson's rule is an extension of the local vacuum level alignment, whereby the Fermi levels of the two materials align while maintaining the difference in natural electron affinity at the interface. Consequently, the vacuum levels experience band bending at the interface. The conduction band and valence band, defined in relation to the vacuum level, also bend, while the difference in electron affinity between the two materials is maintained fixed. However, this method does not account for any charge redistribution at the interface between the two materials. In cases where the two materials possess significantly different electronegativities or lattice spacings, this approximation may substantially deviate from the actual band alignment.

An alternative method, attributed to Kraut,[31,32] combines various measurements to consider charge redistribution across the interface of two materials. Initially, the binding energy of strong core (E_{CL}) levels and the VBM (E_V) are determined by XPS for both materials in vacuum. Subsequently a thin film of one material is deposited onto the other, ensuring that photoelectrons from the lower layer can still escape and be detected during XPS measurements. This facilitates an interface-sensitive measurement where peaks from both materials are resolved. By measuring the separation between core levels in the two materials (ΔE_{CL}), it becomes possible to directly ascertain the valence band offset (ΔE_V) between them, as illustrated schematically in Figure 6.8 and eqn (6.1):

$$\Delta E_V = \left(E_{CL}^{B} - E_{V}^{B}\right) - \left(E_{CL}^{A} - E_{V}^{A}\right) + \Delta E_{CL} \tag{6.1}$$

Here A and B denote bulk materials A and B, and $\Delta E_{CL} = \left(E_{CL}^{A} - E_{CL}^{B}\right)$ is the difference in the core levels at the interface, measured from a thin sample of material B onto material A. This approach operates under the assumption of an abrupt interface, meaning that there is no gradual change in the band structure across the interface, and therefore does not fully account for band bending. The interface is often gradual, and this can cause the band bending to be different from what is predicted by the Kraut method. Also, the accuracy of the results can be affected by factors such as the thickness of the films and the presence of defects at the interface. Nevertheless, it is often a good approximation for thin films.

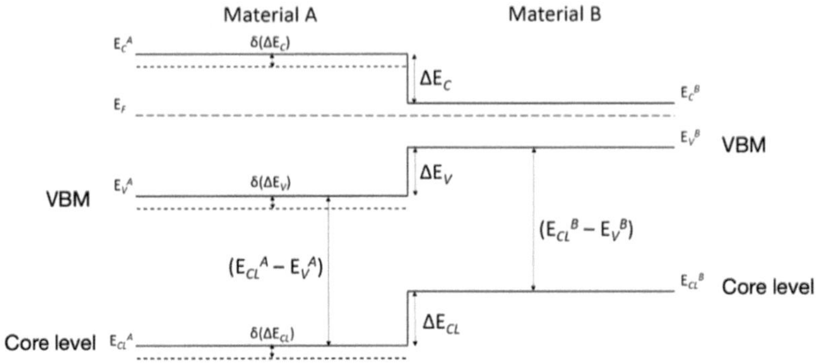

Figure 6.8 Schematic showing PES parameters for the band alignments measured by the Kraut method. The symbol δ refers to the interface of materials A and B. The CBM is inferred from separate knowledge of the band gap. Reproduced from ref. 27, https://doi.org/10.1021/acsaem.0c01477, under the terms of the CC BY 4.0 license, https://creativecommons.org/licenses/by/4.0/.

A drawback to this method lies in sample preparation, which necessitates measurements from three distinct samples: materials A and B, and the interface which must be sufficiently thin. As depicted in Figure 6.9, the inelastic mean free path (IMFP) of a photoelectron is dependent on its kinetic energy, and in accordance with the Beer–Lambert law, 95% of the signal originates within three IMFPs of the surface. Consequently, the XPS sampling depth is approximately 10 nm, necessitating a film thinner than this for the conventional laboratory based XPS application of the Kraut method. An alternative X-ray source, as exemplified by recent investigations into Sb_2Se_3 PV absorbers,[27] offers a solution to overcome this limitation. This source is synchrotron-based hard X-rays and as discussed in Chapter 2, in HAXPES the IMFP extends significantly beyond the range probed by a standard laboratory-based X-ray source, thus enabling probing of buried interfaces.

Antimony selenide (Sb_2Se_3) is a stable binary compound made up of safe and earth-abundant elements, it has a direct band gap of 1.18 eV and a very high absorption coefficient, $>10^5$ cm^{-1}. Its 2D nanoribbon structure allows for very effective carrier transport if the correct orientation is achieved. It is an emerging PV absorber that has attracted considerable recent interest and its device performance has progressed rapidly since first being used in a solar cell, reaching nearly 10% in 2019.[33] The choice of window layer in Sb_2Se_3 devices is a complex one, with no clear consensus on the best material. CdS and TiO_2 are both commonly used, but there are pros and cons to each material, and the role of band alignment in influencing device performance is not fully understood. As indicated in the schematic shown in Figure 6.5, the band alignment between window layer and absorber layer determines the energy barrier for electrons to be transferred from the absorber to the electrode. The band alignment can also affect the recombination of electrons and holes in the device, which can lead to lower device efficiency.

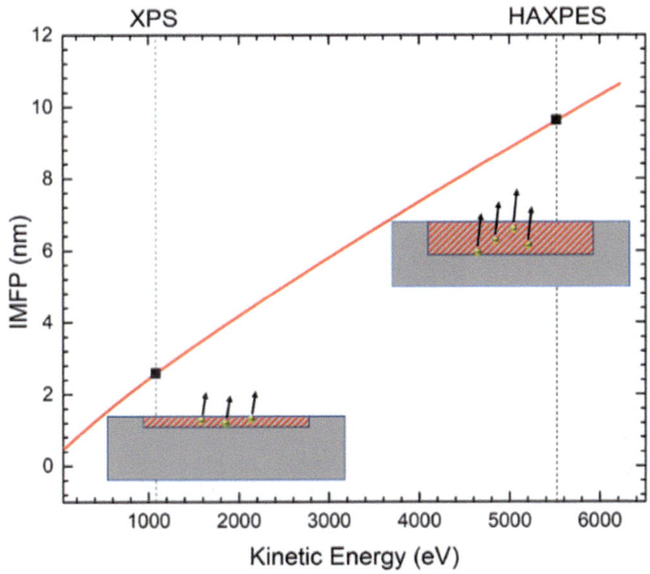

Figure 6.9 The inelastic mean free path of photoelectrons in the PV absorber Sb_2Se_3 with respect to kinetic energy. Black points show the kinetic energy of an electron escaping the Sb 3d orbital for both conventional Al Kα XPS and HAXPES (assuming a photon energy of 5921 eV) and the inset sketches illustrate the relative probing depths of the two techniques. Reproduced from ref. 27, https://doi.org/10.1021/acsaem.0c01477, under the terms of the CC BY 4.0 license, https://creativecommons.org/licenses/by/4.0/.

Recent advances in Sb_2Se_3 device performance have been achieved using close space sublimation (CSS) or vapour transport deposition. These techniques produce large grain sizes with good, preferred orientation for carrier transport. However, they also limit the thickness of the films that can be deposited while still achieving good coverage. This is important for the validity of the Kraut method measurements, which require good coverage.

At a thickness of about 20 nm, good coverage can be achieved by CSS. This thickness is also within the sampling depth of HAXPES, using hard X-rays with an energy of about 6000 eV. This allows photoelectrons with a mean free path of more than 9 nm and a probing depth of more than 27 nm. Using HAXPES, the band offset between a realistic device-relevant layer of Sb_2Se_3 and a window layer can be directly measured. Recently, Shiel *et al.*[27] have used synchrotron based HAXPES to measure the band offsets between Sb_2Se_3 absorber layer and window layers of CdS and TiO_2 of device level thicknesses and correlated their measurements with actual device performance characteristics (Figure 6.10). They deposited about 20 nm films of Sb_2Se_3 films by CSS on RF sputtered substrate films of CdS (thickness 80 nm) and TiO_2 (thickness 60 nm). The valence band offsets between Sb_2Se_3 and CdS and between Sb_2Se_3 and TiO_2 were obtained using several characteristic Sb_2Se_3 core level

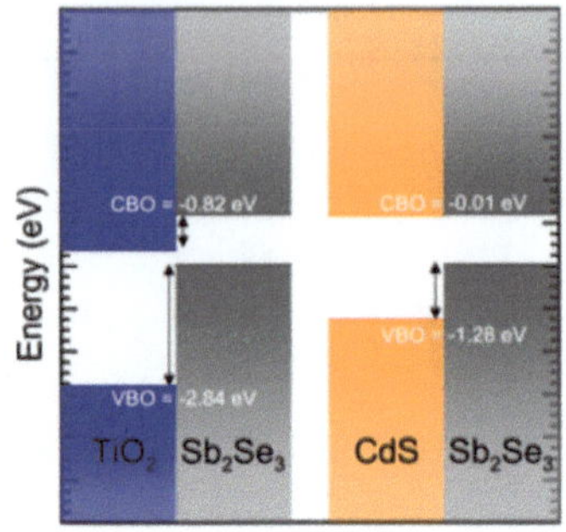

Figure 6.10 Band offsets measured *via* the Kraut method using HAXPES. Reproduced from ref. 27, https://doi.org/10.1021/acsaem.0c01477, under the terms of the CC BY 4.0 license, https://creativecommons.org/licenses/by/4.0/.

peaks (Sb 3d, Sb 4d, and Se 3d) and the valence band edge. For the window layer underneath the absorber layer, the most intense core level peak of Cd 3d and Ti 2p were measured with respect to their respective valence band edge. An average of the VBOs was then calculated from the different core levels, and using separately derived band gaps, the CBOs and the complete band alignment was determined as shown in Figure 6.10. Their results using HAXPES showed a CBO for Sb_2Se_3/CdS of −0.01 eV, but a significantly different CBO of −0.74 eV for Sb_2Se_3/TiO_2. These results suggested that CdS has an optimal band alignment with Sb_2Se_3, while TiO_2-based devices are probably limited by a cliff-like offset that leads to recombination and a limited built-in voltage. This is especially relevant considering the evidence of detrimental intermixing at the CdS/Sb_2Se_3 interface. Some groups have preferred to use TiO_2 as a window layer, despite the unfavourable band alignment. This has also inspired the use of an interdiffusion blocking interlayer and optimising band alignments as discussed by Wand *et al.*[34] Harnessing advantageous band alignments while preventing interdiffusion could provide a platform for pushing the efficiencies of Sb_2Se_3 to the next level.

HAXPES, with its greater probing depth, facilitates investigations of buried interfaces beneath thicker films and, with careful planning, enables the measurement of a full solar cell structure. In a recent publication by Svanström *et al.*[35] a novel methodology employing HAXPES is introduced, enabling the direct measurement of the interfacial photovoltage generated within a complete perovskite based solar cell device during its operation. Perovskite materials, particularly lead halide perovskites, have achieved remarkable power conversion efficiencies exceeding 25%, placing them on par with silicon solar cells.[36] However, the commercialisation of perovskite solar cells has faced obstacles due to their limited long-term stability, which has been partially attributed to the dynamics occurring at the device interfaces.[37] XPS has been used to study X-ray induced decomposition of perovskites in ultra-high vacuum.[38] This study focused on the degradation of $CH_3NH_3PbI_3$ perovskite films induced by X-rays using XPS conducted within an *in situ* ultrahigh vacuum environment. The findings reveal that in pristine methylammonium lead iodine, Pb^{2+} was present without any initial trace of Pb^0. However, after a few hours of exposure to soft X-rays, a distinct Pb^0 signal emerged, suggesting that the $CH_3NH_3PbI_3$ perovskite structure undergoes a decomposition process, resulting in the formation of metallic Pb. Furthermore, it was observed that the nitrogen content experienced a significant decrease within the first hour of X-ray exposure. Perovskite instability under X-ray illumination requires careful measurement setup and device structure. Some organic and hybrid-perovskite materials may experience degradation or instability when exposed to the vacuum conditions typically required for conventional PES measurements. The vacuum exposure can lead to structural and compositional alterations, impacting the electrical characteristics and efficiency of these materials. Moreover, they demonstrate sensitivity to both oxygen and moisture.[39] To alleviate the challenges associated with conducting PES assessments under ultrahigh vacuum conditions, the growing availability of near-ambient pressure PES presents an intriguing development.[40]

Svanström and colleagues[35] present an innovative technique for directly measuring the band alignment and variations in the interfacial electric field within a fully operational lead halide perovskite solar cell structure under operating conditions using HAXPES. Their *operando* measurements extend upon prior laboratory-based XPS studies focusing on the interface and bulk electronic properties of complete Au/spiro-MeOTAD/perovskite/TiO_2 perovskite solar cells designed in a tapered cross-section architecture,[41] where it was determined that most of the photovoltage was generated at the interface with the spiro-MeOTAD layer. Svanström *et al.* elaborate on the essential design considerations necessary for both the solar cell devices and the measurement setup, and they present findings for the perovskite layer, hole transport material, and gold layers at the back contact of the solar cell.

The experiments were conducted at the GALAXIES beamline, situated within the SOLEIL synchrotron, utilising a photon energy of 3000 eV. This choice of photon energy resulted in an inelastic mean free path (IMFP) of approximately 5 nm and a probing depth of about 15 nm. The spectrometer consisted of a Scienta Omicron EW4000 HAXPES hemispherical analyser. In the solar cell architecture, as depicted in Figure 6.11a, the average thickness of gold was 7 nm, with a 10 nm polymeric hole transport layer (HTL) P3 sandwiched between the $Cs_{0.17}FA_{0.83}PbI_3$ perovskite and the gold contact. The HTL layer acts as a buffer between the metal electrode and perovskite to prevent undesirable interaction between the perovskite's Pb component and the gold contact. The inset in Figure 6.11a illustrates the cell mounted on a sample plate designed for external voltage bias measurements at the GALAXIES beamline. In this configuration, the TiO_2/FTO substrate served as the grounding point, while the gold electrode was connected to an external circuit, enabling voltage measurement or application during PES measurements. It is important to note that the illumination setup ensured adequate illumination of the sample for charge separation but did not replicate the solar spectrum.

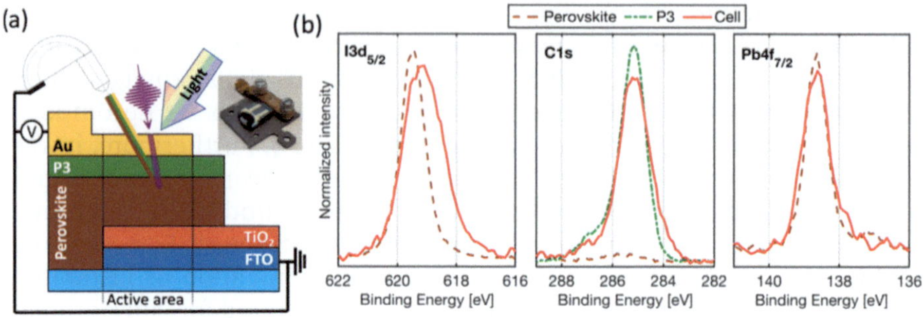

Figure 6.11 (a) Diagram illustrating the cell architecture, areas analysed by photoelectrons, and the electrical connections in the solar cell. The inset image displays the cell mounted on the plate and linked to the electrode. (b) Core level spectra for I $3d_{5/2}$, C 1s, and Pb $4f_{7/2}$ obtained from the pristine perovskite, a reference P3 sample, and the $Cs_{0.17}FA_{0.83}PbI_3$/P3/Au interface within the solar cell device. Reproduced from ref. 35, https://doi.org/10.1021/acsami.2c17527, under the terms of the CC BY 4.0 license, https://creativecommons.org/licenses/by/4.0/.

Figure 6.11b shows the core level spectra for I $3d_{5/2}$, C 1s, and Pb $4f_{7/2}$. These spectra were collected from the perovskite (bare) reference film, the P3 reference film (representing the hole conductor), and from the $Cs_{0.17}FA_{0.83}PbI_3$/P3/Au interface within the solar cell. For the Pb $4f_{7/2}$ core level, only one contribution was observed at 138.6 eV binding energy for both the interface and the perovskite reference, originating from Pb^{2+} within the perovskite. The C 1s core level displayed a single peak at approximately 285.1 eV for the cell, attributed to carbon within the P3 polymer. In contrast, the I $3d_{5/2}$ core level exhibited notable differences between the bare perovskite layer and its incorporation into the device structure. Specifically, the I $3d_{5/2}$ spectra appeared significantly broader in the measurements taken on the device structure when compared to those on the uncovered perovskite layer, indicating the presence of an additional iodide compound within the device structure, possibly due to iodide diffusion into the P3 HTL. Consequently, the authors recommended using the Pb 4f core level over I 3d as a representative signal from the perovskite layer.

Figure 6.12 presents the time evolution of the core levels (Pb $4f_{7/2}$, I $3d_{5/2}$, C 1s, and Au $4f_{7/2}$) at the $Cs_{0.17}FA_{0.83}PbI_3$/P3/Au interface within the solar cell. The measurements were performed under short-circuit conditions in the dark (top row) and at open circuit with illumination (middle row). The bottom row displays the average spectra for each condition. When the cell was short-circuited, the binding energy of all core levels remained relatively constant. However, under open-circuit conditions during illumination, shifts in the core levels towards higher binding energies were observed compared to short-circuit conditions. This shift was attributed to the generation of a photovoltage, resulting in a separation of the Fermi level between the grounded front contact (FTO) and the gold electrode at the back contact.

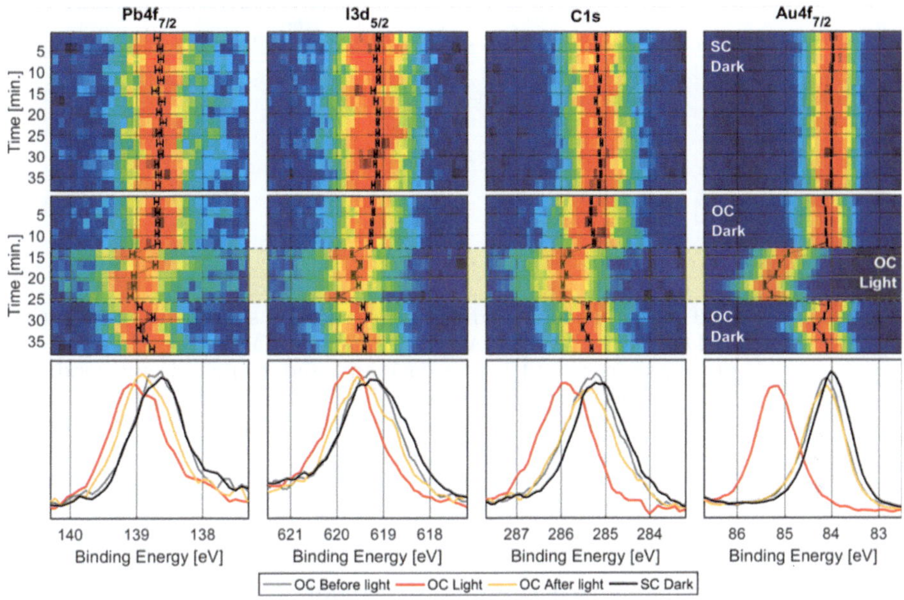

Figure 6.12 Top row: intensity *versus* binding energy and time mapping of the $Cs_{0.17}FA_{0.83}PbI_3$/P3/Au interface in the solar cell device under short circuit conditions in the absence of light. Middle row: corresponding mapping during open circuit conditions with and without light. Bottom row: average core level spectra obtained during short circuit conditions in the dark and open circuit conditions before, during, and after illumination. Reproduced from ref. 35, https://doi.org/10.1021/acsami.2c17527, under the terms of the CC BY 4.0 license, https://creativecommons.org/licenses/by/4.0/.

Figure 6.13 illustrates the shift in binding energy of the Pb 4f, C 1s, and Au 4f core levels over time. The photovoltage generated by the illuminated cell was measured externally using a voltmeter, and these voltage values were overlaid in Figure 6.13b for comparison. The shift in the Au 4f signal closely correlated with the voltage generated by the cell, suggesting that the shift in the Au 4f core level resulted from the photovoltage generated by the cell. Similarly, the shifts observed in the C 1s and Pb 4f core levels were attributed to the internal photovoltage generated within the device stack, specifically within the HTM and perovskite layer. However, it is worth noting that the magnitude and behaviour of the shift varied among the core levels, with average shifts of 1.19 ± 0.19 eV, 0.76 ± 0.07 eV, and 0.33 ± 0.16 eV for the Au 4f, C 1s, and Pb 4f core levels, respectively.

The core level positions are sensitive to changes in internal electric fields. When light illuminated the solar cell, electrons and holes were excited to the conduction and valence bands, leading to the build-up of quasi-Fermi levels for electrons and holes within the perovskite. Holes were extracted from the perovskite to the HTM and then to the gold electrode over the interface. Consequently, the Fermi level of the gold electrode was expected to adjust to

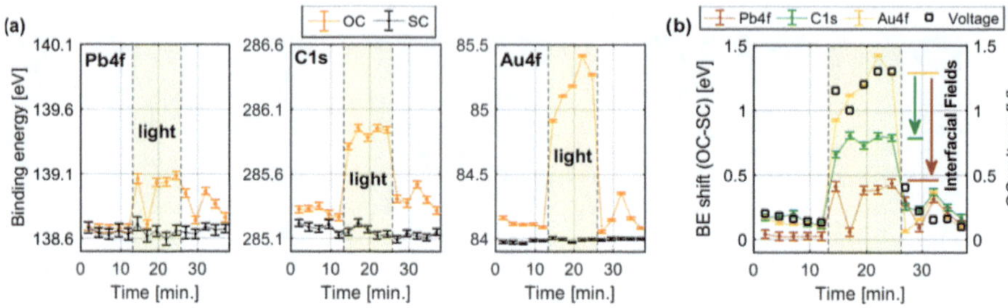

Figure 6.13 (a) Time-dependent binding energy variations of Pb $4f_{7/2}$, I $3d_{5/2}$, C 1s, and Au $4f_{7/2}$ at the $Cs_{0.17}FA_{0.83}PbI_3$/P3/Au interface in the solar cell device during short circuit in the dark and under open circuit conditions with and without light. (b) The shift in binding energy during open circuit relative to the average binding energy during short circuit. Reproduced from ref. 35, https://doi.org/10.1021/acsami.2c17527, under the terms of the CC BY 4.0 license, https://creativecommons.org/licenses/by/4.0/.

match the quasi-Fermi level of the holes in the perovskite. In this open-circuit, illuminated state, hole accumulation generated a photovoltage through light-induced charge redistribution. As a result, the internal electric fields differed from the equilibrium state in the dark, leading to shifts in the core levels of various materials. These shifts provided a means to detect absolute changes in potential between different layers.

The total electric field change across the device corresponded to the 1.19 eV shift in the Au 4f core level, reflecting an open-circuit voltage of approximately 1.2 V. The electric field change from Au up to the P3 layer, calculated as the difference between the shifts in Au 4f and C 1s (0.43 eV), corresponded to an internal photovoltage of 0.43 V. Similarly, the photovoltage between the P3 layer and the perovskite, reflected by the binding energy shifts of 0.77 and 0.33 eV, was determined to be 0.43 V. Consequently, it was determined that approximately 70% of the open-circuit voltage was generated at the back interfaces, evenly distributed between the perovskite/HTM (0.43 V) and HTM/Au interfaces (0.43 V).

These core level measurements were used to estimate the band alignment at the $Cs_{0.17}FA_{0.83}PbI_3$/P3/Au interfaces under different conditions: short-circuit in the dark and open-circuit with illumination. Reference measurements of the valence band for pure P3 and perovskite samples were used, along with core level to valence band maximum (VBM) binding energy differences, to determine the band alignment in the buried interfaces using the method given by Kraut.[31] The resulting band diagrams (Figure 6.14) provided insights into the energy alignment at the back contacts of the solar cell, both in the dark at short circuit and under illumination at open circuit. It is important to note that these band diagrams represent average band positions and do not account for band bending which could be present at these interfaces and change between the two cases. This study clearly demonstrates the photovoltage generation and measurement across the device and shows that

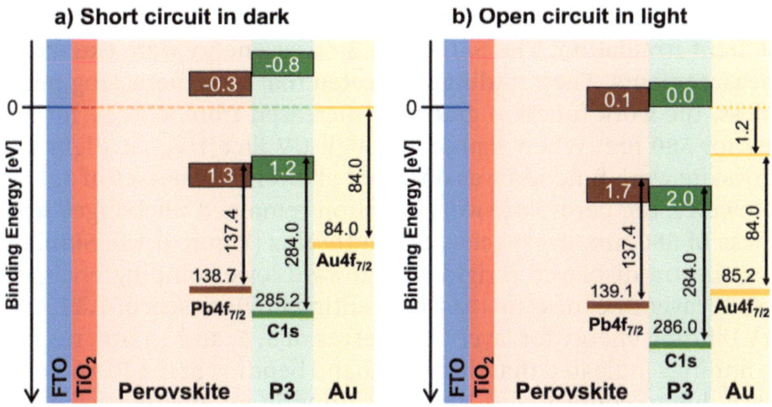

Figure 6.14 Schematic alignment of valence bands determined by the difference between core-level and VBM binding energies, along with the band gap, for the solar cells under short circuit conditions in the dark (a) and open circuit during illumination, generating a voltage of 1.2 V (b). The positions of core levels, valence bands, and conduction bands are depicted proportionately in the figure, reflecting a system where charge redistribution primarily occurs at the interfaces, leaving the bulk material of each layer unchanged. In panel (a), the dashed line signifies the Fermi level at 0 eV binding energy, as FTO is grounded to the spectrometer. In panel (b), the Fermi level of FTO is at 0 eV binding energy since the FTO is still grounded, and the Fermi level of Au is at 1.2 eV binding energy, representing the 1.2 V open-circuit voltage. Reproduced from ref. 35, https://doi.org/10.1021/acsami.2c17527, under the terms of the CC BY 4.0 license, https://creativecommons.org/licenses/by/4.0/.

contrary to the conventional assumption that interfacial energy-level alignment remains fixed once the interface forms, significant energy-level realignment occurs at the interfaces during photoexcitation.

Related studies[42–44] employed laboratory-based measurements of work function and valence band of ITO/HTL/Perovskite stacks to demonstrate significant energy-level realignment during photoexcitation at the interfaces between organic semiconductors and metal-halide perovskites. To illustrate the effect of visible-light-induced changes in interfacial energy level alignment, the study utilised poly-[bis-(4-phenyl)-(2,4,6-trimethylphenyl)-amin] (PTAA) as the HTL. PTAA is widely used in efficient p–i–n perovskite solar cells.[43] Initially, they investigated the bare PTAA film deposited onto indium-tin-oxide (ITO) using UPS which was conducted using a monochromated helium discharge lamp (photon energy of 21.22 eV) in an ultrahigh vacuum system. In the dark and under illumination (white light with ≈1.5 sun intensity), the HTL exhibited a relatively constant work function (Φ) of 4.37 eV and a HOMO level onset at 0.68 eV binding energy.

To determine the band alignment at the PTAA/perovskite interface, $FA_{0.85}Cs_{0.15}PbI_3$ films of varying thicknesses (ranging from ≈3 to ≈580 nm) were deposited onto PTAA/ITO substrates, and UPS measurements were conducted for each perovskite thickness, both in the dark and under 1.5 sun visible light irradiation. The SEC and VB onset energy were extracted from the measurements. Their findings indicated that with increasing perovskite thickness, the work function gradually increased from 4.35 eV for 3 nm to 4.79 eV for 580 nm. When exposed to high UV flux (UV_H), a slight increase in perovskite work function was observed at layer thicknesses of 3, 5, and 10 nm. However, the perovskite work function remained unchanged at a layer thickness of 580 nm, irrespective of the UV flux (Figure 6.15). Similarly, the valence spectra displayed a rigid shift toward lower binding energy as the perovskite layer thickness increased, resulting in VB onsets of 1.48, 1.39, and 1.42 eV binding energy for layer thicknesses of 3, 5, and 10 nm, respectively. These findings indicated that there was band bending at the PTAA/perovskite interface due to contact with PTAA, resulting in an energy level alignment corresponding to a type-II heterojunction with a significant energy offset of ≈0.80 eV for hole extraction from the perovskite to the PTAA (Figure 6.15c).

Under 1.5 sun visible light illumination, a consistent increase in the sample work function by ≈ 0.43 eV was determined for all thicknesses. This shift in the secondary electron cutoff was quantitatively identical to the shift in the valence band, raising the perovskite VB onset toward the PTAA HOMO

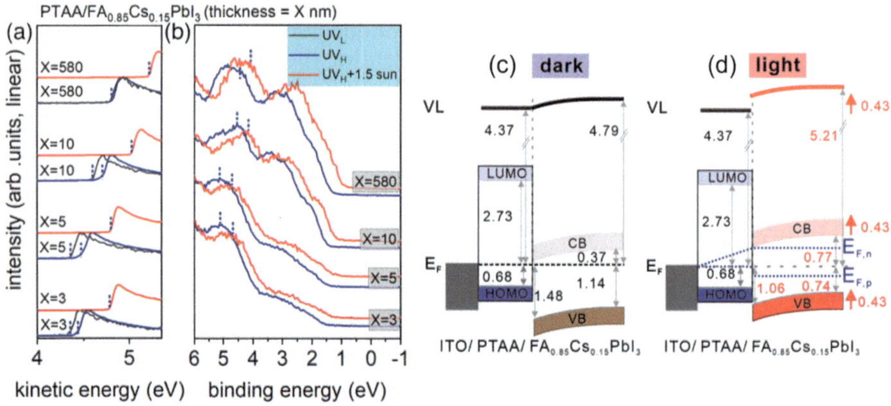

Figure 6.15 UPS spectra of $ITO/PTAA/FA_{0.85}Cs_{0.15}PbI_3$ stacks with varied $FA_{0.85}Cs_{0.15}PbI_3$ film thickness ranging from 3 to 580 nm. (a) SEC and (b) valence region. UV_L (grey lines) and UV_H (blue lines) denote the UV flux at low and high intensity, respectively. Schematic energy level diagrams of $ITO/PTAA/FA_{0.85}Cs_{0.15}PbI_3$ stacks in (c) dark and under (d) 1.5 sun white light irradiation. VL refers to the vacuum level. The indicated energy values are given with respect to E_F set to zero. The blue dashed lines in (b) and (c) represent the quasi-Fermi levels $E_{F,n}$ and $E_{F,p}$ for electrons and holes, respectively; the red arrows and numbers indicate the illumination-induced change of the perovskite energy levels with respect to E_F of the conductive ITO substrate. Reproduced from ref. 43, https://doi.org/10.1002/solr.202101065, under the terms of the CC BY 4.0 license, https://creativecommons.org/licenses/by/4.0/.

level onset, yielding an offset of about 0.38 eV (Figure 6.15d). Furthermore, it is noteworthy that upon deactivation of the light source, all energy levels returned to their initial states, strongly suggesting that these reversible shifts have an electrostatic origin. It is important to emphasise that the energy levels of PTAA remained unchanged during light exposure, as discussed earlier. The fundamental mechanism at play revolved around the generation of photoexcited carriers within the perovskite layer and their subsequent transfer to the PTAA HTL. This process leads to the accumulation of electrons within the perovskite layer. Notably, the shifts in energy levels induced by illumination in the thin perovskite films support the conclusion that a significant portion of the accumulated negative charge within the perovskite layer is situated near the PTAA interface. This observation is further reinforced by the fact that even high levels of UV flux induced noticeable shifts at the PTAA/perovskite interface, but this effect diminishes as the perovskite thickness reaches 580 nm.

An alternative approach to determine energy band alignment at heterointerfaces in real-world solar cell samples, and where band bending can also be determined, is through depth profiling achieved by meticulous Ar^+ ion sputtering, as recently highlighted by Gansukh *et al.*[45] This approach is suitable where the materials are not susceptible to damage either from the ion sputtering or the X-ray beam. In their research, they harnessed UPS and XPS to investigate the junction between CdS and CZTS, as well as the alloy Ag-CZTS. The incorporation of silver (Ag) into CZTS has been proven to mitigate antisite defects within the bulk material, consequently enhancing efficiency. The thin films of photovoltaic (PV) absorbers CZTS and ACZTS were fabricated *via* pulsed laser deposition (PLD) onto soda-lime glass substrates, followed by the deposition of CdS through a chemical bath process. Subsequently, the CdS film underwent precise sputtering while concurrently recording photoemission data, traversing from the bulk CdS layer through the interface and into the bulk absorber films. This study underscores the utilisation of synchrotron radiation in PV research. The UPS and XPS measurements were conducted at the Matline end-station situated within the ASTRID storage ring facility. Photon energy values were selected to optimise the cross-section, ensuring the most effective measurement of both valence band and core levels. Specifically, for UPS spectra, excitation energies of 60 and 130 eV were applied. Meanwhile, for XPS core level spectra, varying excitation energies were employed: 60 and 130 eV for Cd 4d and Zn 3d levels, 400 eV for Ag 3d levels, and 220 eV for S 2p levels.

Figure 6.16 illustrates a series of valence band spectra for the two heterostructure samples. The outcomes reveal the ability to distinguish individual valence band edges in the spectra, portraying the interface as a composite of features related to the absorber and CdS overlayer. Figure 6.17 presents the valence band spectra from three distinct depths. The valence band maximum of CdS is found to reside at −2.1 eV below the Fermi level, while that of CZTS is at −0.46 eV. Disregarding band bending effects, they deduced natural band offsets of 1.64 eV for CdS/CZTS and 1.52 eV for CdS/

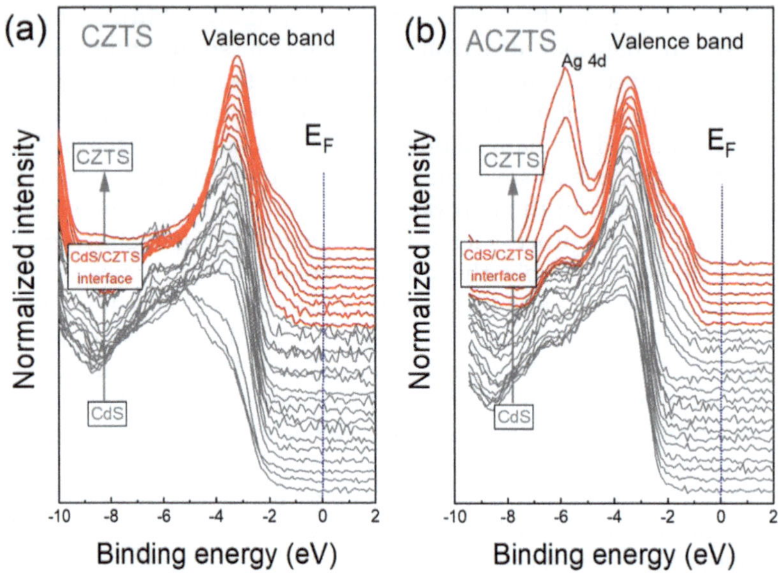

Figure 6.16 UPS spectra taken at photon energy 60 eV at different sputtering depths across (a) the CdS/CZTS and (b) CdS/ACZTS heterostructures. Red plots indicate the interface regions and E_F denotes the Fermi level. Reproduced from ref. 45, https://doi.org/10.1038/s41598-020-73828-0, under the terms of the CC BY 4.0 license, https://creativecommons.org/licenses/by/4.0/.

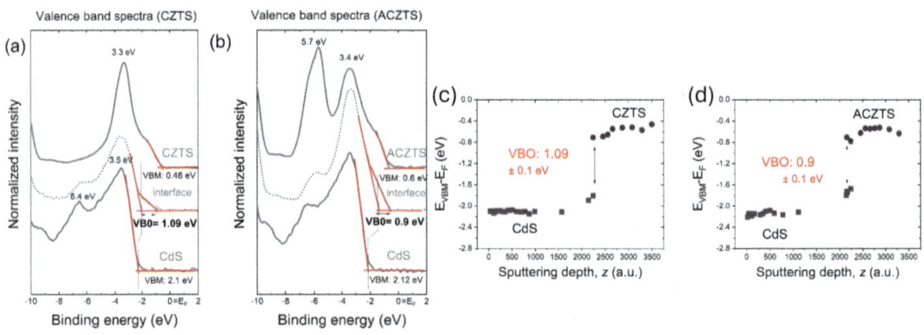

Figure 6.17 (a) and (b) PES spectra of CdS, interface and CZTS/ACZTS. The VBO was determined from the difference between the VBM positions at the interface. (c) and (d) Evolution of the VBM as a function of sputtering depth for CdS/CZTS and CdS/ACZTS. Reproduced from ref. 45, https://doi.org/10.1038/s41598-020-73828-0, under the terms of the CC BY 4.0 license, https://creativecommons.org/licenses/by/4.0/.

ACZTS. With the introduction of Ag through alloying, the VBM of CZTS shifts lower by 0.14 eV, a phenomenon consistent with predictions from DFT calculations. This shift arises from the hybridisation of deeper-lying Ag 4d states with S 3p states.

As the thickness of the top layer diminishes, band bending at the interface influences the local charge distribution, resulting in a shift of the CdS VBM. Consequently, values of 1.09 eV and 0.9 eV were determined for the valence band offsets (VBO) at the interface. Figure 6.17c and d provides a comprehensive depiction of the evolution of VBM positions, illustrating the band bending at the interface and the VBO between CdS and the two absorbers. These findings were substantiated by core-level measurements and reveal a substantial shift of core-level peaks and the CdS valence band edge to lower energy (exceeding 0.4 eV) upon the removal of the CdS layer, stemming from band bending and charge redistribution at the interface. Their study demonstrates a distinctive cliff-like band alignment at both CdS/CZTS and CdS/Ag-CZTS heterointerfaces. The applied experimental approach showcased the ability to conduct nanometre-scale depth profiling of the electronic structure at the p–n junction between the absorber and window layers.

6.5 Energy Band Alignment Studies in Organic and Hybrid-perovskite PV

In recent times, there has been a significant surge of interest in organic photovoltaic cells (OPVs) due to their appealing characteristics, including lightweight construction, flexibility, cost-effectiveness, and eco-friendliness. Unlike conventional solar technologies, OPVs have demonstrated great potential for applications in building-integrated energy generation systems, such as windows and greenhouses. This is attributed to their ability to efficiently absorb light in the near-infrared spectrum while maintaining semi-transparency and, crucially, possessing a neutral optical density in the visible range. Notably, the emergence of ladder-type non-fullerene acceptors (NFAs) has ushered in a new era for OPVs, achieving power conversion efficiencies (PCEs) of approximately 18% in non-transparent cells.[46,47]

XPS has been instrumental in examining various aspects of OPVs. One area of focus has been interface engineering, where XPS is employed to both characterise and modify the chemical composition and energy level alignment at interfaces, including organic–organic and organic–metal interfaces. This strategic approach aims to enhance charge transport efficiency and minimise losses within OPVs. Furthermore, XPS has played a crucial role in the realm of surface modification of organic materials. It has been particularly valuable in studying the effects of surface treatments such as chemical doping or the application of self-assembled monolayers. These investigations delve into how such treatments influence the electronic structure of organic semiconductors, ultimately aiding in the design of organic materials that exhibit improved charge transport properties and photon absorption. Insertion of interfacial layers between the active layer and the TCO in OPV cells plays a vital role in device performance, as demonstrated by Song *et al.*[48] who used standard XPS measurements to determine the work function and HOMO energies of SAM modified glass/ITO surfaces. Moreover, the susceptibility of organic solar cells to degradation over time due to environmental factors, such as moisture and oxygen, has been a subject of concern. In this regard, XPS has been employed effectively to monitor the chemical changes occurring at the surface of organic materials during aging or exposure to environmental stressors. These studies provide crucial insights into the mechanisms of degradation, offering opportunities for enhancing the stability and durability of organic solar cells.[49,50]

Most contemporary research on OPVs focuses on solution-processed bulk heterojunction cells. In these cells, the photoactive layer, composed of donor and acceptor materials, is intricately blended at the nanoscale, facilitating the dispersion of interfaces within the active layer at optimal diffusion distances. This design maintains the necessary layer thickness for efficient light absorption. OPV architectures can be categorised as either conventional or inverted, depending on the orientation of the electrodes. Charge-carrier transport within the device is facilitated by hole-transport and electron-transport interfacial layers (ETLs) on opposite sides of the active layer. In a conventional stack, a typical HTL might involve PEDOT:PSS, often paired with an ITO anode, while a typical electron-transport layer could feature materials like PFN-Br or Bathocuproine (BCP), typically paired with a metal cathode (such as silver, gold, or aluminium, Figure 6.18). This architecture typically includes a transparent substrate, like glass or flexible plastic, followed by an ITO layer serving as a transparent conductive electrode. The HTL resides between the ITO and the active layer, followed by the photoactive layer, the ETL, and ultimately, a metal electrode. Energy level alignment within this architecture plays a pivotal role in efficient charge carrier generation and collection. The HTL's work function must be lower than the ionisation potential of the active layer, while the ETL's work function should be higher than the ionisation potential of the active layer. This ensures that holes are attracted to the HTL and electrons to the ETL, allowing them to be efficiently transported to their respective electrodes.[51]

The photoactive layer comprises an acceptor material with a narrow energy band gap, signifying a small difference between its HOMO and LUMO levels. This characteristic is strategically employed to promote efficient hole transfer from the acceptor to the donor and facilitate electron transfer from the donor polymer to the acceptor's LUMO. Metal oxides, such as TiO_2, ZnO, MoO_3, NiO, and others, are frequently employed in OPV devices, favoured for their proficiency as either charge transport layers or charge extraction layers (see references given in ref. 52). An advantage lies in the ability to tune their properties, such as bandgap, electronic structure, and work function, by manipulating their composition and processing conditions. For example, the work function of these metal oxides can be tailored for the VBM to align effectively with the HOMO levels of the organic photoactive donor materials. This alignment streamlines hole extraction and mitigates energy losses at the interface between the organic material and the HTL. The precise alignment of energy levels results in a reduced energy barrier for hole transport, minimising resistive losses. Additionally, metal oxides can be engineered to

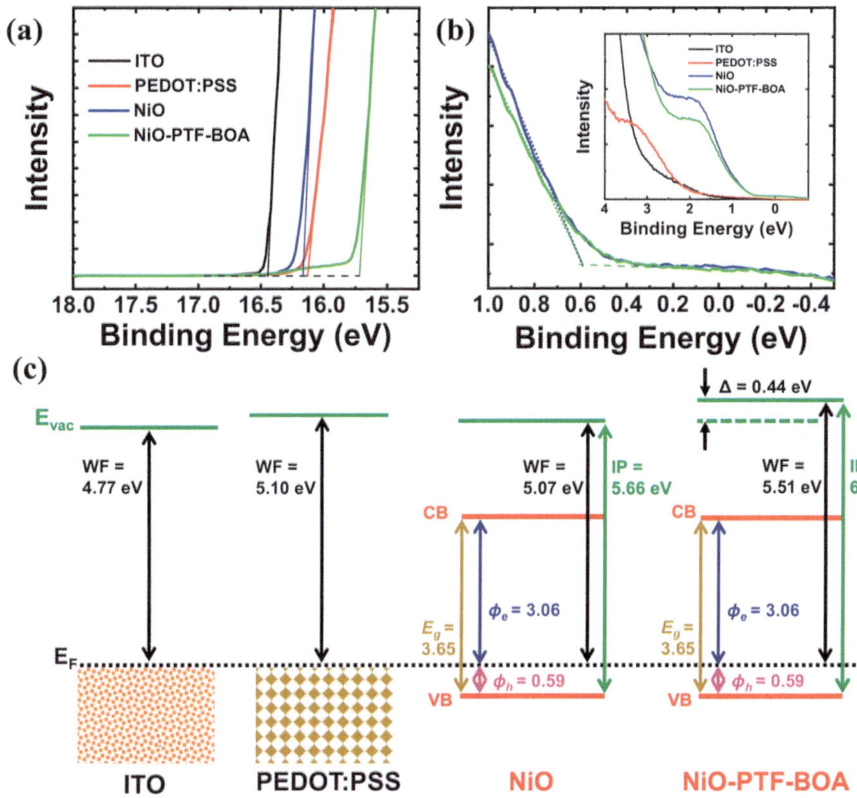

Figure 6.18 a) Secondary electron cut-off region and b) valence band edge obtained from the UPS spectra of PEDOT:PSS, NiO, and NiO-PTF-BOA treated layers. (c) Energy band diagrams of PEDOT:PSS, NiO, and NiO-PTF-BOA layers obtained from the measurements. Reproduced from ref. 53, https://doi.org/10.1002/admi.202201274, under the terms of the CC BY 4.0 license, https://creativecommons.org/licenses/by/4.0/.

exhibit exceptional optical transparency in the visible and near-infrared portions of the electromagnetic spectrum. This transparency is of paramount importance as it permits incident sunlight to permeate the HTL and penetrate the organic active layer for absorption.

In a recent study by Tran *et al.*,[53] UPS was employed to assess the ionisation potential and work function of various hole-transport layers in non-fullerene organic solar cells (OSCs) featuring the PBDB-T-2F:Y6 active material. The investigation included a comparative analysis of the commonly used PEDOT:PSS HTL with NiO and NiO subjected to surface modification using 4-(trifluoromethyl)benzoic acid (PTF-BOA). The intrinsic NiO nanoparticle layer, in its unaltered state without any surface treatment, exhibited a lower work function (WF) value of 5.1 eV, which is a typical value for solution-processed NiO. Conversely, the NiO layer treated with PTF-BOA, denoted as NiO-PTF-BOA, displayed an elevated WF value of 5.5 eV (Figure 6.19).

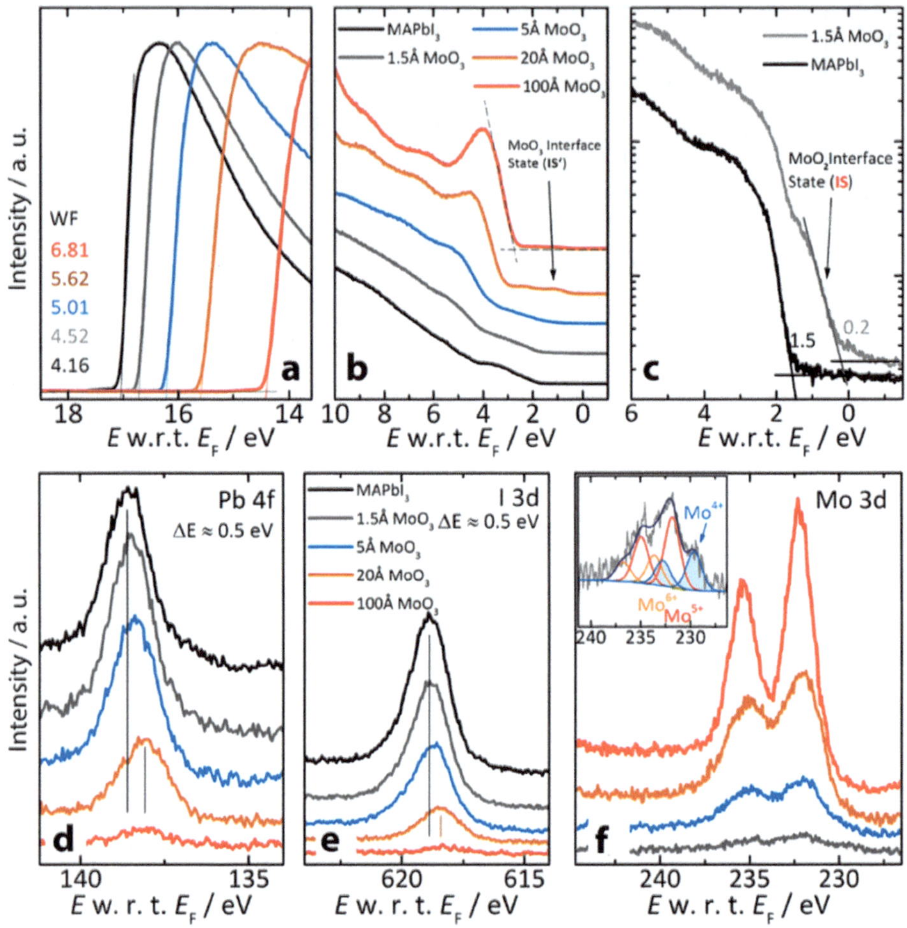

Figure 6.19 (a)–(c) UPS and (d)–(f) XPS photoemission spectra of $MAPbI_3$ on TiO_2 with MoO_3 incrementally evaporated on top. (a) SEC and work function (WF), (b) valence band region and (c) close-up of valence band region in semilog plot, emphasising the molybdenum oxide interface state IS (see also Figure 6.3a). (d) Pb 4f, (e) I 3d, and (f) Mo 3d core level spectra. The inset in f depicts a lineshape analysis of the Mo 3d core level spectrum of the 1.5 Å MoO_3 film, showing the Mo^{4+} component (shaded, blue). Reproduced from ref. 58 with permission from American Chemical Society, Copyright 2016.

Further analysis using impedance spectroscopy revealed that PBDB-T-2F:Y6 solar cells incorporating the PTF-BOA-treated NiO HTL exhibited reduced recombination rates and shorter hole transport times. Consequently, these solar cells achieved an impressive efficiency of 16.36%, marking the highest reported efficiency for OSCs utilising NiO as the HTL.

The use of XPS was further demonstrated by measurements of core-level spectra, specifically examining the Ni $2p_{3/2}$ and O 1s peaks derived from both

NiO and NiO-PTF-BOA. Following a careful peak-fitting analysis, they elucidated that the increase in work function can be attributed to the interaction of the organic layer's carboxyl group, COOH, with the NiO surface. This interaction leads to the attachment and subsequent dissociation of COOH on the NiO surface, resulting in the formation of NiOOH. This NiOOH layer, in turn, constitutes a dipole layer that contributes to the enhancement of the work function. Steirer *et al.* showed that the WF of solution based NiO_x can be strongly modified by an oxygen plasma treatment. An increase from 4.75 eV to 5.32 eV upon O_2-plasma treatment has been observed which was attributed to an introduction of nickel oxyhydroxide surface species inducing a large interface dipole.

As illustrated earlier, it is evident that the work function and electronic structure of transition metal oxides can be adjusted through surface modifications as well as by the introduction of defects, or altering the oxidation state of cations, all with the aim of achieving optimal energy-level alignment. This capability allows these oxides to position molecular energy levels in proximity to the Fermi level for many organic semiconductors, a feature that facilitates efficient charge transfer as demonstrated in studies such as those by Greiner *et al.*[54] In their studies, XPS core-level and valence-band photoemission spectroscopy were employed to characterise a wide range of oxide samples and organic layers to correlate the electronic structures of the oxides and the organic molecules. The investigated oxides encompassed a diverse range of electronic properties and work functions, and included n-type, p-type wide band gap semiconductors and metallic oxides. These oxides were paired with different organic semiconductors, and the energy-level alignment was determined by assessing the HOMO levels of the various organic materials. Notably, while various potential correlations were explored, such as connections between the organic HOMO energy and the oxide's VBM, CBM, and work function, the only relationship found to influence energy alignment was the energy difference between the oxide's work function and the ionisation energy of the organic layer. This remarkable versatility empowers transition metal oxides with a wide range of applications in the field of organic electronics. In a recent study Ma *et al.*[55] used XPS to find that oxidation state and work function of the MoO_x buffer layer can be increased through a post-air annealing approach. This enhancement results in a more robust driving force for charge transport and open-circuit voltage, leading to improved photovoltaic performance in polymer solar cells (PSCs) that utilise the post-air annealed MoO_x as the hole transport layer.

The concept of charge transport layers (HTL and ETL), which has been a cornerstone in the realm of organic photovoltaic devices, has recently found exciting applications in perovskite PV research. This innovative crossover has brought new insights and opportunities for optimising perovskite solar cell performance. Utilising the concepts of energy-level alignment and charge transport inspired by organic PV technology, significant progress has been made in enhancing perovskite materials and interfaces, and XPS has played an important role in these studies, as recently reviewed in the use of photoemission in the characterisation of halide perovskites.[56] Metal oxides as charge transfer layers play a crucial role in the continuous endeavour to utilise the remarkable photovoltaic characteristics of perovskite materials, ultimately aiming for efficient and economically viable solar energy conversion. Most metal oxides are characterised as semiconductors possessing a substantial bandgap spanning several eVs. Consequently, the electronic properties, encompassing attributes like the work function, positions of the conduction band and valence band, become especially crucial for governing the flow of charges within these substances. The charge-injection and extraction behaviours of metal oxides are greatly influenced by interfacial phenomena and the subsequent alignment of energy levels in organic and hybrid perovskite devices.

MoO_3, as well as V_2O_5 and WO_3, are categorised as n-type semiconductors and exhibit electronic states that are situated at significant energy depths. To illustrate, MoO_3 possesses an electron affinity, work function, and ionisation energy of 6.7 eV, 6.9 eV, and 9.5 eV, respectively. When these materials are thermally evaporated to form films, they become highly n-doped due to the presence of oxygen vacancies, causing a substantial shift of the Fermi level very close to the conduction band. The density of these oxygen vacancies can vary considerably depending on the specific thermal evaporation conditions, leading to the formation of gap states in proximity to the valence band edge. Numerous reports have provided evidence that interlayers of MoO_3, V_2O_5, and WO_3 facilitate processes such as hole-injection and hole-extraction. Hole injection occurs at the interface between the metal oxide and the organic layer, driven by an electron transfer from the organic HOMO to the CB of the metal oxide. Similarly, for hole-extraction, electrons residing in the CB of the metal oxide migrate towards the organic interface, where they recombine with photo-generated holes.[57]

Schulz *et al.*[58,59] conducted a study to examine the impact of utilising high work function MoO_3 as a hole extraction layer in a methylammonium lead iodide ($MAPbI_3$) perovskite absorber photovoltaic material. Molybdenum trioxide (MoO_3) with its high work function has been effectively employed as a material for both hole injection and extraction in organic electronics, including organic solar cells, where it additionally contributes to the creation of a substantial built-in voltage. In these devices, the efficient extraction of holes is attributed to the direct transfer of electrons from the exceptionally low-lying conduction band of n-type MoO_3 (with an electron affinity, EA, approximately 6.7 eV) into the HOMO of the donor molecules within the adjacent active layer. In the study reported by Schulz *et al.*,[58,59] photoemission experiments were performed within a custom-built ultra-high vacuum chamber, which allowed for *in situ* sample transfer between the measurement position and the MoO_3 deposition compartment. The samples under examination consisted of $MAPbI_3$ layered on TiO_2-covered FTO glass substrates. UPS spectra were collected using He I (21.22 eV) and He II (40.8 eV) excitation sources, at a nominal spectral resolution below 0.15 eV. XPS core level spectra were acquired using Al Kα radiation (1486.7 eV) with a modest resolution of approximately 0.8 eV. This particular study serves as an illustration of how the determination of electronic structure, band alignment, and chemical composition at an interface can be accomplished through the combined utilisation of laboratory-based UPS and XPS techniques.

The photoemission spectra of the $MAPbI_3$ layer, both in its bare form and with incremental thicknesses of top-deposited MoO_3, are displayed in Figure 6.19. The work function of the film, as determined from the SEC in the UPS spectra presented in Figure 6.19a, exhibited a steady increase with the thickness of the oxide. When a 100 Å thick layer of MoO_3 was applied, the vacuum level was measured at 6.8 eV above the Fermi level (E_F). In the valence band region of the film, clear evidence pointed to the development of a MoO_3 film, with the VBM, indicated by the dotted line in Figure 6.19b situated 2.8 eV below E_F, and the oxide ionisation energy, I_E, at 2.8 + 6.8 = 9.6 eV. The 3 eV band gap oxide was n-doped, and its CBM was positioned just 0.2 eV above E_F. The emergence of surface/interface states (IS) within the oxide gap, located near E_F, was clearly discernible in the 20 Å spectrum (Figure 6.19b), representing a characteristic feature of sub-stoichiometric evaporated MoO_3. The surface/interface state was more distinctly observed when comparing semi-logarithmic plots of the valence band region of the $MAPbI_3$ film in its pristine state and when covered with a submonolayer of MoO_3 with a nominal thickness of 1.5 Å. In this comparison, an interface state density with an onset at 0.2 eV below E_F was evident (referred to as IS in Figure 6.19c). The presence of these surface/interface states was attributed to the chemical reactions occurring within the MoO_3 film at the precise interface with $MAPbI_3$.

Additional insights into the energy level alignment at the perovskite/oxide interface were obtained from core-level XPS data. As increasing quantities of MoO_3 were deposited, there was a noticeable shift in both the iodine and lead core levels (I $3d_{5/2}$ and Pb $4f_{7/2}$) of the perovskite layer toward lower binding energies, as depicted in Figure 6.19d and e. This shift amounted to a total of 0.5 eV after the deposition of 20 Å of MoO_3. This observed shift strongly suggested the presence of MoO_3-induced band bending within the perovskite film, consistent with the deposition of the high work function oxide. The spin–orbit split components of the Mo 3d core level, on the other hand, did not exhibit any significant shift, as shown in Figure 6.19f. However, when the Mo 3d lineshape was deconvoluted into components corresponding to different Mo oxidation states, a distinct component, Mo^{4+}, indicative of sub-stoichiometric molybdenum oxide was revealed (inset Figure 6.19f). Notably, this Mo^{4+} contribution was only discernible at exceptionally low film thicknesses (1.5 and 5 Å, as shown in Figure 6.19f), suggesting that the significant reduction to molybdenum dioxide (MoO_2) was confined to the interface between $MAPbI_3$ and the oxide layer and involved the perovskite layer.

To suppress this interface chemical reaction, a 35 Å thick buffer layer of the HTM spiro-MeOTAD was evaporated from a Knudson cell in the vacuum chamber on top of the perovskite film, to protect the absorber from the subsequently evaporated MoO_3. The core level analysis revealed the same shift of the Pb 4f and I 3d levels observed for the system without spiro-MeOTAD (data not shown here). However, in contrast to the previous case, the Mo3d core level did not exhibit any significant Mo^{4+} component, indicating that the spiro-MeOTAD layer prevented the chemical reaction at the perovskite/molybdenum oxide interface, and thus the formation of MoO_2. The HOMO onset of the organic HTM was at ~1.0 eV below E_F prior to MoO_3 deposition. The subsequently evaporated oxide layer p-dopes the spiro-MeOTAD, and the HTM HOMO onset shifted to 0.6 eV below E_F.

The combined results from the PES measurements were consolidated into energy level diagrams illustrating the interface, both with and without the spiro-MeOTAD buffer layer (Figure 6.20). The shift observed in the Pb 4f

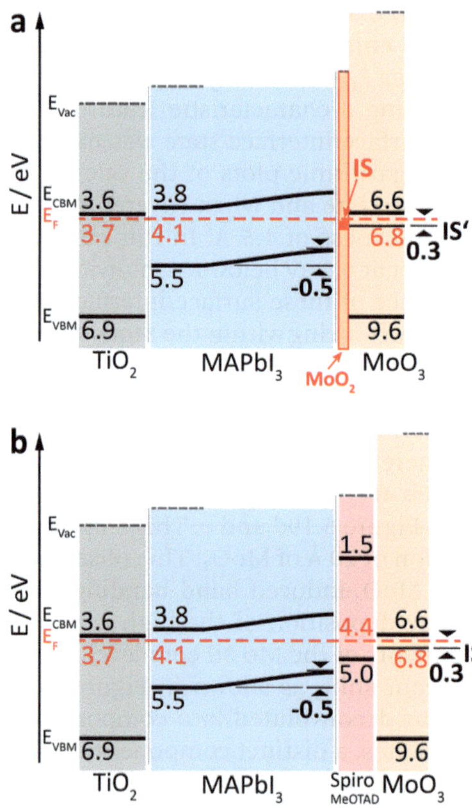

Figure 6.20 Schematic energy level diagram of a $MAPbI_3$ perovskite layer with (a) MoO_3 in immediate contact showing the emergence of an interface state (IS), and (b) with a spiro-MeOTAD interlayer. In both cases, a 0.5 eV band bending is induced in the perovskite layer making the film more intrinsic at the interface. For each layer, the Fermi level position (E_F, dotted line), work function (in red), and electron affinity and ionisation energy (in black) are indicated in eV. Reproduced from ref. 58 with permission from American Chemical Society, Copyright 2016.

and I 3d core levels remained consistent in the absence and presence of the spiro-MeOTAD buffer layer, signifying a 0.5 eV band bending effect induced within the perovskite layer due to the deposition of MoO_3, from the electrostatic potential imposed by the high work function of the molybdenum oxide layer. It's worth noting that determining the extent of band bending within the perovskite layer was not possible in these laboratory-based experiments, but in principle could be determined using HAXPES to probe different regions of the buried perovskite.

To assess the respective impacts of MoO_3 and spiro-MeOTAD/MoO_3 interlayers on actual device characteristics, the researchers incorporated copper(II) phthalocyanine (CuPc) as a hole transport layer, which was evaporated atop the designed hole extraction layers. Their observations indicated that the glass/FTO/TiO_2/$MAPbI_3$/MoO_3/CuPc/Au devices with a bare MoO_3 interlayer exhibited significantly diminished photovoltaic activity. This outcome was attributed to the presence MoO_2 at the interface. The introduction of a protective buffer layer, such as an extremely thin film of spiro-MeOTAD in the Glass/FTO/TiO_2/$MAPbI_3$/u-spiro-MeOTAD/MoO_3/CuPc/Au device, effectively inhibited the chemical reaction between the oxide and the absorber while preserving the band bending within the perovskite. Despite both device configurations exhibiting unconventional impairments in device characteristics, this study serves as an illustrative example of the pivotal role played by UPS/XPS techniques in the design of hybrid perovskite-based electronics. The authors have thoroughly reviewed the interface physics of halide perovskites.[60] They conclude that PES/IPES represent the preferred techniques for assessing the energy levels of the VBM and CBM, as well as the DOS of electronic states located on the surface of semiconductors. Furthermore, by extracting the vacuum level from these measurements, it becomes possible to identify crucial parameters such as ionisation energy and electron affinity. These parameters play a leading role in the determination of both chemical and optoelectronic properties at the interface of a device, as well as in understanding carrier transfer rates across that interface.

6.6 Conclusions

In the realm of solar photovoltaics (PV), a profound understanding of electronic band alignment at interfaces and the factors influencing it is paramount for achieving efficient energy conversion. PES is an advanced and mature analytical technique, and it has emerged as a powerful and indispensable tool for precisely characterising band alignments across various solar cell technologies. These encompass traditional cells, emerging perovskite-based systems, and organic PV devices. Through intricate analyses of core-level spectra and valence band structures, PES furnishes invaluable insights into energy-level positions, interface modifications, and material properties that govern charge carrier dynamics and, consequently, device performance.

One of the fundamental contributions of XPS lies in its unique ability to provide critical information, including the secondary electron cut-off, the work function of materials, and the onset of the valence band for a material. These measurements offer a straightforward means of visualising the natural band alignment between materials, often serving as a pivotal starting point in solar PV design. To illustrate the impact of this capability, consider the studies on SnS systems, where XPS measurements elucidate why certain materials fail to function effectively when paired with conventional window materials. Furthermore, when combined with sophisticated DFT calculations, XPS allows researchers to delve deeper into the composition and orbital hybridisation making up the valence band. As exemplified by the SnS systems, XPS can elucidate the pivotal role played by lone pair electrons, characteristic of specific elements in certain PV materials. These electrons introduce additional electronic states, leading to shifts in energy levels that significantly influence band alignments.

XPS also offers a precise determination of band offsets at heterojunctions or interfaces within a PV stack. This capability can be achieved directly or through methods like the Kraut method, and it can even exploit the surface sensitivity of XPS to tune the photon energy for studying buried interfaces. Understanding band bending and the effects of charge redistribution at heterostructures is crucial for optimising device performance. Furthermore, it plays an essential role in characterising the electronic structure of PV materials by examining core-level spectra. It is possible to extract valuable information about chemical composition, oxidation states, and defects, all of which have profound effects on energy-level alignments. XPS is particularly adept at monitoring chemical changes occurring at the surface of PV materials during aging or exposure to environmental stressors, offering insights into material degradation mechanisms.

In the context of conventional solar cells, XPS has been pivotal in elucidating band alignments at heterojunction interfaces, thereby enabling the enhancement of charge separation and collection efficiency. Precise measurements of energy levels within semiconductor layers and interface regions have enabled the tailoring of device architecture to minimise energy losses and optimise overall efficiency. XPS has also played a critical role in characterising surface passivation layers, which are instrumental in reducing recombination losses and improving solar cell efficiency.

Perovskite solar cells, with their remarkable rise in efficiency and affordability, have captivated the PV research community. XPS has proven indispensable in unravelling the intricate interplay of energy levels within perovskite materials. Extensive investigations into perovskite absorbers and their interfaces with charge transport layers have yielded insights into band alignments, influencing charge carrier extraction and recombination processes. This understanding has not only facilitated the design of more efficient perovskite solar cells but also addressed stability issues that have hindered their commercialisation.

In the realm of organic photovoltaics (OPVs), XPS has been pivotal in elucidating energy-level alignments at the interfaces between organic materials and between organic and metal layers. By characterising and modifying the chemical composition of these interfaces, it has been possible to enhance charge transport and minimise losses, resulting in improved OPV device performance. Additionally, XPS has played a crucial role in studying the effects of surface treatments, such as chemical doping and self-assembled monolayers, on the electronic structure of organic semiconductors. This knowledge has paved the way for the design of organic materials with superior charge transport properties and increased photon absorption.

In summary, XPS stands as an invaluable and versatile tool for characterising band alignment in various solar PV technologies. Whether applied to traditional silicon-based cells, burgeoning perovskite solar cells, or organic PV systems, XPS consistently delivers critical insights that drive advancements in efficiency, stability, and overall performance. As the field of solar energy continues its evolution, XPS will remain a fundamental asset for researchers and engineers striving to develop more efficient and sustainable photovoltaic technologies. By unravelling the intricacies of energy-level alignment, XPS contributes significantly to the ongoing pursuit of clean and renewable energy sources that will power our future.

References

1. V. Shutthanandan, M. Nandasiri, J. Zheng, M. H. Engelhard, W. Xu, S. Thevuthasan and V. Murugesan, *J. Electron Spectrosc. Relat. Phenom.*, 2019, **231**, 2.
2. S. Iida, M. Terashima, K. Mamiya, H. Chang, S. Sasaki, A. Ono, T. Kimoto and T. Miyayama, *J. Vac. Sci. Technol. B*, 2021, **39**, 044001.
3. M. I. Nandasiri, L. E. Camacho-Forero, A. M. Schwarz, V. Shutthanandan, S. Thevuthasan, P. B. Balbuena, K. T. Mueller and V. Murugesan, *Chem. Mater.*, 2017, **29**, 4728.
4. W. Xue, Q. Yan, G. Xu, L. Suo, Y. Chen, C. Wang, C.-A. Wang and J. Li, *Nano Energy*, 2017, **38**, 12.
5. K. N. Wood, K. X. Steirer, S. E. Hafner, C. Ban, S. Santhanagopalan, S. Lee and G. Teeter, *Nat. Commun.*, 2018, **9**, 2490.
6. C. P. Muzzillo, S. Glynn, G. Teeter and L. M. Mansfield, *ACS Appl. Energy Mater.*, 2022, **5**, 11328.
7. T. D. Veal, D. O. Scanlon, R. Kostecki and E. Arca, *J. Phys. Energy*, 2021, **3**, 032001.
8. M. A. Isaacs, J. Davies-Jones, P. R. Davies, S. Guan, R. Lee, D. J. Morgan and R. Palgrave, *Mater. Chem. Front.*, 2021, **5**, 7931.
9. J. Hu, P. Chen, D. Luo, D. Wang, N. Chen, S. Yang, Z. Fu, M. Yu, L. Li, R. Zhu and Z. Lu, *Commun. Mater.*, 2022, **3**, 39.
10. Z. Wu, M. Jiang, Z. Liu, A. Jamshaid, L. K. Ono and Y. Qi, *Adv. Energy Mater.*, 2020, **10**, 1903696.

11. C. Das, M. Wussler, T. Hellmann, T. Mayer and W. Jaegermann, *Phys. Chem. Chem. Phys.*, 2018, **20**, 17180.
12. J. P. C. Baena, L. Steier, W. T. Tress, M. Saliba, S. Neutzner, T. Matsui, F. Giordano, T. J. Jacobsson, A. R. S. Kandada, S. M. Zakeeruddin, A. Petrozza, A. Abate, M. K. Nazeeruddin, M. Grätzel and A. Hagfeldt, *Energy Environ. Sci.*, 2015, **8**, 2928.
13. H. Fan, Z. Yang, X. Ren, M. Yin, F. Gao and S. Liu, *AIP Adv.*, 2016, **6**, 015314.
14. C. Das, M. Kedia, W. Zuo, C. Mortan, M. Kot, J. I. Flege and M. Saliba, *Sol. RRL*, 2022, **6**, 2200348.
15. Q. Wang, J. Yang, A. Gerlach, F. Schreiber and S. Duhm, *J. Phys. Mater.*, 2022, **5**, 044010.
16. S. Olthof, *Adv. Opt. Mater.*, 2021, **9**, 2100227.
17. T. Schultz, P. Bärmann, E. Longhi, R. Meena, Y. Geerts, Y. Gogotsi, S. Barlow, S. R. Marder, T. Petit and N. Koch, *Phys. Rev. Mater.*, 2023, **7**, 045002.
18. R. S. Hall, D. Lamb and S. J. C. Irvine, *Energy Sci. Eng.*, 2021, **9**, 606.
19. J. H. Lee, D. Shin, R. Rhee, S. Yun, K. M. Yeom, D. H. Chun, S. Lee, D. Kim, Y. Yi, J. H. Noh and J. H. Park, *J. Phys. Chem. Lett.*, 2019, **10**, 21.
20. A. Kahn, *Mater. Horiz.*, 2016, **3**, 7.
21. R. Schlaf, *Calibration of Photoemission Spectra and Work Function Determination Tutorial*, University of South Florida. http://rsl.eng.usf.edu/Documents/Tutorials/PEScalibration.pdf. Accessed August 2023.
22. T. J. Whittles, *PhD thesis*, University of Liverpool, 2015, DOI: 10.17638/03010978.
23. T. J. Whittles, L. A. Burton, J. M. Skelton, A. Walsh, T. D. Veal and V. R. Dhanak, *Chem. Mater.*, 2016, **28**, 3718.
24. L. A. Burton, T. J. Whittles, D. Hesp, W. M. Linhart, J. M. Skelton, B. Hou, R. F. Webster, G. O'Dowd, C. Reece, D. Cherns, D. J. Fermín, T. D. Veal, V. R. Dhanak and A. Walsh, *J. Mater. Chem. A*, 2016, **4**, 1312.
25. A. Walsh, D. J. Payne, R. G. Egdell and G. W. Watson, *Chem. Soc. Rev.*, 2011, **40**, 4455.
26. M. J. Smiles, J. M. Skelton, H. Shiel, L. A. H. Jones, J. E. N. Swallow, H. J. Edwards, P. A. E. Murgatroyd, T. J. Featherstone, P. K. Thakur, T.-L. Lee, V. R. Dhanak and T. D. Veal, *J. Mater. Chem. A*, 2021, **9**, 22440.
27. H. Shiel, O. S. Hutter, L. J. Phillips, J. E. N. Swallow, L. A. H. Jones, T. J. Featherstone, M. J. Smiles, P. K. Thakur, T.-L. Lee, V. R. Dhanak, J. D. Major and T. D. Veal, *ACS Appl. Energy Mater.*, 2020, **3**, 11617.
28. T. J. Whittles, T. D. Veal, C. N. Savory, A. W. Welch, F. W. de Souza Lucas, J. T. Gibbon, M. Birkett, R. J. Potter, D. O. Scanlon, A. Zakutayev and V. R. Dhanak, *ACS Appl. Mater. Interfaces*, 2017, **9**, 41916.
29. T. J. Whittles, T. D. Veal, C. N. Savory, P. J. Yates, P. A. E. Murgatroyd, J. T. Gibbon, M. Birkett, R. J. Potter, J. D. Major, K. Durose, D. O. Scanlon and V. R. Dhanak, *ACS Appl. Mater. Interfaces*, 2019, **11**, 27033.
30. M. Wuttig, C.-F. Schön, M. Schumacher, J. Robertson, P. Golub, E. Bousquet, C. Gatti and J.-Y. Raty, *Adv. Funct. Mater.*, 2022, **32**, 2110166.

31. E. A. Kraut, R. W. Grant, J. R. Waldrop and S. P. Kowalczyk, *Phys. Rev. Lett.*, 1980, **44**, 1620.
32. J. R. Waldrop, R. W. Grant, S. P. Kowalczyk and E. A. Kraut, *J. Vac. Sci. Technol. A*, 1985, **3**, 835.
33. (a) A. Mavlonov, T. Razykov, F. Raziq, J. Gan, J. Chantana, Y. Kawano, T. Nishimura, H. Wei, A. Zakutayev, T. Minemoto, X. Zu, S. Li and L. Qiao, *Sol. Energy*, 2020, **201**, 227. (b) Z. Duan, X. Liang, Y. Feng, H. Ma, B. Liang, Y. Wang, S. Luo, S. Wang, R. E. Schropp, Y. Mai and Z. Li, *Adv. Mater.*, 2022, **34**, 2202969.
34. Y. Wang, S. Ji and B. Shin, *J. Phys. Energy*, 2022, **4**, 044002.
35. S. Svanström, A. G. Fernández, T. Sloboda, T. J. Jacobsson, F. Zhang, F. O. L. Johansson, D. Kühn, D. Céolin, J.-P. Rueff, L. Sun, K. Aitola, H. Rensmo and U. B. Cappel, *ACS Appl. Mater. Interfaces*, 2023, **15**, 12485.
36. NREL Best Research-Cell Efficiency Chart, https://www.nrel.gov/pv/cell-efficiency.html. Accessed September 2023.
37. S. Shao and M. A. Loi, *Adv. Mater. Interfaces*, 2020, **7**, 1901469.
38. W.-C. Lin, W.-C. Lo, J.-X. Li, Y.-K. Wang, J.-F. Tang and Z.-Y. Fong, *npj Mater. Degrad.*, 2021, **5**, 13.
39. B. Han, C. Wu, J. Liu, Z. Sun, X. Gu and C. Shen, *Nanomaterials*, 2023, **13**, 1983.
40. K. J. C.-R. Chun, A. S. Walton, D. J. Lewis, A. Tedstone, P. O'Brien, A. G. Thomas and W. R. Flavell, *Chem. Commun.*, 2017, **53**, 5231.
41. M. Wussler, T. Mayer, C. Das, E. Mankel, T. Hellmann, C. Prabowo, I. Zimmermann, M. K. Nazeeruddin and W. Jaegermann, *Adv. Funct. Mater.*, 2020, **30**, 1910679.
42. M. Stolterfoht, C. M. Wolff, J. A. Márquez, S. Zhang, C. J. Hages, D. Rothhardt, S. Albrecht, P. L. Burn, P. Meredith, T. Unold and D. Neher, *Nat. Energy*, 2018, **3**, 847.
43. F. Zu, M. Roß, L. Frohloff, D. Shin, N. Tessler, S. Albrecht and N. Koch, *Sol. RRL*, 2022, **6**, 2101065.
44. F. Zu, J. H. Warby, M. Stolterfoht, J. Li, D. Shin, E. Unger and N. Koch, *Phys. Rev. Lett.*, 2021, **127**, 246401.
45. M. Gansukh, Z. Li, M. E. Rodriguez, S. Engberg, F. M. A. Martinho, S. L. Mariño, E. Stamate, J. Schou, O. Hansen and S. Canulescu, *Sci. Rep.*, 2020, **10**, 18388.
46. M. Zhang, L. Zhu, G. Zhou, T. Hao, C. Qiu, Z. Zhao, Q. Hu, B. W. Larson, H. Zhu, Z. Ma, Z. Tang, W. Feng, Y. Zhang, T. P. Russell and F. Liu, *Nat. Commun.*, 2021, **12**, 309.
47. Y. Li, X. Huang, K. Ding, H. K. M. Sheriff Jr, L. Ye, H. Liu, C.-Z. Li, H. Ade and S. R. Forrest, *Nat. Commun.*, 2021, **12**, 5419.
48. C. K. Song, A. C. White, L. Zeng, B. J. Leever, M. D. Clark, J. D. Emery, S. J. Lou, A. Timalsina, L. X. Chen, M. J. Bedzyk and T. J. Marks, *ACS Appl. Mater. Interfaces*, 2013, **5**, 9224.
49. M. Wang, X. Cai, Y. Huang, K. Zheng and F. Li, *Front. Phys.*, 2020, **8**, 578317.
50. J. Kettle, Z. Ding, M. Horie and G. C. Smith, *Org. Electron.*, 2016, **39**, 222.

51. J. Zhang, H. S. Tan, X. Guo, A. Facchetti and H. Yan, *Nat. Energy*, 2018, **3**, 720.
52. (a) A. K. K. Kyaw, X. W. Sun, C. Y. Jiang, G. Q. Lo, D. W. Zhao and D. L. Kwong, *Appl. Phys. Lett.*, 2008, **93**, 221107. (b) C. Anrango-Camacho, K. Pavón-Ipiales, B. A. Frontana-Uribe and A. Palma-Cando, *Nanomaterials*, 2022, **12**, 443. (c) X. Hu, L. Chen and Y. Chen, *J. Phys. Chem. C*, 2014, **118**, 9930.
53. H. N. Tran, H. Lee, C. B. Park, N. V. Krishna, F. T. A. Wibowo, S.-Y. Jang, J. Y. Kim and S. Cho, *Adv. Mater. Interfaces*, 2022, **9**, 2201274.
54. M. T. Greiner, M. G. Helander, W.-M. Tang, Z.-B. Wang, J. Qiu and Z.-H. Lu, *Nat. Mater.*, 2012, **11**, 76.
55. P. Ma, J. Xu, C. Wang, C. Wang, F. Meng, Y. Xie and S. Wen, *RSC Adv.*, 2021, **11**, 35141.
56. S. Béchu, M. Ralaiarisoa, A. Etcheberry and P. Schulz, *Adv. Energy Mater.*, 2020, **10**, 1904007.
57. K. Zilberberg, J. Meyer and T. Riedl, *J. Mater. Chem. C*, 2013, **1**, 4796.
58. P. Schulz, J. O. Tiepelt, J. A. Christians, I. Levine, E. Edri, E. M. Sanehira, G. Hodes, D. Cahen and A. Kahn, *ACS Appl. Mater. Interfaces*, 2016, **8**, 31491.
59. P. Schulz, *ACS Energy Lett.*, 2018, **3**, 1287.
60. P. Schulz, D. Cahen and A. Kahn, *Chem. Rev.*, 2019, **119**, 3349.

Subject Index

References to figures are given in *italic* type. References to tables are indicated by **bold** type.